*Peter Bodrogi and Tran Quoc Khanh*

**Illumination, Color and Imaging**

# Wiley-SID Series in Display Technology

Series Editor:
**Anthony C. Lowe**

Consultant Editor:
**Michael A. Kriss**

Display Systems: Design and Applications
**Lindsay W. MacDonald and Anthony C. Lowe (Eds.)**

Electronic Display Measurement: Concepts, Techniques, and Instrumentation
**Peter A. Keller**

Reflective Liquid Crystal Displays
**Shin-Tson Wu and Deng-Ke Yang**

Colour Engineering: Achieving Device Independent Colour
**Phil Green and Lindsay MacDonald (Eds.)**

Display Interfaces: Fundamentals and Standards
**Robert L. Myers**

Digital Image Display: Algorithms and Implementation
**Gheorghe Berbecel**

Flexible Flat Panel Displays
**Gregory Crawford (Ed.)**

Polarization Engineering for LCD Projection
**Michael G. Robinson, Jianmin Chen, and Gary D. Sharp**

Fundamentals of Liquid Crystal Devices
**Deng-Ke Yang and Shin-Tson Wu**

Introduction to Microdisplays
**David Armitage, Ian Underwood, and Shin-Tson Wu**

Mobile Displays: Technology and Applications
**Achintya K. Bhowmik, Zili Li, and Philip Bos (Eds.)**

Photoalignment of Liquid Crystalline Materials: Physics and Applications
**Vladimir G. Chigrinov, Vladimir M. Kozenkov and Hoi-Sing Kwok**

Projection Displays, Second Edition
**Matthew S. Brennesholtz and Edward H. Stupp**

Introduction to Flat Panel Displays
**Jiun-Haw Lee, David N. Liu and Shin-Tson Wu**

LCD Backlights
**Shunsuke Kobayashi, Shigeo Mikoshiba and Sungkyoo Lim (Eds.)**

Liquid Crystal Displays: Addressing Schemes and Electro-Optical Effects, Second Edition
**Ernst Lueder**

Transflective Liquid Crystal Displays
**Zhibing Ge and Shin-Tson Wu**

Liquid Crystal Displays: Fundamental Physics and Technology
**Robert H. Chen**

3D Displays
**Ernst Lueder**

OLED Display Fundamentals and Applications
**Takatoshi Tsujimura**

Illumination, Color and Imaging: Evaluation and Optimization of Visual Displays
**Tran Quoc Khanh and Peter Bodrogi**

*Peter Bodrogi and Tran Quoc Khanh*

# Illumination, Color and Imaging

Evaluation and Optimization of Visual Displays

WILEY-VCH

WILEY-VCH Verlag GmbH & Co. KGaA

**The Authors**

**Dr. Peter Bodrogi**
TU Darmstadt
Laboratory of Lighting Technology
Darmstadt, Germany
bodrogi@lichttechnik.tu-darmstadt.de

**Prof. Tran Quoc Khanh**
TU Darmstadt
Laboratory of Lighting Technology
Darmstadt, Germany
khanh@lichttechnik.tu-darmstadt.de

**The Series Editor**

**Tony Lowe**
Lambent Consultancy
Braishfield, UK
lambentconsultants.com

All books published by **Wiley-VCH** are carefully produced. Nevertheless, authors, editors, and publisher do not warrant the information contained in these books, including this book, to be free of errors. Readers are advised to keep in mind that statements, data, illustrations, procedural details or other items may inadvertently be inaccurate.

**Library of Congress Card No.:** applied for

**British Library Cataloguing-in-Publication Data**
A catalogue record for this book is available from the British Library.

**Bibliographic information published by the Deutsche Nationalbibliothek**
The Deutsche Nationalbibliothek lists this publication in the Deutsche Nationalbibliografie; detailed bibliographic data are available on the Internet at http://dnb.d-nb.de.

© 2012 Wiley-VCH Verlag & Co. KGaA, Boschstr. 12, 69469 Weinheim, Germany

All rights reserved (including those of translation into other languages). No part of this book may be reproduced in any form – by photoprinting, microfilm, or any other means – nor transmitted or translated into a machine language without written permission from the publishers. Registered names, trademarks, etc. used in this book, even when not specifically marked as such, are not to be considered unprotected by law.

**Print ISBN:** 978-3-527-41040-8
**ePDF ISBN:** 978-3-527-65075-0
**ePub ISBN:** 978-3-527-65074-3
**mobi ISBN:** 978-3-527-65073-6
**oBook ISBN:** 978-3-527-65072-9

**Cover Design**  Spieszdesign, Neu-Ulm
**Typesetting**  Thomson Digital, Noida, India
**Printing and Binding**  Markono Print Media Pte Ltd, Singapore

Printed on acid-free paper

*To Prof. János Schanda, for his research and teaching in the domains of color science, colorimetry, photometry and visual technologies*

# Contents

**Series Editor's Foreword** *XIII*
**Preface** *XV*
**About the Authors** *XXI*

**1 Color Vision and Self-Luminous Visual Technologies** *1*
1.1 Color Vision Features and the Optimization of Modern Self-Luminous Visual Technologies *2*
1.1.1 From Photoreceptor Structure to Colorimetry *2*
1.1.2 Spatial and Temporal Contrast Sensitivity *6*
1.1.3 Color Appearance Perception *12*
1.1.4 Color Difference Perception *15*
1.1.5 Cognitive, Preferred, Harmonic, and Emotional Color *17*
1.1.6 Interindividual Variability of Color Vision *18*
1.2 Color Vision-Related Technological Features of Modern Self-Luminous (Nonprinting) Visual Technologies *18*
1.3 Perceptual, Cognitive, and Emotional Features of the Visual System and the Corresponding Technological Challenge *20*
References *23*

**2 Colorimetric and Color Appearance-Based Characterization of Displays** *25*
2.1 Characterization Models and Visual Artifacts in General *25*
2.1.1 Tone Curve Models and Phosphor Matrices *26*
2.1.2 Measured Color Characteristics, sRGB, and Other Characterization Models *27*
2.1.3 Additivity and Independence of the Color Channels *35*
2.1.4 Multidimensional Phosphor Matrices and Other Methods *35*
2.1.5 Spatial Uniformity and Spatial Independence *39*
2.1.6 Viewing Direction Uniformity *45*
2.1.7 Other Visual Artifacts *46*
2.1.8 The Viewing Environment: Viewing Conditions and Modes *48*

| | | |
|---|---|---|
| 2.1.9 | Application of CIELAB, CIELUV, and CIECAM02 to Self-Luminous Displays  *49* | |
| 2.2 | Characterization Models and Visual Artifacts of the Different Display Technologies  *51* | |
| 2.2.1 | Modern Applications of the Different Display Technologies  *52* | |
| 2.2.2 | Special Characterization Models of the Different Displays  *53* | |
| 2.2.2.1 | CRT  *53* | |
| 2.2.2.2 | PDP  *55* | |
| 2.2.2.3 | Various LCD Technologies and Their Viewing Direction Uniformity  *60* | |
| 2.2.2.4 | Head-Mounted Displays and Head-Up Displays  *67* | |
| 2.2.2.5 | Projectors Including DMD and LCD  *68* | |
| 2.2.2.6 | OLEDs  *71* | |
| 2.3 | Display Light Source Technologies  *72* | |
| 2.3.1 | Projector Light Sources  *73* | |
| 2.3.2 | Backlight Sources  *75* | |
| 2.3.3 | Color Filters, Local Dimming, and High Dynamic Range Imaging  *79* | |
| 2.4 | Color Appearance of Large Viewing Angle Displays  *81* | |
| 2.4.1 | Color Appearance Differences between Small and Large Color Stimuli  *81* | |
| 2.4.1.1 | Color Appearance of an Immersive Color Stimulus on a PDP  *82* | |
| 2.4.1.2 | Xiao *et al.*'s Experiment on the Appearance of a Self-Luminous 50° Color Stimulus on an LCD  *87* | |
| 2.4.2 | Mathematical Modeling of the Color Size Effect  *87* | |
| | References  *91* | |
| **3** | **Ergonomic, Memory-Based, and Preference-Based Enhancement of Color Displays**  *97* | |
| 3.1 | Ergonomic Guidelines for Displays  *97* | |
| 3.2 | Objectives of Color Image Reproduction  *105* | |
| 3.3 | Ergonomic Design of Color Displays: Optimal Use of Chromaticity Contrast  *107* | |
| 3.3.1 | Principles of Ergonomic Color Design  *107* | |
| 3.3.2 | Legibility, Conspicuity, and Visual Search  *108* | |
| 3.3.3 | Chromaticity Contrast for Optimal Search Performance  *111* | |
| 3.3.4 | Chromaticity and Luminance Contrast Preference  *123* | |
| 3.4 | Long-Term Memory Colors, Intercultural Differences, and Their Use to Evaluate and Improve Color Image Quality  *134* | |
| 3.4.1 | Long-Term Memory Colors for Familiar Objects  *135* | |
| 3.4.2 | Intercultural Differences of Long-Term Memory Colors  *139* | |
| 3.4.3 | Increasing Color Quality by Memory Colors  *141* | |
| 3.5 | Color Image Preference for White Point, Local Contrast, Global Contrast, Hue, and Chroma  *142* | |
| 3.5.1 | Apparatus and Method to Obtain a Color Image Preference Data Set  *143* | |
| 3.5.2 | Image Transforms of Color Image Preference  *144* | |
| 3.5.3 | Preferred White Point  *144* | |
| 3.5.4 | Preferred Local Contrast  *147* | |

| | | |
|---|---|---|
| 3.5.5 | Preferred Global Contrast *147* | |
| 3.5.6 | Preferred Hue and Chroma *150* | |
| 3.6 | Age-Dependent Method for Preference-Based Color Image Enhancement with Color Image Descriptors *151* | |
| | References *156* | |

**4 Color Management and Image Quality Improvement for Cinema Film and TV Production** *161*

| | | |
|---|---|---|
| 4.1 | Workflow in Cinema Film and TV Production Today – Components and Systems *161* | |
| 4.1.1 | Workflow *161* | |
| 4.1.2 | Structure of Color Management in Today's Cinema and TV Technology *164* | |
| 4.1.3 | Color Management Solutions *165* | |
| 4.2 | Components of the Cinema Production Chain *166* | |
| 4.2.1 | Camera Technology in Overview *166* | |
| 4.2.2 | Postproduction Systems *174* | |
| 4.2.3 | CIELAB and CIEDE 2000 Color Difference Formulas Under the Viewing Conditions of TV and Cinema Production *176* | |
| 4.2.3.1 | Procedure of the Visual Experiment *178* | |
| 4.2.3.2 | Experimental Results *181* | |
| 4.2.4 | Applications of the CIECAM02 Color Appearance Model in the Digital Image Processing System for Motion Picture Films *184* | |
| 4.3 | Color Gamut Differences *191* | |
| 4.4 | Exploiting the Spatial–Temporal Characteristics of Color Vision for Digital TV, Cinema, and Camera Development *195* | |
| 4.4.1 | Spatial and Temporal Characteristics in TV and Cinema Production *195* | |
| 4.4.2 | Optimization of the Resolution of Digital Motion Picture Cameras *199* | |
| 4.4.3 | Perceptual and Image Quality Aspects of Compressed Motion Pictures *205* | |
| 4.4.3.1 | Necessity of Motion Picture Compression *205* | |
| 4.4.3.2 | Methods of Image Quality Evaluation *205* | |
| 4.4.3.3 | The Image Quality Experiment *207* | |
| 4.4.4 | Perception-Oriented Development of Watermarking Algorithms for the Protection of Digital Motion Picture Films *214* | |
| 4.4.4.1 | Motivation and Aims of Watermarking Development *214* | |
| 4.4.4.2 | Requirements for Watermarking Technology *216* | |
| 4.4.4.3 | Experiment to Test Watermark Implementations *217* | |
| 4.5 | Optimum Spectral Power Distributions for Cinematographic Light Sources and Their Color Rendering Properties *223* | |
| 4.6 | Visually Evoked Emotions in Color Motion Pictures *229* | |
| 4.6.1 | Technical Parameters, Psychological Factors, and Visually Evoked Emotions *229* | |
| 4.6.2 | Emotional Clusters: Modeling Emotional Strength *231* | |
| | References *233* | |

| 5 | **Pixel Architectures for Displays of Three- and Multi-Color Primaries** *237* |
| --- | --- |
| 5.1 | Optimization Principles for Three- and Multi-Primary Color Displays to Obtain a Large Color Gamut *238* |
| 5.1.1 | Target Color Sets *240* |
| 5.1.2 | Factors of Optimization *244* |
| 5.1.2.1 | Color Gamut Volume *244* |
| 5.1.2.2 | Quantization Efficiency *244* |
| 5.1.2.3 | Number of Color Primaries *245* |
| 5.1.2.4 | White Point *245* |
| 5.1.2.5 | Technological Constraints *246* |
| 5.1.2.6 | *P/W* Ratio *247* |
| 5.1.2.7 | Roundness *249* |
| 5.1.2.8 | RGB Tone Scales and Display Black Point *250* |
| 5.2 | Large-Gamut Primary Colors and Their Gamut in Color Appearance Space *250* |
| 5.2.1 | Optimum Color Primaries *251* |
| 5.2.2 | Optimum Color Gamuts in Color Appearance Space *252* |
| 5.3 | Optimization Principles of Subpixel Architectures for Multi-Primary Color Displays *257* |
| 5.3.1 | The Color Fringe Artifact *258* |
| 5.3.2 | Optimization Principles *259* |
| 5.3.2.1 | Minimum Color Fringe Artifact *259* |
| 5.3.2.2 | Modulation Transfer Function *260* |
| 5.3.2.3 | Isotropy *260* |
| 5.3.2.4 | Luminance Resolution *261* |
| 5.3.2.5 | High Aperture Ratio *261* |
| 5.4 | Three- and Multi-Primary Subpixel Architectures and Color Image Rendering Methods *262* |
| 5.4.1 | Three-Primary Architectures *262* |
| 5.4.2 | Multi-Primary Architectures *264* |
| 5.4.3 | Color Image Rendering Methods *268* |
|  | Acknowledgment *270* |
|  | References *271* |
|  |  |
| 6 | **Improving the Color Quality of Indoor Light Sources** *273* |
| 6.1 | Introduction to Color Rendering and Color Quality *273* |
| 6.2 | Optimization for Indoor Light Sources to Provide a Visual Environment of High Color Rendering *276* |
| 6.2.1 | Visual Color Fidelity Experiments *276* |
| 6.2.2 | Color Rendering Prediction Methods *282* |
| 6.2.2.1 | Deficits of the Current Color Rendering Index *282* |
| 6.2.2.2 | Proposals to Redefine the Color Rendering Index *285* |
| 6.3 | Optimization of Indoor Light Sources to Provide Color Harmony in the Visual Environment *286* |
| 6.3.1 | Visual Color Harmony Experiments *287* |

| | | |
|---|---|---|
| 6.3.2 | Szab et al.'s Mathematical Model to Predict Color Harmony | 287 |
| 6.3.3 | A Computational Method to Predict Color Harmony Rendering | 289 |
| 6.4 | Principal Components of Light Source Color Quality | 293 |
| 6.4.1 | Factors Influencing Color Quality | 293 |
| 6.4.2 | Experimental Method to Assess the Properties of Color Quality | 296 |
| 6.4.3 | Modeling Color Quality: Four-Factor Model | 302 |
| 6.4.4 | Principal Components of Color Quality for Three Indoor Light Sources | 303 |
| 6.5 | Assessment of Complex Indoor Scenes Under Different Light Sources | 304 |
| 6.5.1 | Psychological Relationship between Color Difference Scales and Color Rendering Scales | 305 |
| 6.5.2 | Brightness in Complex Indoor Scenes in Association with Color Gamut, Rendering, and Harmony: A Computational Example | 311 |
| 6.5.3 | Whiteness Perception and Light Source Chromaticity | 316 |
| 6.6 | Effect of Interobserver Variability of Color Vision on the Color Quality of Light Sources | 318 |
| 6.6.1 | Variations of Color Vision Mechanisms | 319 |
| 6.6.2 | Effect of Variability on Color Quality | 320 |
| 6.6.2.1 | Variability of the Visual Ratings of Color Quality | 321 |
| 6.6.2.2 | Variability of Perceived Color Differences and the Color Rendering Index | 321 |
| 6.6.2.3 | Variability of Similarity Ratings | 322 |
| 6.6.3 | Relevance of Variability for Light Source Design | 324 |
| | Acknowledgments | 324 |
| | References | 324 |
| | | |
| **7** | **Emerging Visual Technologies** | **329** |
| 7.1 | Emerging Display Technologies | 329 |
| 7.1.1 | Flexible Displays | 329 |
| 7.1.2 | Laser and LED Displays | 330 |
| 7.1.3 | Color Gamut Extension for Multi-Primary Displays | 334 |
| 7.2 | Emerging Technologies for Indoor Light Sources | 339 |
| 7.2.1 | Tunable LED Lamps for Accent Lighting | 339 |
| 7.2.2 | Optimization for Brightness and Circadian Rhythm | 341 |
| 7.2.3 | Accentuation of Different Aspects of Color Quality | 347 |
| 7.2.4 | Using New Phosphor Blends | 348 |
| 7.2.5 | Implications of Color Constancy for Light Source Design | 354 |
| 7.3 | Summary and Outlook | 357 |
| | Acknowledgments | 360 |
| | References | 360 |

**Index** 363

## Series Editor's Foreword

Display manufacturers spend a great deal of time and resource improving the visual characteristics of their display products. Such improvements encompass resolution, contrast, color gamut, viewing angle, and switching speed. Yet the manner in which displays are used is often haphazard, with too little attention being paid to the orientation of the display to sources of ambient illumination, to the ambient illuminance, or to the hue of the illuminant. How much better their visual experience would be if users or those responsible for display use within an organization had more knowledge of all these factors and applied them appropriately. How much more effectively could manufacturers and product developers use their resources if they paid greater attention to the realistic limits imposed by the human visual system and by the gamut of the majority of colors we experience in real life. Too often, marketing statements enter the realm of improbability with claims of massive color gamuts and contrast ratios achievable only under dark room conditions.

This latest book in the series is written by two respected experts in the field of display evaluation and optimization. It addresses the issues I have outlined above and a great deal more. It is a very complete book. In fact, the authors have provided such a complete description of its contents in the preface that I shall not comment further on it in detail here.

There are, however, some general comments I would make. Many, perhaps most of those, who have made measurements on displays they are researching will have been solely interested in the temporal and contrast characteristics of their particular display. That is all well and good; such measurements are the fundamental basis of characterizing displays. However, what this book reveals is the complexity and richness of the stages of development that follow and that, in the authors' own words, emphasize how to use the features of the human visual system to meet today's technological challenges. Those challenges include familiar elements such as the colorimetric and color appearance-based characterization and calibration of color monitors and color management in digital TV and cinema applications. However, they also include the less familiar optimization of pixel and subpixel architectures for displays of more than three primary colors, the concepts of color conspicuity, color memory, and color preference-based enhancement of color displays for visual ergonomics and pleasing image rendering. I am among those becoming familiar with visual changes that are related to the aging process, but new to me was a quantitative treatment of cultural differences. The last of the challenges the book addresses is perhaps better considered as an opportunity. It concerns the ability to optimize the spectral power distribution of modern light sources that can be used either as indoor illuminants or as display backlights.

This book contains a significant amount of previously unpublished material. A much needed and very up-to-date work, it will provide great benefit and vital guidance to an extremely wide and diverse audience that includes but is definitely not limited to those involved in the development of image capture and display devices and systems, light sources and illumination systems, and image optimization, processing, and production software.

*Braishfield, United Kingdom*  *Anthony C. Lowe*
*Series Editor*

# Preface

This book is a monograph about how to exploit the knowledge of the human color information processing system in order to design usable, ergonomic, and pleasing information displays, entertainment displays, or a high-quality visual environment. For the designer of modern self-luminous visual technologies including displays and light sources for general lighting, optimization principles derived from the human visual system are presented. This book has arisen from the need for a specialist text that brings together these principles derived from a comprehensive view of human color information processing from retinal photoreceptors to cognition, preference, harmony, and emotions arising in the visual brain with the recent amazing developments of display technology and general indoor light source technology. In this sense, this book is not a textbook on human vision, colorimetry, color science, display technology, or light source technology. Instead, the emphasis is on how to use the features of the human visual system to meet today's technological challenges including the colorimetric and color appearance-based characterization and calibration of color monitors, color management in digital TV and cinema, optimization of pixel and subpixel architectures for displays of three or more primary colors, color conspicuity, color memory, and color preference-based enhancement of color displays for visual ergonomics and pleasing image rendering, also concerning cultural and age differences, and last but not least the optimization of spectral power distributions of modern light sources used to illuminate an indoor scene or an image rendering pixel architecture as a backlight.

Concerning the intended audience of this book, researchers and engineers of display and camera development (cameras, monitors, televisions, projectors, and head-mounted displays) may be concerned, for example, lighting engineers who develop novel light sources, researchers and engineers who develop color image optimization algorithms, software developers involved in color image processing, engineers of imaging and display systems, scientists involved in color vision research, designers of human interfaces and systems, application software developers for special effects in digital cinema postproduction, designers of lighting environments, postgraduate students in these domains, and anyone implementing a color management system. The material of this monograph can also be taken as a background reading for master's degrees in color image science and for researchers and design scientists, physicists, and engineers in the field of imaging technologies and their applications as well as university students in this field. The book may also be interesting for professionals working on software development for media and entertainment, video and film production, indoor architecture, and social aspects of home media technology as well as for graphics students and web developers.

Throughout the book, the term "self-luminous visual technologies" is used in the context of imaging technologies and illuminating technologies but printing technologies are excluded. Printing technologies and conventional photography represent a huge domain of knowledge that is out of the scope of this book. The issues of outdoor light sources such as street lighting or automotive lighting address the very complex mechanisms of human visual performance in the mesopic (twilight) luminance range; hence, these issues are also out of scope. In this book, the term "imaging technologies" is intended to mean all technologies that capture, digitalize, transmit, compress, transform, or display spectral, temporal, and spectral distributions of light, while the term "illuminating technologies" refers to all light source technologies used to illuminate reflecting or translucent objects to provide a visual environment consisting of the illuminated colored objects optimal for the user. The term "illuminating technologies" also covers the design of light sources used in digital or analog projectors or in backlit display technologies.

The book is organized into seven chapters. Chapter 1 is an introduction to color vision and self-luminous visual technologies. The question is what technology and which technological component is a specific feature of color vision relevant for and why. These features include retinal photoreceptor structure, spatial and temporal contrast sensitivity, color appearance perception, color difference perception, legibility, visibility, and conspicuity of colored objects, cognitive, preferred, harmonic, and emotional color, and the interindividual variability of color vision. Specific problems, features, and optimization potentials arising from the characteristics of color vision are described that are relevant for each technology including digital film and TV, cameras, color monitors, head-mounted displays, digital signage displays and large tiled displays, microdisplays, projectors, light sources of display backlighting, and general indoor illumination. At the end, Chapter 1 contains a table summarizing the perceptual, cognitive, and emotional features of the visual system and the corresponding technological challenge with links to specific sections later in the monograph.

Chapter 2 deals with the colorimetric and color appearance-based characterization of displays starting with a general description of display characterization models such as tone curve models, phosphor matrices, sRGB, and other characterization models. The additivity or independence of the monitor's color channels is an important criterion for an efficient characterization model. Multidimensional phosphor matrices and other methods are presented to reduce the colorimetric error arising from color channel interdependence. Methods are presented to test and ensure the spatial uniformity of the display to achieve accurate colors in every point. Also, the color predicted at a specific point should not depend on the color of other positions on the screen according to the important criterion of spatial independence. Methods to predict spatial interdependence are also described and the concept of viewing direction uniformity is presented that is especially important for liquid crystal displays. A paragraph is devoted to the miscellaneous visual artifacts, that is, the visually disturbing patterns arising from the imperfectness of display technology. The effect of the viewing environment including viewing conditions, viewing modes, and ambient light is described to be able to apply CIELAB, CIELUV, and CIECAM02 to a self-luminous display. Specific characterization models are described for the specific display technologies. Different projector light sources and backlighting light sources including LEDs are compared with relevance to the use of color filters, their white points, local dimming, and high dynamic range imaging. Finally, Chapter 2 also deals with the color appearance difference between small and large color stimuli, the so-called color size effect, and its mathematical modeling. Specifically, the color appearance of large color stimuli (e.g., 60–100° on a PDP) is different from small to medium size colors (i.e., below 20°). This effect is accounted for by an extension of CIELAB for the specific viewing condition of large self-luminous displays.

Chapter 3 deals with the ergonomic, memory-based, and preference-based enhancement of color displays. Ergonomic guidelines of visual displays and the objectives of color image reproduction are summarized. The principles of ergonomic color design are described for color displays to support effective work with the user interface appearing on the display based on the relationship among legibility, conspicuity, and visual search. A method of optimal use of chromaticity contrast to optimize search performance is presented together with the issues of chromaticity contrast preference and luminance contrast preference for young and elderly display users. In Chapter 3, long-term memory colors of familiar objects are located in color space and their intercultural differences are pointed out. A method to obtain a color image preference data set and a preference-based color image enhancement method are presented containing color image transforms that influence color image preference including the preferred white point, local contrast, global contrast, hue, and chroma.

Chapter 4 deals with the issues of color management and image quality improvement for cinema film and TV production. The components and systems of color management workflows in today's cinema film and TV production are described together with the components of the cinema production chain. An overview of camera technology and postproduction systems is given and the applicability of CIELAB and CIEDE2000 color difference formulas under the viewing conditions of TV and cinema production is dealt with. It is described how to apply the CIECAM02 color appearance model in the digital image processing system for motion picture films. Color gamut differences among cinema motion picture digital cameras, HDTV CRT monitors, film projectors, and DLP projectors are pointed out. It is shown how to exploit the spatial–temporal characteristics of color vision for digital TV, cinema, and camera development including how to optimize the resolution of digital motion picture cameras and how to compress motion pictures without impairing their perceived image quality. Methods of image quality evaluation and an image quality experiment are described. The important issue of watermarking algorithms for the protection of digital motion picture films is dealt with in detail. This is one of the most typical applications of human visual principles to advance display technology described in this book. The next issue of Chapter 4 concerns the optimum spectral power distributions for cinematographic light sources to optimize their color image rendering properties. Finally, the interesting question of visually evoked emotions in color motion pictures is dealt with. The question is how the technological parameters of video sequences influence or strengthen those parts of human emotions that are evoked by the visual appearance of the movie.

Chapter 5 deals with the different pixel architectures for self-luminous displays with three or more primary colors. To optimize the color gamut of the display, several factors are considered including the target colors to be covered by the optimized color gamut, color quantization, the number of primary colors, the white point, the issues of virtual primaries and technological constraints, and also the visually acceptable luminance ratio between a primary color and the white point. Several sets of optimum primary colors are presented together with the shape of their optimum color gamuts in color appearance space. In Chapter 5, a set of principles derived from human spatial color vision are also described to optimize the subpixel architectures of modern displays with three to seven primary colors including the requirements of minimal color fringe error, good modulation transfer function, isotropy, good luminance resolution, high aperture ratio, and large color gamut. Examples of actual subpixel architectures and color image rendering methods are also shown.

Chapter 6 deals with the optimization of color quality for indoor light sources of general lighting. The issues of color rendering and color quality are introduced including the psychological dimensions of color quality and their metrics such as the metrics used to quantify color fidelity. Visual color fidelity experiments are also described together with a set of color rendering prediction methods to be used for both conventional light sources and solid-state light sources such as LED lamps. Visual color harmony experiments, mathematical methods to predict the color harmony of different color combinations, and computational methods of color harmony rendering represent an interesting special case of color quality evaluation completed by several other factors of color quality such as perceived brightness, visual clarity, color discrimination capability, and color preference. Chapter 6 also shows the result of a principal component analysis of the latter factors followed by a description of a so-called "acceptability" experiment that deals with realistic colored test objects illuminated by different light sources of different color rendering properties of various color distributions. Finally, the effect of interobserver variability on the color quality of light sources is discussed.

Chapter 7 deals with today's emerging visual technologies including flexible displays, lasers, and LED displays with LED lifetime considerations. Color gamut extension algorithms for multi-primary displays are also described together with the temperature dependence of their color gamut by the example of a four-primary color sequential (RGCB) model LED display consisting of colored chip LEDs. Red and cyan colored chip LEDs were replaced by red and cyan phosphor-converted LEDs and the model computation was repeated. Chapter 7 also deals with the emerging technologies for indoor light sources including tunable LED lamps for accent lighting and a possible co-optimization of LED spectral power distributions for brightness and circadian rhythm. Additional issues addressed in Chapter 7 include the accentuation of different aspects of color quality, the use of new phosphor blends, and the implications of color constancy for light source design. Finally, a summary of the whole book and an outlook for future research is given.

This book contains material from various sources including the authors' articles previously published in *Color Research and Application, Displays*, the German journal *Licht*, the *Journal of Electronic Imaging*, Proceedings of AIC, CGIV, and CIE conferences, the German journal *FKT (TV and Cinema Technology)*, and the authors' lecture qualification theses. This material has been organized and is now presented in a consistent and more readable way because the material has been reviewed very thoroughly and then reformulated. The authors' original ideas have been reconsidered, refined, and further explained to include several new insights from the lighting engineer's point of view, also in the view of numerous recent literature items including patent publications. Complex interdependences across the material have been pointed out. Thus, this book provides a more detailed, more comprehensive, more thorough, and more systematic treatment of the subject than the original articles. In addition to this, the book contains numerous new ideas and a lot of new material published in the sections of this monograph for the first time. To obtain this latter material, we gratefully acknowledge the help from the coworkers of the Laboratory of Lighting Technology of the Technische Universität Darmstadt, especially Mr. Marvin Böll, Mr. Stefan Brückner, Ms. Nathalie Krause, Mr. Wjatscheslaw Pepler, and Mr. Quang Vinh Trinh, in no particular order. The authors would like to thank the colleagues and the diploma students of the company Arnold & Richter (Munich, Germany) for the cooperation during the development of the film scanner, film recorder, and the digital cinema camera with all related research and development aspects, especially Mr. Franz Kraus, Dr. Johannes Steurer, Dr. Achim Oehler, Dr. Peter Geissler, Mr. Michael Koppetz, Mr. Joachim Holzinger, Mr. Harald Brendel, Mr. Christian Bueckstuemmer, Ms. Doreen Wunderlich, Mr. Alexander Vollstaedt, Dr. Sebastian Kunkel, Mr. Ole Gonschorek, Mr. Andreas Kraushaar, Mr. Constantin Seiler, Ms. Christina Hacker, and Mr. Nils Haferkemper.

*P. Bodrogi*
*T.Q. Khanh*

## About the Authors

**Peter Bodrogi** is a senior research fellow at the Laboratory of Lighting Technology of the Technische Universität Darmstadt in Darmstadt, Germany. He graduated in Physics from the Loránd Eötvös University of Budapest, Hungary. He obtained his PhD degree in Information Technology from the University of Pannonia in Hungary. He has co-authored numerous scientific publications and invented patents about color vision and self-luminant display technology. He has received several scientific awards including a Research Fellowship of the Alexander von Humboldt Foundation, Germany, and the Walsh-Weston Award, Great Britain. He has been member of several Technical Committees of the International Commission of Illumination (CIE).

**Tran Quoc Khanh** is University Professor and Head of the Laboratory of Lighting Technology at the Technische Universität Darmstadt in Darmstadt, Germany. He graduated in Optical Technologies, obtained his PhD degree in Lighting Engineering, and his degree of lecture qualification (habilitation) for his thesis in Colorimetry and Colour Image Processing from the Technische Universität Ilmenau, Germany. He has gathered industrial experience as a project manager by ARRI Cine Technik in Munich, Germany. He has been the organizer of the well-known series of international symposia for automotive lighting (ISAL) in Darmstadt, Germany, and is a member of several Technical Committees of the International Commission of Illumination (CIE).

# 1
# Color Vision and Self-Luminous Visual Technologies

Color vision is a complicated phenomenon triggered by visible radiation from the observer's environment imaged by the eye on the retina and interpreted by the human visual brain [1]. A visual display device constitutes an interface between a supplier of electronic information (e.g., a television channel or a computer) and the human observer (e.g., a person watching TV or a computer user) receiving the information stream converted into light. The characteristics of the human component of this interface (i.e., the features of the human visual system such as visual acuity, dynamic luminance range, temporal sensitivity, color vision, visual cognition, color preference, color harmony, and visually evoked emotions) cannot be changed as they are determined by biological evolution.

Therefore, to obtain an attractive and usable interface, the hardware and software features of the display device (e.g., size, resolution, luminance, contrast, color gamut, frame rate, image stability, and built-in image processing algorithms) should be optimized to fit the capabilities of human vision and visual cognition. Accordingly, in this chapter, the most relevant characteristics of human vision – especially those of color vision – are introduced with special respect to today's different display technologies.

The other aim of this chapter is to present a basic overview of some essential concepts of colorimetry [2] and color science [3–5]. Colorimetry and color science provide a set of numerical scales for the different dimensions of color perception (so-called correlates for, for example, the perceived lightness or saturation of a color stimulus). These numerical correlates can be computed from the result of physical light measurement such as the spatial and spectral light power distributions of the display. Using these numerical correlates, the display can be evaluated and optimized systematically by measuring the spectral and spatial power distributions of their radiation – without cumbersome and time-consuming direct visual evaluations.

*Illumination, Color and Imaging: Evaluation and Optimization of Visual Displays*, First Edition.
Peter Bodrogi and Tran Quoc Khanh.
© 2012 Wiley-VCH Verlag GmbH & Co. KGaA. Published 2012 by Wiley-VCH Verlag GmbH & Co. KGaA.

## 1.1
## Color Vision Features and the Optimization of Modern Self-Luminous Visual Technologies

This section summarizes the most important features of color vision for the evaluation and optimization of self-luminous color displays including the photoreceptor structure of the retina, the spatial and temporal contrast sensitivity of the human visual system, color appearance and color difference perception, the components of visual performance and ergonomics (legibility, visibility, and conspicuity of colored objects), and certain features arising at a later stage of human visual information processing such as cognitive, preferred, harmonic, and emotional color phenomena. The important issue of interindividual variability of color vision will also be dealt with in this section.

### 1.1.1
### From Photoreceptor Structure to Colorimetry

Human color vision is trichromatic [1]. This feature has its origin in the retinal photoreceptor structure consisting of three types of photoreceptors that are active at daytime light intensity levels: the L-, M-, and S-cones. Rods constitute a further type of retinal photoreceptors but as they are responsible for nighttime vision and partially for twilight viewing conditions, they are out of the scope of this book. Displays should ensure a high enough general luminance level (e.g., higher than 50–100 cd/m$^2$, depending on the chromaticity of the stimulus) for the three types of cones to operate in an optimum state for the best possible perception of colors. Generally, above a luminance of about 100 cd/m$^2$, rods produce no signal for further neural processing and it is possible to predict the matching and the appearance of colors from the cone signals only.

L-, M-, and S-cones constitute a characteristic retinal cone mosaic. The central (rod-free) part of the cone mosaic can be seen Figure 1.1.

As can be seen from Figure 1.1, the inner area of the central part (subtending a visual angle of about 0.3° or 100 μm) is free of S-cones resulting in the so-called small-field tritanopia, that is, the insensitivity to bluish light for very small central viewing fields. There are on average 1.5 times as many L-cones as M-cones in this region of the retina [1]. L- and M-cones represent 93% of all cones, while S-cones represent the rest (7%).

Spectral sensitivities of the three types of cones [1] are depicted in Figure 1.2, while a more extensive database of the characteristic functions describing human color vision can be found on the Web [1)]. These cone sensitivities were measured at the cornea of the eye; hence, they include the filtering effect of the ocular media and the central yellow pigment on the retina (so-called macular pigment). Sensitivity curves

---

1) Web Database of the Color & Vision Research Laboratory, Institute of Ophthalmology, University College London, London, UK, www.cvrl.org

## 1.1 Color Vision Features and the Optimization of Modern Self-Luminous Visual Technologies

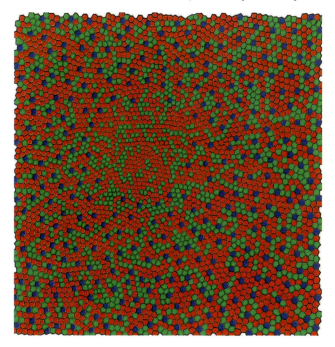

**Figure 1.1** The cone mosaic of the rod-free inner fovea, that is, the central part of the retina subtending about 1°, that is, about 300 μm. Red dots: long-wavelength sensitive cone photoreceptors (L-cones). Green dots: middle-wavelength sensitive cones (M-cones). Blue dots: short-wavelength sensitive cones (S-cones). Source: Figure 1.1 from Sharpe, L.T., Stockman, A., Jägle, H., and Nathans, J. (1999) Opsin genes, cone photopigments, color vision and color blindness, in Ref. [1], pp. 3–51. Reproduced with permission from Cambridge University Press.

were adjusted to the average relative numbers of the L-, M-, and S-cones, that is, 56, 37, and 7%, respectively.

As can be seen from Figure 1.2, the spectral bands of the L-, M-, and S-cones provide three initial color signals like the CCD or CMOS array of a digital camera. From these initial color signals, the retina computes two chromatic signals (or chromatic channels), $L-M$ (red–green opponent channel) and $S-(L+M)$ (yellow–blue opponent channel), and one achromatic signal, $L+M$. The latter signal is called luminance signal or luminance channel. As can be seen from Figure 1.2, the maxima of the L-, M-, and S-sensitivity curves in Figure 1.2 occur at 566, 541, and 441 nm, respectively [1]. Note that these spectral sensitivity curves are expressed in quantal units. To express them in energy units, the logarithm of the wavelength should be added to each value and the curve renormalized [1].

For stimuli subtending a visual angle of 1–4°, the spectral sensitivity of the luminance channel is usually approximated by the $V(\lambda)$ function, the spectral luminous efficiency function for photopic vision also defining the CIE standard photometric observer for photopic vision (the basis of photometry) [2]. The $V(\lambda)$

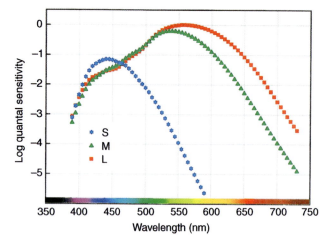

**Figure 1.2** Spectral sensitivities of the three types of cones measured in quantal units (to obtain energy units, add log($\lambda$) to each value and renormalize [1]) as measured at the cornea of the eye, thus containing the filtering effect of the ocular media and the macular pigment. Sensitivities adjusted to average relative numbers of L-, M-, and S-cones (i.e., 56, 37, and 7%, respectively). Source: Figure 1.1 from Sharpe, L.T., Stockman, A., Jägle, H., and Nathans, J. (1999) Opsin genes, cone photopigments, color vision and colorblindness, in Ref. [1], pp. 3–51. Reproduced with permission from Cambridge University Press.

function seriously underestimates the spectral sensitivity of the luminance channel at short wavelengths[1].

Due to historical reasons, the spectral sensitivities of the three types of cones (Figure 1.2) are currently not widely used to characterize the radiation (so-called color stimulus) reaching the human eye and resulting in color perceptions. Instead of that, for color stimuli subtending a visual angle of 1–4°, the so-called color matching functions of the CIE 1931 standard colorimetric observer [2] are applied, while interindividual variability cannot be neglected (see Section 1.1.6). These color matching functions are denoted by $\bar{x}(\lambda)$, $\bar{y}(\lambda)$, $\bar{z}(\lambda)$ and constitute the basis of standard colorimetry. At this point, we would like to direct the attention of the interested reader to the recent updates of photometry and colorimetry[1] [6].

To describe the color matching of more extended stimuli, that is, for visual angles greater than 4° (e.g., 10°), the so-called CIE 1964 standard colorimetric observer is recommended [2]. These color matching functions are denoted by $\bar{x}_{10}(\lambda)$, $\bar{y}_{10}(\lambda)$, $\bar{z}_{10}(\lambda)$. Latter functions are compared with the $\bar{x}(\lambda)$, $\bar{y}(\lambda)$, $\bar{z}(\lambda)$ functions in Figure 1.3.

The aim of colorimetry is to predict which spectral power distributions result in the same color appearance (so-called matching colors) in a single (standard) viewing condition, that is, directly juxtaposed 2° stimuli imaged to the central retina for an average observer of normal color vision. In this sense, two matching colors have the same so-called *XYZ* tristimulus values. *XYZ* tristimulus values are recommended to be the basis of CIE colorimetry [2].

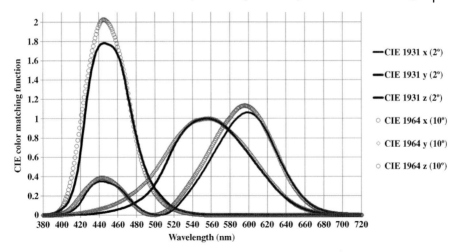

**Figure 1.3** Black curves: color matching functions of the CIE 1931 standard colorimetric observer [2][1)] denoted by $\bar{x}(\lambda)$, $\bar{y}(\lambda)$, $\bar{z}(\lambda)$ intended to describe the matching of color stimuli subtending a visual angle of 1–4°. Open gray circles: color matching functions of the CIE 1964 standard colorimetric observer [2][1)] denoted by $\bar{x}_{10}(\lambda)$, $\bar{y}_{10}(\lambda)$, $\bar{z}_{10}(\lambda)$ intended to describe the matching of color stimuli subtending greater than 4°.

To compute the XYZ tristimulus values, the spectral radiance distribution of the color stimulus $L(\lambda)$ measured by a spectroradiometer on a color patch (a color sample reflecting the light from a light source or a self-luminous light emitting surface) shall be multiplied by one of the three color matching functions ($\bar{x}(\lambda)$, $\bar{y}(\lambda)$, $\bar{z}(\lambda)$), integrated in the entire visible spectrum (360–830 nm), and multiplied by a constant $k$ (see Equation 1.1).

$$X = k \int_{360\,\text{nm}}^{830\,\text{nm}} L(\lambda)\bar{x}(\lambda)\,d\lambda$$

$$Y = k \int_{360\,\text{nm}}^{830\,\text{nm}} L(\lambda)\bar{y}(\lambda)\,d\lambda \quad (1.1)$$

$$Z = k \int_{360\,\text{nm}}^{830\,\text{nm}} L(\lambda)\bar{z}(\lambda)\,d\lambda$$

For reflecting color samples, the spectral radiance of the stimulus ($L(\lambda)$) is equal to the spectral reflectance ($R(\lambda)$) of the sample multiplied by the spectral irradiance from the light source illuminating the reflecting sample ($E(\lambda)$). Equation 1.2 expresses this for diffusely reflecting materials.

$$L(\lambda) = \frac{R(\lambda)E(\lambda)}{\pi} \quad (1.2)$$

The value of $k$ is computed according to Equation 1.3 [2].

$$k = \frac{100}{\int_{360\,nm}^{830\,nm} L(\lambda)\bar{y}(\lambda)d\lambda} \tag{1.3}$$

As can be seen from Equation 1.3, for reflecting color samples, the constant $k$ is chosen so that $Y = 100$ for ideal white objects with $R(\lambda) \equiv 1$.

For self-luminous objects (such as self-luminous displays), the value of $k$ can be chosen to be 683 lm/W [2]. Then the value of $Y$ will be equal to the luminance of the self-luminous object. In case of a self-luminous display, the peak white of the display is often visible as a background or as a white frame around an image. In this case, it makes sense to compute the relative tristimulus values of the color stimulus appearing on the self-luminous display by dividing every tristimulus value of any color stimulus ($X$, $Y$, and $Z$) by the $Y$ value of peak white (i.e., by peak white luminance) and multiplying by 100. The CIECAM02 color appearance model anticipates such relative tristimulus values (see Section 2.1.9).

For color stimuli with visual angles greater than $4°$, the tristimulus values $X_{10}$, $Y_{10}$, and $Z_{10}$ can be computed substituting $\bar{x}(\lambda)$, $\bar{y}(\lambda)$, $\bar{z}(\lambda)$ by $\bar{x}_{10}(\lambda)$, $\bar{y}_{10}(\lambda)$, $\bar{z}_{10}(\lambda)$ in Equation 1.1. As can be seen from Figure 1.3, the two sets of color matching functions, that is, $\bar{x}(\lambda)$, $\bar{y}(\lambda)$, $\bar{z}(\lambda)$ and $\bar{x}_{10}(\lambda)$, $\bar{y}_{10}(\lambda)$, $\bar{z}_{10}(\lambda)$, differ significantly. The consequence is that two matching color stimuli subtending a visual angle of, for example, $1°$ generally will not match if their size is increased to, for example, $10°$.

The so-called chromaticity coordinates ($x$, $y$, $z$) are defined by Equation 1.4.

$$x = \frac{X}{X+Y+Z}, \quad y = \frac{Y}{X+Y+Z}, \quad z = \frac{Z}{X+Y+Z} \tag{1.4}$$

The diagram of the chromaticity coordinates $x$, $y$ is called the CIE 1931 chromaticity diagram or the CIE ($x$, $y$) chromaticity diagram [2]. Figure 1.4 illustrates how color perception changes across the $x$, $y$ diagram.

As can be seen from Figure 1.4, chromaticities are located inside the curved boundary of quasi-monochromatic radiations of different wavelengths (so-called spectral locus) and the purple line. White tones are positioned in the middle range of the diagram with increasing saturation toward the spectral locus. Perceived hue changes (purple, red, yellow, green, cyan, and blue) when going around the region of white tones in the middle of the $x$, $y$ diagram.

### 1.1.2
### Spatial and Temporal Contrast Sensitivity

The user of the display would like to discern visual objects such as letters, numbers, or symbols from their background and perceive the fine spatial structure of objects, for example, analyze the colored textures of different objects in a photorealistic image, discern a thin colored line of a diagram with colored background, or recognize a complex Asian letter based on its composition of tiny strokes. To be able to do so, the user needs an appropriate display hardware and image rendering software respecting the spatial frequency characteristics of the achromatic (L + M) and chromatic (L − M, S − (L + M)) channels of the human visual system.

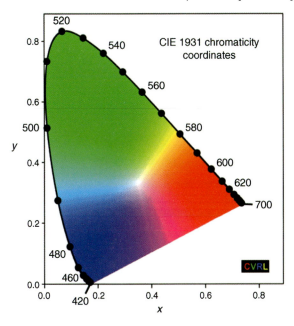

**Figure 1.4** Illustration of how color perception changes across the CIE (x, y) chromaticity diagram [2]. The curved boundary of colors with three-digit numbers (wavelengths in nanometer units) represents the locus of monochromatic (i.e., most saturated) radiation. Source: Figure 7 from Ref. [7]. Reproduced with permission from Wiley-VCH Verlag GmbH & Co. KGaA.

To understand these spatial frequency characteristics, it is essential to learn how the human visual system analyzes the spatial structures of the retinal image. L-, M-, and S-cone signals are processed by different cell types of the retina including the so-called ganglion cells. Ganglion cells process the signals from several cones located inside their receptive fields. Receptive fields of ganglion cells are built to be able to amplify the spatial contrasts (i.e., edges) of the image in the following way.

Every receptive field has a circular center and a concentric circular surround. Stimulation of the center and the surround exhibits opposite firing reactions of the ganglion cell: it is firing when the stimulus is in the center ("on-center cell"), while it is inhibited when the stimulus is in the surround. The other type of ganglion cell ("off-center cell") is inhibited when the stimulus is in the center and firing when the stimulus is in the surround. This way, spatially changing stimuli (contrasts or edges) increase firing, while spatially homogeneous stimuli generate only a minor response (see Figure 1.5).

On the human retina, achromatic contrast (i.e., spatial changes of the L + M signal) is detected according to the principle of Figure 1.5. Similar receptive field structures produce the chromatic signals for chromatic contrast, that is, spatial changes of the L − M or S − (L + M) signals. But in this case, the spectral sensitivity of the center differs from the spectral sensitivity of the surround due to the different combinations of the L-, M-, and S-cones in the center and in the surround. This receptive field structure is called double opponent because there is a spatial

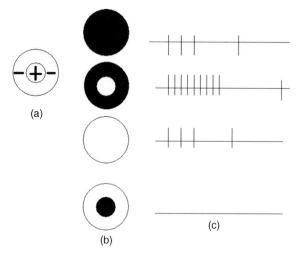

**Figure 1.5** (a) Schematic representation of the receptive field of an "on-center" ganglion cell: +, center; −, surround. (b) Black: no light; white: light stimulus; from top to bottom: (1) no light over the whole receptive field; (2) contrast – light on the center, no light on the surround; (3) light over the whole receptive field; (4) light on the surround. (c) Firing rate, from top to bottom: weak, strong, weak, no response [8].

opponency (center/surround) and a cone opponency (L/M or S/(L + M)). Table 1.1 summarizes its possible cone signal combinations.

To produce chromatic signals for homogeneous color patches, a so-called single-opponent receptive field (with cone signal opponency but no spatial opponency) is responsible. In this kind of receptive field, the center and the surround (containing the combinations of Table 1.1) overlap in space [9].

It is the size and sensitivity of the receptive fields and the spatial aberrations of the eye media (cornea, lens, and vitreous humor) that determine the spatial frequency characteristics of the achromatic and chromatic channels [8]. In practical applications including self-luminous displays, the basic question is how much achromatic or

**Table 1.1** Possible combinations of L-, M-, and S-cone signals in the center and in the surround of the receptive field structure producing the signals of the chromatic channels to detect chromatic contrast (chromatic edges).

| Chromatic channel | Cell type | Center | Surround |
|---|---|---|---|
| L − M | On-center | +L | −M |
| L − M | Off-center | −L | +M |
| L − M | On-center | +M | −L |
| L − M | Off-center | −M | +L |
| S − (L + M) | On-center | +S | −(L + M) |
| S − (L + M) | Off-center | +(L + M) | −S |

chromatic contrast is needed to detect a visual object of a given size corresponding to a given spatial frequency (see Sections 3.3 and 4.4). Size is usually expressed in degrees of visual angle, while spatial frequency is expressed in cycles per degree (cpd) units. For example, 10 cpd means that there are 10 pairs of thin black and white lines within a degree of visual angle.

Contrast (C) can be measured either by the contrast ratio, that is, the signal value (e.g., L + M or L − M) of the object ($S_O$) divided by the signal value of its background ($S_B$) (i.e., $S_O/S_B$), or by the so-called Michelson contrast $(S_O - S_B)/(S_O + S_B)$. Contrast sensitivity (CS) is defined as the reciprocal value of the threshold value of contrast needed to detect the object. Achromatic contrast sensitivity is a band-pass function of spatial frequency increasing up to about 3–5 cpd and then decreasing toward high spatial frequencies. For about 40 cpd (corresponding to a visual object of about 1 arcmin) or above, achromatic contrast sensitivity is equal to zero. This means that it is no use increasing the contrast (even up to infinity, that is, black on white) if the object is smaller than about 1 arcmin. This is the absolute limit of (foveal) visual acuity. Below this limit, generally more contrast is needed for higher spatial frequencies to be able to detect an object, as can be seen from Figure 1.6.

In Figure 1.6, the spatial frequency of the pattern increases from top to bottom and contrast increases from left to right. For each spatial frequency, there is a horizontal threshold position where the pattern can just be detected. These visual threshold positions correspond to the achromatic contrast sensitivity function plotted in Figure 1.7.

As can be seen from Figure 1.7, achromatic contrast sensitivity is higher for higher retinal illuminance levels (e.g., 2200 Td) because at such a high level, the visual system is operating in its optimum (i.e., truly photopic) state of adaptation. The conventional unit of retinal illuminance is the troland (Td), the product of photopic luminance in $cd/m^2$ and the pupil area in $mm^2$. Replacing the grayscale sinusoidal pattern of Figure 1.6 by pure chromatic transition patterns (without achromatic contrast), the contrast sensitivity of the chromatic channels becomes visible. An example can be seen in Figure 3.19b. The latter example shows a combination of L − M contrast and S − (L + M) contrast without any achromatic contrast. Chromatic contrast sensitivity functions of the L − M and S − (L + M) channels are compared with the achromatic contrast sensitivity function (at a high retinal illuminance level) in Figure 1.8.

As can be seen from Figure 1.8, while the L + M (luminance) contrast sensitivity function exhibits band-pass nature, chromatic functions are low-pass functions. Chromatic contrast sensitivity is limited to a narrow spatial frequency range up to 8 cpd. Even for lower spatial frequencies, chromatic contrast sensitivity is low compared to achromatic contrast sensitivity (see Section 3.3). This knowledge is exploited to develop image and video compression algorithms (e.g., JPEG, MPEG) for digital still and motion images and the dataflow in digital TV and cinema (see Section 4.4.3). The low contrast sensitivity of the S − (L + M) channel can be used to watermark video sequences without noticing it visually (see Section 4.4.4).

Concerning temporal contrast sensitivity, increasing the temporal frequency (measured in Hz units) of temporally modulated stimuli, first flicker is perceived

**Figure 1.6** Demonstration of achromatic contrast sensitivity (so-called Campbell–Robson contrast sensitivity chart). The spatial frequency of the pattern increases from top to bottom. Achromatic contrast increases from left to right. For each spatial frequency, there is a horizontal threshold position where the pattern can just be detected corresponding to the achromatic contrast sensitivity function plotted in Figure 1.7. Try to reconstruct this position as a function of spatial frequency visually and draw the contrast sensitivity function of your own eye [8].

and then, for higher temporal frequencies, a constant stimulus appears. The transition point between the two is called critical flicker frequency (CFF) playing an important role in the visual ergonomics of displays (see Section 3.1). The temporal contrast sensitivity of the achromatic (luminance) channel exhibits band-pass nature, while the temporal contrast sensitivity of the chromatic channels is a low-pass function (see Figure 1.9).

As can be seen from Figure 1.9, the temporal contrast sensitivity of the chromatic channels is much less than the temporal contrast sensitivity of the achromatic channel. Critical flicker frequency of the chromatic channels is equal to about 6–7 Hz, while the critical flicker frequency of the achromatic channel is equal to about

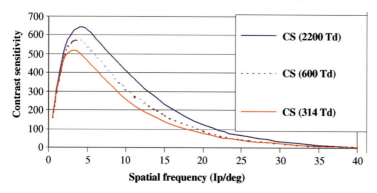

**Figure 1.7** Achromatic contrast sensitivity functions for different values of retinal illuminance level (in Td units). Retinal illuminance corresponds to the luminance of the stimulus (in cd/m$^2$) scaled by the pupil area (in mm$^2$). Abscissa: spatial frequency in cpd units; ordinate: achromatic contrast sensitivity (relative units) [8].

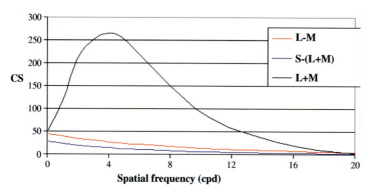

**Figure 1.8** Chromatic contrast sensitivity functions of the L − M and S − (L + M) channels compared with the achromatic contrast sensitivity function (at a high retinal illuminance level). Abscissa: spatial frequency in cpd units; ordinate: contrast sensitivity (relative units) [8, 10].

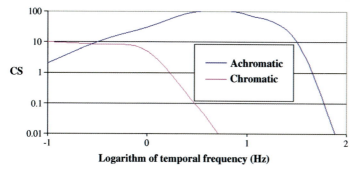

**Figure 1.9** Temporal contrast sensitivity functions of the achromatic (luminance) and chromatic channels. Abscissa: temporal frequency of the altering stimulus in Hz units; ordinate: contrast sensitivity (relative units) [8].

50–70 Hz depending on luminance level and stimulus eccentricity. This knowledge is essential to measure the $V(\lambda)$ function to be able to "switch off" the influence of the more sluggish chromatic channels. Modern displays and film projectors use high frame rates to avoid any flicker artifact even for higher luminance levels and peripheral perception (see Section 4.4.1).

### 1.1.3
**Color Appearance Perception**

The description of color stimuli in the system of tristimulus values ($X$, $Y$, and $Z$) results in a nonuniform and nonsystematic representation of the color perceptions corresponding to these stimuli. More specifically, the relevant psychological attributes of perceived colors (i.e., perceived lightness, brightness, redness–greenness, yellowness–blueness, hue, chroma, saturation, and colorfulness) cannot be expressed in terms of $XYZ$ values directly. To model color perception, numerical correlates have to be derived from the $XYZ$ values of the stimulus for each attribute (as mentioned at the beginning of this chapter).

Hue is the attribute of a visual sensation according to which a color stimulus appears to be similar to the perceived colors red, yellow, green, and blue, or a combination of two of them [11]. Brightness is the attribute of a color stimulus according to which it appears to emit more or less light [11]. Lightness is the brightness of a color stimulus judged relative to the brightness of a similarly illuminated reference white (appearing white or highly transmitting) [3].

Colorfulness is the attribute of a color stimulus according to which the stimulus appears to exhibit more or less chromatic color. For a given chromaticity, colorfulness generally increases with luminance [12]. In an indoor environment, observers tend to assess the *chroma* of surface colors. The perceived attribute chroma refers to the colorfulness of the color stimulus judged in proportion to the brightness of the reference white [3].

Saturation is the colorfulness of a stimulus judged in proportion to its own brightness [11]. A perceived color can be very saturated without exhibiting a high level of chroma. For example, a deep red sour cherry is quite saturated but it exhibits less chroma because the sour cherry is colorful compared to its (low) own brightness but it is not so colorful in comparison to the brightness of the reference white. Figure 1.10 illustrates the three perceived attributes, hue, chroma, and lightness.

The numerical scales modeling the above attributes of color perception (so-called numerical correlates) should be perceptually uniform. This means that equal differences of their scales should correspond to equal perceptual differences. Otherwise, they are not useful for practice. If the above-mentioned numerical correlates are computed, then the color stimuli can be arranged in a three-dimensional space, the so-called color space.

In a color space, the three perpendicular axes and certain angles and distances carry psychologically relevant meanings related to the perceived color attributes. Hence, these color spaces are very useful tools of color display design and evaluation, including all aspects of color perception, cognition, preference, and emotion. For

## 1.1 Color Vision Features and the Optimization of Modern Self-Luminous Visual Technologies

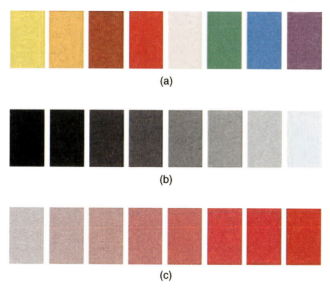

**Figure 1.10** Illustration of three attributes of perceived color: (a) changing hue, (b) changing lightness, and (c) changing chroma. Reproduction of Figure 1 from Ref. [13] with permission from *Color Research and Application*.

example, preferred or ergonomic colors of a display user interface can be easily represented and understood if they are specified in such a color space. A schematic illustration of the structure of a color space can be seen in Figure 1.11.

As can be seen from Figure 1.11, lightness increases from black to white from the bottom to the top along the gray lightness scale in the middle of color space. At every

**Figure 1.11** Schematic illustration of the general structure of color space. Lightness increases from black to white from the bottom to the top along the gray lightness scale in the middle. Chroma increases from the gray scale toward the outer colors of high chroma. The perceptual attribute of hue varies when rotating the image plane around the gray axis in space.

lightness level, chroma increases from the gray scale toward the most saturated outer colors. The perceptual attribute of hue varies when rotating the image plane around the gray axis in space.

CIE colorimetry recommends two such coordinate systems, CIELAB and CIELUV, the so-called CIE 1976 uniform color spaces [2]. Computations of the approximate numerical correlates of the perceived color attributes in these two uniform color spaces start from the XYZ values of the color stimulus and the XYZ values of a specified reference white color stimulus ($X_n$, $Y_n$, $Z_n$).

In many cases, the reference white is an object color, that is, the perfect reflecting diffuser illuminated by the same light source as the test object. The application of color spaces to self-luminous displays is described in Section 2.1.9. Although CIELAB and CIELUV represent standard practice today, their defining equations are repeated below. The CIELAB color space is defined by Equation 1.5.

$$L^* = 116f(Y/Y_n) - 16$$
$$a^* = 500[f(X/X_n) - f(Y/Y_n)] \quad (1.5)$$
$$b^* = 200[f(Y/Y_n) - f(Z/Z_n)]$$

In Equation 1.5, the function $f$ is defined by Equation 1.6.

$$f(u) = u^{1/3} \qquad \text{if} \quad u > (24/116)^3$$
$$f(u) = (841/108)u + (16/116) \quad \text{if} \quad u \leq (24/116)^3 \quad (1.6)$$

In CIELAB, output quantities (approximate correlates of the perceived attributes of color) include $L^*$ (CIE 1976 lightness of Equation 1.5), CIELAB chroma ($C^*_{ab}$), and CIELAB hue angle ($h_{ab}$). The quantities $a^*$ and $b^*$ in Equation 1.5 can be interpreted as rough correlates of perceived redness–greenness (red for positive values of $a^*$) and perceived yellowness—blueness (yellow for positive values of $b^*$). $L^*$, $a^*$, and $b^*$ constitute the three orthogonal axes of CIELAB color space; compare with the illustration of color space shown in Figure 1.11. Equation 1.7 shows how to calculate $C^*_{ab}$ and $h_{ab}$ from $a^*$ and $b^*$.

$$C^*_{ab} = \sqrt{a^{*2} + b^{*2}}$$
$$h_{ab} = \arctan(b^*/a^*) \quad (1.7)$$

Similar quantities are defined in the other color space, CIELUV, as well. The value of $L^*$ of the CIELUV color space is identical to the value of $L^*$ of the CIELAB color space. The rectangular coordinates $u^*$ and $v^*$ are computed by Equation 1.8.

$$u^* = 13L^*(u' - u'_n)$$
$$v^* = 13L^*(v' - v'_n) \quad (1.8)$$

In Equation 1.8, the $u'$, $v'$ values of the so-called CIE 1976 uniform chromaticity scale diagram (UCS diagram or $u'$, $v'$ diagram) are defined starting from the chromaticity coordinates $x$, $y$ defined by Equation 1.4 (see Equation 1.9). The subscript $n$ in Equation 1.8 refers to the reference white.

$$u' = 4x/(-2x + 12y + 3)$$
$$v' = 9y/(-2x + 12y + 3) \quad (1.9)$$

In the CIELUV color space ($L^*$, $u^*$, $v^*$), CIELUV chroma ($C^*_{uv}$) and CIELUV hue angle ($h_{uv}$) are defined by substituting $a^*$ and $b^*$ by $u^*$ and $v^*$ in Equation 1.7, respectively. In addition to these, CIELUV also defines a numerical correlate of perceived *saturation*, $s_{uv}$, according to the underlying perceptually uniform UCS chromaticity diagram. This is shown in Equation 1.10.

$$s_{uv} = 13\sqrt{(u'-u'_n)^2 + (v'-v'_n)^2} \tag{1.10}$$

It is important to read the notes of the CIE publication [2] carefully stating (among others) that CIELUV and CIELAB are "intended to apply to comparisons of differences between object colors of the same size and shape, viewed in identical white to middle-gray surroundings by an observer photopically adapted to a field of chromaticity not too different from that of average daylight." To compare the color appearance of color stimuli viewed in different viewing conditions including tungsten light/daylight, average/dark/dim surround luminance levels, or different backgrounds, so-called color appearance models shall be used such as the CIECAM02 color appearance model [14] (see also Section 2.1.9). The CIECAM02 model computes numerical correlates for all perceived attributes of color. CIECAM02 correlates represent an improved model of color perception compared to CIELAB or CIELUV.

### 1.1.4
### Color Difference Perception

A disadvantage of the CIE ($x$, $y$) chromaticity diagram [2] is that it is perceptually not uniform. In Figure 1.4, observe that a distance in the green region of the diagram represents a less change of perceived chromaticness than the same distance in the blue–purple region. The so-called MacAdam ellipses [15] quantify this effect (see Figure 1.12). Roughly speaking, perceived chromaticity differences are hardly noticeable inside the ellipse (for a more precise definition of the MacAdam ellipses, see Ref. [15]). Note that the ellipses of Figure 1.12 are magnified 10 times.

As can be seen from Figure 1.12, MacAdam ellipses are large in the green region of the CIE ($x$, $y$) chromaticity diagram while they are small in blue–purple region and the orientation of the ellipses also changes. To overcome these difficulties, the $x$ and $y$ axes were distorted so as to make identical circles from the MacAdam ellipses and this resulted in the $u'$, $v'$ diagram of Equation 1.9.

The $u'$, $v'$ diagram is perceptually uniform (at least approximately, in the sense that equal distances represent equal changes of perceived chromaticity in any part of the diagram) if the relative luminance difference of the two color stimuli is small, for example, $\Delta Y < 0.5$. This means that the $u'$, $v'$ diagram is useful to evaluate differences of perceived chromaticness without lightness differences.

Perceived total color differences between two color stimuli ($\Delta E^*_{ab}$ and $\Delta E^*_{uv}$) are modeled by the Euclidean distances between them. Euclidean distances shall be computed in the rectangular CIELAB ($L^*$, $a^*$, $b^*$) and CIELUV ($L^*$, $u^*$, $v^*$) color spaces. Lightness, chroma, and hue angle differences of two color stimuli ($\Delta L^*$, $\Delta C^*_{ab}$, and $\Delta h_{ab}$) can be computed by subtracting the lightness, chroma, and hue angle values of the two color stimuli.

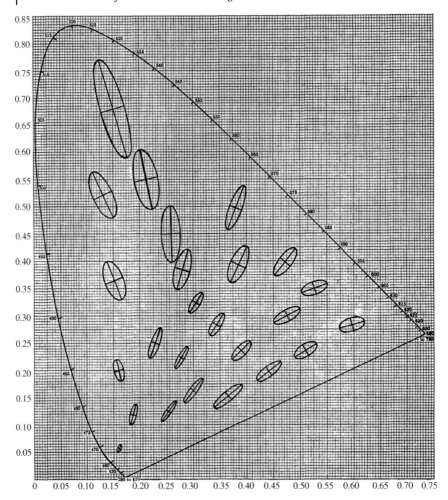

**Figure 1.12** MacAdam ellipses [15] in the CIE (x, y) chromaticity diagram. Abscissa: chromaticity coordinate x; ordinate: chromaticity coordinate y. Roughly speaking, perceived chromaticity differences are not noticeable inside the ellipses. For a more precise definition of the MacAdam ellipses, see Ref. [15]. Ellipses are magnified 10 times. Reproduced from Ref. [15] with permission from the *Journal of the Optical Society of America*.

Hue differences ($\Delta H^*_{ab}$) must not be confused with hue angle differences ($\Delta h_{ab}$). Hue differences include the fact that the same hue change results in a large color difference for large chroma and in a small color difference for small chroma (i.e., in the neighborhood of the CIELAB or CIELUV $L^*$ axis). CIELAB hue difference is defined by Equation 1.11. In Equation 1.11, $\Delta H^*_{ab}$ has the same sign as $\Delta h_{ab}$.

$$\Delta H^*_{ab} = \sqrt{(\Delta E^*_{ab})^2 - (\Delta L^*)^2 - (\Delta C^*_{ab})^2} \tag{1.11}$$

However, CIELAB and CIELUV color differences exhibit perceptual nonuniformities depending on the region of color space (e.g., reddish or bluish colors), color difference magnitude (small, medium, or large color differences), and miscellaneous viewing parameters including sample separation, texture, and background color [16]. The CIEDE2000 total color difference formula corrects the nonuniformity of CIELAB for small color differences under a well-defined set of reference conditions [17]. The CIEDE2000 formula introduces weighting functions for the hue, chroma, and lightness components of CIELAB total color difference and a factor to account for hue–chroma interaction.

Recently, uniform color spaces based on the CIECAM02 color appearance model were introduced [18] to describe small (CIECAM02-SCD) and large color differences (CIECAM02-LCD). An intermediate space (CIECAM02-UCS) was also introduced. Recently, the superior performance of CIECAM02-UCS was corroborated in visual experiments on color rendering [19, 20] (see Section 6.2.1).

### 1.1.5
### Cognitive, Preferred, Harmonic, and Emotional Color

Color perceptions undergo further processing in the visual brain, giving rise to cognitive, esthetic, emotional, and memory-related color phenomena. These effects can be exploited to enhance the usability and image quality of visual displays. Perceived color is classified into color categories described by color names such as yellow, orange, brown, red, pink, green, blue, purple, white, gray, or black. This categorization is the basic process of color cognition. The distinction between perception and cognition is that while perception refers to immediate mapping of objects or events of the real world into the brain, cognition refers to subsequent higher order processes of semantic and verbal classification of the perceptions or to the mental imagery of the same objects or events [21, 22].

Long-term memory colors of familiar objects (e.g., blue sky, green grass, skin, tan skin, or yellow banana) represent a further type of cognitive color [13]. The color quality of pictorial images on a display can be enhanced by shifting the actual image colors toward these long-term memory colors (Section 3.4). Cognitive color is also relevant in visual ergonomics (Sections 3.1–3.3) because it improves visual search performance due to the control of visual attention "filtering out" unattended visual objects or events [23]. Color is an effective code when used as a cue or alerting signal or a method of grouping similar items or separating items [24].

The esthetic aspect of color is related to the pleasing or preferred appearance of stand-alone color patches, pictorial color images, or combinations of color patches. The latter aspect (i.e., esthetic value or preference of color combinations) is called color harmony (Section 6.3). As an example, more or less harmonic combinations of watercolors can be seen in Figure 6.19. Color can also evoke very strong emotions often in combination with other visual and nonvisual factors of still or motion images and these emotions can be enhanced by dedicated video processing algorithms (Section 4.6).

## 1.1.6
**Interindividual Variability of Color Vision**

In the previous sections, interindividual differences of color perception were neglected and a hypothetic average observer, the CIE 1931 standard colorimetric observer [2], was considered. In reality, however, there are observers with anomalous or deficient color vision. According to the observer's genotype, spectral sensitivity maxima of the L-, M-, and S-cones can be shifted by up to 4 nm. Some observers have less L-, M-, or S-cones and exhibit protanomalous, deuteranomalous, or tritanomalous color vision, respectively. If one of the cone types is completely missing, then they are called protanope, deuteranope, or tritanope observers. The interesting domain of visual displays and deficient color vision is out of the scope of this book.

Even within the limits of normal trichromatic color vision, there is a large variability of retinal mosaics especially concerning the ratios of the L- and M-cones varying between 0.4 and 13 [25]. The postreceptoral mechanisms of color vision are very adaptable and – at least in principle – able to counterbalance this variability of photoreceptor mosaics. There are, however, in turn large variations among the subjects at the later stages of neural color signal processing including the perception of color differences (Section 6.6), color cognition, preference (Sections 3.5 and 3.6), harmony, long-term memory colors (Section 3.4.2), and visually evoked emotions.

## 1.2
**Color Vision-Related Technological Features of Modern Self-Luminous (Nonprinting) Visual Technologies**

In this section, specific features of the different display technologies (and cameras) are described related to specific color vision characteristics, for every relevant type of display technology separately. For digital film and high-resolution digital TV, image quality is a main issue, especially the accurate reproduction of color appearance. To do so, the set of displayable colors (the so-called color gamut) has to be optimized to cover the most important colors (see Sections 5.1 and 5.2). Color resolution should be high enough to be able to render continuous color shadings (see Section 2.2.2.5).

Image quality can be further improved if spatial resolution is increased by subpixel rendering and by reducing the extent of spatial color artifacts at the same time (see Sections 5.3 and 5.4). High dynamic range (HDR) imaging means the emergence of highlights (Section 2.3.3) on the display enhancing the emotional effect of the motion picture (Section 4.6). One important aim of color management (Section 4.1.3) is that the displayed colors have the same color appearance across different displays, for example, on a proof monitor, in an analog cinema, and in a digital cinema (see also Section 3.2). To do so, the colorimetric characterization of the display (Chapter 2) has to be carried out on one hand and a color appearance model accounting for the adaptation of the human visual system has to be applied on the other hand. These two components have to be built into the display's hardware and/or software (in the so-called color management system) converting digital electronic signals into visible radiation for the observer.

It is also important to evaluate the color differences between a proof monitor and the actual appearance in the cinema or on the TV (Section 4.2.3). To reduce the huge amount of data for digital TV and cinema, image compression without loss of visually perceptible spatial, color, and motion information is necessary. This can be done by exploiting the knowledge about achromatic and chromatic contrast sensitivity of the human visual system (Section 4.4). Note that, for digital cinema, often very specific viewing conditions apply, including dim viewing conditions and large viewing angles (Section 2.4) influencing color appearance.

Motion picture theaters evoke special emotions visually and this feature requires a specific film-like color appearance (Sections 4.2.4 and 4.6). Long-term human color memory and color image preference can also be considered to provide pleasing images, possibly also depending on the intended group of observers, for example, depending on the cultural background or on the age of the observers (Sections 3.4, 3.5, 3.6).

In a camera, the colorimetric, spatial, and temporal resolution of the sensor array and the lens determine the quality of the captured image (Section 4.4.2). The colorimetric characterization of the camera is equally important to be able to transform the raw image consisting of the sensor signals into a device-independent format, for example, *XYZ* values at each pixel. A color appearance model helps apply further corrections such as adjusting the white balance or the tone characteristics of the image. It is essential to apply visually error-free image compression to reduce the bandwidth of video data transmission.

Color monitors represent similar features to digital film and TV except that the viewing conditions and the aims of use are different. Color monitors are usually viewed in light office environments where ambient light cannot be neglected and this has to be taken into account when applying a color appearance model (Section 2.1.8). The size of monitors is usually less; hence, the color size effect (Section 2.4) can be neglected.

Instead of film-like color image appearance and visually evoked emotions, visual ergonomics plays a more important role (Sections 3.1 and 3.3) for color monitors because instead of entertainment purposes, color monitors are used as a component of a computer workplace or for infotainment. Thus, the basic concepts of visual ergonomics, visibility, legibility, readability, visual attention, and visual search features become the substantial factors of display hardware and software design.

Head-mounted displays (HMDs, Section 2.2.2.4) often provide an immersive visual environment that can be either a projection of the real world or an artificial visual world. As HMDs often visualize three dimensions, an important requirement is to reduce parallax artifacts arising from the imperfect representation of depth information. As immersion means very large viewing angles, the color size effect is also relevant (Section 2.4). In head-up displays, additional visual information is superimposed upon the directly viewed image of the real world; hence, it is essential to match the luminance of this superposition to the actual luminance level of the real-world image (Section 2.2.2.4).

Digital signage displays and large tiled displays cover large areas (Section 2.4) on indoor or outdoor walls of buildings and provide visual information for numerous users simultaneously. Results from visual ergonomics are necessary to ensure the legibility of displayed information. Removing flicker, jitter, and ambient light reflections is also substantial.

For projectors, the image is viewed in a dim or dark environment and this has to be considered in the color appearance model. The spectral power distributions of projector light sources have to be matched to the spectral transmission of the color filters of the projector to achieve a large color gamut (Section 2.3.3). Alternatively, for LED projectors, the peak wavelengths of the LEDs have to be chosen in a similar way.

Light sources of display backlighting (Section 2.3.2) should provide a spatially uniform illumination of the color filter mosaic of the display (Section 2.1.5). Again, to achieve a large color gamut, the spectral power distribution of the backlight should match the spectral transmission of the color filters (Section 2.3.3). Another criterion of co-optimizing backlight spectra and filter transmissions is that the primary colors of the display should be bright enough compared to the brightness of the white point (Section 5.1).

For the light sources of indoor illumination, specific visual requirements apply (see Chapter 6). The reason is that they illuminate a room with usually white walls and several reflecting colored objects inside the room. Figure 6.17 shows a so-called tabletop arrangement of colored objects intended to model this situation. First of all, the light source itself should provide an appropriate white tone (visible on the white standard in Figure 6.17), for example, warm white for home illumination in Western countries and cool or cold white for office environments.

In addition to this, the colors of the reflecting objects should be rendered by the light source in an appropriate way. Reflected colors should not be undersaturated or oversaturated and they should exhibit a natural hue similar to the usual color appearance of each object under daylight or tungsten light. Even if the white tone of the light source itself is acceptable, that is, it has no strange tints such as a greenish or reddish shade, the reflecting colors of the objects can be rendered poorly if certain spectral ranges are missing from the spectral power distribution of the light source (see Section 6.2).

Besides this so-called color rendering or color fidelity property of the light source, there exist several other aspects of color quality including the color harmony among the different colored objects (see Section 6.3). Other color quality aspects are dealt with in Section 6.4 including visual clarity, continuous color transitions, color preference, and the rendering of long-term memory colors by the light source (see also Section 3.4.1).

## 1.3
### Perceptual, Cognitive, and Emotional Features of the Visual System and the Corresponding Technological Challenge

The starting point of the optimization of visual display technologies and indoor light sources is the analysis of the human user's characteristics, including the properties of the human visual system relevant for the most important visual task (e.g., work with a computer user interface or entertainment) on the display. In a second step, the technological challenge is that the visual display or the indoor light source should be designed to achieve the best perceptual, cognitive, emotional, or preference-based image appearance.

For self-luminous displays, the user's characteristics include the user's age, cultural background, and personality together with his or her spatial and color vision features as well as cognitive, emotional, and image preference characteristics. The task analysis should consider the mode of observation (e.g., still images or motion pictures), the surround luminance level (dark, dim, average, or bright), and the type of the user's task, for example, surveillance, monitoring, textual input on a user interface, programming, web browsing with extensive visual search, or watching still images or motion pictures for entertainment or infotainment purposes and appraising their spatiotemporal color appearance.

The next step in the optimization is considering the crucial visual mechanisms involved in the task, for example, the achromatic (luminance) channel of the visual system for the reading task. The final step is the optimization of the temporal, spatial, and colorimetric technological properties of the display in order to fulfill the requirements posed by the visual system for good image quality. For example, the design of new subpixel architectures (Section 5.3) can apply a set of design principles derived from the characteristics of the human visual system and novel types of subpixel architectures can be invented (Section 5.4).

Provided that the display is used for observers of normal color vision working in a well-lit office environment using predominantly still images, the color gamut of the new display can be co-optimized with its good spatial resolution in accordance with the properties of the retinal mosaic where the image of the display is projected by the user's eye lens. Thus, a huge amount of information stored in the computer memory can be mapped onto the user's brain very efficiently via the optical radiation emitted by the display and detected by the retina constituting an integral part of the visual brain.

For indoor light sources, the optimization workflow differs from the self-luminous display's workflow in the following aspects. In this case, human visual mechanisms should be considered from the point of view of indoor lighting, that is, the color appearance of the reflecting objects in the room, color discrimination among the different reflecting colors of the objects, the perceived color harmony of their combinations, and the fulfillment of the observer's color preference demands in the environment lit by the indoor light source. To optimize the light source, all available light source technologies should be kept in mind with the technological possibilities of tailoring their spatial and spectral power distributions by considering the spectral reflectance curves of the important objects that possibly appear in the indoor environment.

Since the beginning of the twenty-first century, lighting research has focused special attention on the spectral sensitivity of human circadian behavior, that is, the 24 h cycles of human activity synchronized by the "body clock." This circadian rhythm influences work concentration, sleep quality, and well-being of office and industrial workers. Today's technological challenge is to optimize the spectral and spatial power distributions of the light source to stimulate a special type of photoreceptors, the so-called intrinsically photosensitive retinal ganglion cells (ipRGCs, see Section 7.2.2).

Table 1.2 shows a selection of important perceptual, cognitive, and emotional features of the human visual system (examples) together with the challenges of display or light source technology. The corresponding sections of this book are also indicated in Table 1.2.

**Table 1.2** Perceptual, cognitive, and emotional features of the human visual system and challenges of display or light source technology.

| Feature | Technological challenge | |
|---|---|---|
| Trichromatic color vision, color matching | Accurate colorimetric characterization of color displays | Sections 2.1 and 2.2 |
| Chromaticity contrast and visual search | Ergonomic color design of a user interface on a display | Sections 3.3.2 and 3.3.3 |
| Spatial color vision | Ergonomic design of a color display with preferred color contrast for young and elderly users | Sections 3.3.4 and 4.4 |
| Spatiochromatic properties of the retinal mosaic | Optimization of multicolor subpixel architectures on a color display, digital cameras, motion picture compression algorithms, and watermarking | Sections 4.4 and 5.3 |
| Color appearance | Color gamut optimization of a modern multi-primary color display | Section 5.1 |
| Color appearance of large color stimuli | Providing accurate color appearance on a large or immersive color display | Section 2.4 |
| Color difference perception | Evaluation of color differences between soft-proof monitors and digital cinema | Section 4.2.3 |
| Color appearance, color fidelity, chromatic adaptation, color preference, color harmony | Improving the color quality of the lit environment | Chapter 6 |
| Cognitive color | Ergonomic presentation of information on a color display to enhance the user's recognition and visual search characteristics | Section 3.3.1 |
| Long-term color memory | Enhancement of the color quality or perceived naturalness of pictorial color images | Section 3.4 |
| Visually evoked emotions | Enhancement of the strength of the emotional effect in motion images | Section 4.6 |
| Image color quality and preference | Enhancement of the color image quality of pictorial color images | Sections 3.5, 3.6, and 4.4.3 |
| Circadian behavior | Optimize light source according to circadian behavior | Section 7.2.2 |

The corresponding sections/chapters of this book are indicated in the last column.

# References

1. Gegenfurtner, K.R. and Sharpe, L.T. (eds) (1999) *Color Vision: From Genes to Perception*, Cambridge University Press.
2. CIE 015:2004 (2004) *Colorimetry*, 3rd edn, Commission Internationale de l'Éclairage.
3. Hunt, R.W.G. and Pointer, M.R. (2011) *Measuring Colour* (*Wiley-IS&T Series in Imaging Science and Technology*), 4th edn, John Wiley & Sons, Ltd.
4. Wyszecki, G. and Stiles, W.S. (2000) *Color Science: Concepts and Methods, Quantitative Data and Formulae* (*Wiley Series in Pure and Applied Optics*), 2nd edn, John Wiley & Sons, Inc.
5. Fairchild, M.D. (2005) *Color Appearance Models* (*Wiley-IS&T Series in Imaging Science and Technology*), 2nd edn, John Wiley & Sons, Ltd.
6. CIE 170-1:2006 (2006) *Fundamental Chromaticity Diagram with Physiological Axes – Part 1*, Commission Internationale de l'Éclairage.
7. Stockman, A. (2004) Colorimetry, in *The Optics Encyclopedia: Basic Foundations and Practical Applications*, vol. **1** (eds T.G. Brown, K. Creath, H. Kogelnik, M.A. Kriss, J. Schmit, and M.J. Weber), Wiley-VCH Verlag GmbH & Co. KGaA, Berlin, pp. 207–226.
8. Khanh, T.Q. (2004) Physiologische und Psychophysische Aspekte in der Photometrie, Colorimetrie und in der Farbbildverarbeitung (Physiological and psychophysical aspects in photometry, colorimetry and in color image processing). Habilitationsschrift (Lecture qualification thesis), Technische Universitaet Ilmenau, Ilmenau, Germany.
9. Solomon, S.G. and Lennie, P. (2007) The machinery of color vision. *Nat. Rev. Neurosci.*, **8**, 276–286.
10. Nadenau, M. and Kunt, M. (2000) Integration of human color vision models into high quality image compression. Dissertation No. 2296, Ecole Polytechnique Fédérale de Lausanne.
11. CIE S 017/E:2011 (2011) *ILV: International Lighting Vocabulary*, Commission Internationale de l'Éclairage.
12. Hunt, R.W.G. (1977) The specification of colour appearance. I. Concepts and terms. *Color Res. Appl.*, **2** (2), 55–68.
13. Derefeldt, G., Swartling, T., Berggrund, U., and Bodrogi, P. (2004) Cognitive color. *Color Res. Appl.*, **29** (1), 7–19.
14. CIE 159-2004 (2004) *A Color Appearance Model for Color Management Systems: CIECAM02*, Commission Internationale de l'Éclairage.
15. MacAdam, D.L. (1942) Visual sensitivities to color differences in daylight. *J. Opt. Soc. Am.*, **32** (5), 247–274.
16. CIE 101-1993 (1993) *Parametric Effects in Color-Difference Evaluation*, Commission Internationale de l'Éclairage.
17. CIE 142-2001 (2001) *Improvement to Industrial Color-Difference Evaluation*, Commission Internationale de l'Éclairage.
18. Luo, M.R., Cui, G., and Li, Ch. (2006) Uniform color spaces based on CIECAM02 color appearance model. *Color Res. Appl.*, **31**, 320–330.
19. Li, Ch., Luo, M.R., Li, Ch., and Cui, G. (2011) The CRI-CAM02UCS color rendering index. *Color Res. Appl.*, doi: 10.1002/col.20682.
20. Bodrogi, P., Brückner, S., and Khanh, T.Q. (2010) Ordinal scale based description of color rendering. *Color Res. Appl.*, doi: 10.1002/col.20629.
21. Barsalou, L.W. (1999) Perceptual symbol systems. *Behav. Brain Sci.*, **22**, 577–660.
22. Humphreys, G.W. and Bruce, V. (1989) *Visual Cognition: Computational, Experimental, and Neuropsychological Perspectives*, Lawrence Erlbaum Associates, Hove,UK/Hillsdale, NJ.
23. Crick, F. (1994) *The Astonishing Hypothesis*, Simon & Schuster Ltd., London.
24. Krebs, M.J. and Wolf, J.D. (1979) Design principles for the use of color in displays. Proceedings of the Society for Information Display, vol. 20, pp. 10–15.
25. Carroll, J., Neitz, J., and Neitz, M. (2002) Estimates of L:M cone ratio from ERG flicker photometry and genetics. *J. Vis.*, **2**, 531–542.

# 2
# Colorimetric and Color Appearance-Based Characterization of Displays

The aim of this chapter is to explain the components of display characterization models in general and for specific types of displays. A colorimetric characterization model describes how to convert the electronic color signals of an image into colored light of given tristimulus values ($XYZ$) to create accurate color stimuli at every pixel of the display for the human observer of normal color vision. Anomalous color vision is out of the scope of this book. For good image color quality, the characterization model should go beyond colorimetry to ensure the expected appearance of the displayed colors, that is, accurate perceived hue, lightness, and chroma of the colors. Color appearance depends on the viewing conditions of the display including surround luminance level, the spatial structure of the displayed image, and the size (viewing angle) of the stimulus. On large viewing angle displays, color appearance is subject to change compared to standard image size. Good colorimetric characterization of a display also helps avoid visual artifacts, that is, the disturbing visual phenomena arising from the imperfections of the display's color technology.

Within the above-defined framework, Section 2.1 deals with the general description of the components of display characterization models, that is, tone curves and phosphor matrices, certain artifacts that can be partially removed by appropriate characterization models, and the method of applying color appearance models to displays. Section 2.2 describes specific characterization models and visual artifacts of specific display technologies including cathode ray tube (CRT), plasma display panel (PDP), liquid crystal display (LCD), and projectors. Section 2.3 introduces the different projector and display backlighting light sources together with their color filters. Finally, Section 2.4 describes the above-mentioned color size effect, that is, the change of color appearance between standard size and large color stimuli.

## 2.1
### Characterization Models and Visual Artifacts in General

This section deals with characterization models and related visual artifacts in general, such as tone curve models, phosphor matrices, differences between measured and predicted color characteristics, sRGB, and other characterization models for

*Illumination, Color and Imaging: Evaluation and Optimization of Visual Displays*, First Edition.
Peter Bodrogi and Tran Quoc Khanh.
© 2012 Wiley-VCH Verlag GmbH & Co. KGaA. Published 2012 by Wiley-VCH Verlag GmbH & Co. KGaA.

computer-controlled displays. Color additivity depending on the independence of the monitor's color channels is an important criterion for an efficient characterization model. The method of multidimensional phosphor matrices is presented to reduce the colorimetric error arising from the technological violations of color channel independence.

A method is shown to test and ensure the spatial uniformity of the display to achieve an accurate color stimulus at every point. Also, the color predicted at a specific point should not depend on the color at other pixels on the screen – this is the important criterion of spatial independence. Methods to predict spatial interdependence are also described in this section. The concept of viewing direction uniformity is introduced. This is especially important for LCDs. A paragraph is devoted to miscellaneous visual artifacts. Finally, the effect of the viewing environment including viewing conditions, viewing modes, and ambient light is dealt with to show how to apply CIELAB, CIELUV, and the CIECAM02 color appearance model to a self-luminous display.

### 2.1.1
### Tone Curve Models and Phosphor Matrices

To establish a colorimetric characterization model for a display, spectroradiometric or colorimetric measurements have to be carried out. Colorimetric characterization means the prediction of the CIE $X$, $Y$, $Z$ tristimulus values (briefly $XYZ$ values) of the color stimuli displayed on the monitor from their $r$, $g$, $b$ digital color counts (briefly $rgb$ values) of the video memory or graphics card of the computer controlling the display. These $rgb$ values are converted into voltage signals to control the colors at every pixel of the display. If the monitor is not suitable for accurate color display or it is suitable but poorly characterized, then perceptual color distortions take place. Color distortions may lead to incorrect decisions in systems where the correct identification of colors is essential, such as medical, military, traffic surveillance, or computer graphics applications.

To establish the characterization model, spectroradiometric or colorimetric measurements of the display should be conducted. The displays should be switched on well before the measurement for their stabilization. For the measurements themselves, test color patches of various $rgb$ values should be shown on the display in a dark room. The measurement accuracy of the actual colorimeter can be checked, for example, by using high-end calibrated colorimeters or spectroradiometers. Colorimeters can be calibrated before the measurements using a standard lamp and white standards such as a $BaSO_4$ or a PTFE plate. A calibration against filtered light sources with comparable spectral power distributions as the display primary colors (e.g., the CRT phosphors) may provide a better colorimetric accuracy.

After the colorimetric characterization, it is essential to carry out control measurements by using a set of test color stimuli of different $rgb$ values than the ones used for characterization to estimate the colorimetric error caused by possible inaccuracies of the model. There are special devices on the market (with combined hardware and software) for the automatic colorimetric characterization of displays. The colorimeter or spectroradiometer of these devices is controlled by a computer program that also generates the test color patches.

In general, colorimetric characterization models have two components, tone curve models and phosphor matrices. Tone curve models predict the relative luminous output $R$, $G$, and $B$ ($0 \leq R, G, B \leq 1$) of the individual color channels (usually red, green, and blue, but in some modern displays, more than three primary colors are used) from the *rgb* values ranging between 0 and $2^n$, where the exponent $n$ represents the color resolution of each channel. $RGB$ values are predicted in each pixel of the display corresponding to the *rgb* values of that pixel stored in the computer's memory.

In today's self-luminous color technology, the value of $n$ ranges between 6 (low color resolution) and 10 (high color resolution). The number of different colors is equal to $2^{kn}$, where $k$ represents the number of color channels (e.g., three: red, green, and blue), for example, $2^{3 \cdot 8} = 16\ 777\ 216$. The (usually nonlinear) shape of the measured $R(r)$, $G(g)$, $B(b)$ functions varies among the different display technologies; typical shapes include power functions (CRT) or S-shaped curves (LCD). Built-in converters may change the native tone curve of the display; hence, a converted curve (often looking like a CRT tone curve) can be measured.

Phosphor matrices model the additive mixture of light from the (three) color channels by converting $RGB$ values to $XYZ$ values describing the color stimulus. Note that $XYZ$ values can be further converted to so-called numerical correlates of perceived color attributes (hue, chroma, lightness) for the human visual system observing the display. Phosphor matrices usually consist of $3 \times 3$ numbers for a conventional display of three color channels (or three primary colors). The first (or second and third) row of the matrix converts $RGB$ values into $X$ (or $Y$ and $Z$). After the characterization measurement, the second row of the matrix can be given in cd/m$^2$ units. Then, the resulting $Y$ values predict the display's luminance at every pixel.

The term *phosphor matrix* refers to CRT display technology where three phosphors (red, green, and blue) are used to convert the energy of electrons into visible radiation. Other self-luminous color technologies use different methods such as color filter mosaics in LCDs or PDPs or rotating color filter wheels and digital micromirror devices (DMDs) in projectors. The common feature of every self-luminous display technology is that separate color channels are mixed additively at a later stage to display the full color gamut for the user of the display (i.e., a human observer). This additive mixture can be modeled by multiplying the relative output of the separate color channels by a matrix to obtain the $XYZ$ values. In this sense, the term *phosphor matrix* can be retained for non-CRT technologies.

### 2.1.2
**Measured Color Characteristics, sRGB, and Other Characterization Models**

In this section, measured tone curves of different CRT monitors are analyzed and compared with a standard colorimetric characterization model, sRGB [1], and with another model. It will be seen that different settings of CRT monitors such as gain and offset controls and the position and size of the test color patch influence the results of the characterization measurements. Chromaticity errors and total color errors resulting from the use of the sRGB model are estimated by comparing measured tone curves with sRGB tone curves [2].

The standard colorimetric characterization model or standard display color space (sRGB) was proposed as a method for a tolerably accurate color reproduction for not high-end users of CRT-based monitors and also for other types of displays [1]. sRGB is a simple but efficient tool to represent display color in a roughly *device-independent* way if the numerical representation of color appearance is not necessary. It was, for example, used to represent large-scale Internet-based lexical color resources consisting of color terms related to self-luminous colors collected from users across the World Wide Web [3].

It is very interesting to take a closer look at the colorimetric accuracy of applying the sRGB standard to different CRT displays of different white points and phosphor chromaticities. In this section, it will be shown that the most important item is the difference in the white point of the monitors and that the chromaticity difference of the phosphors is of minor importance.

Tone curves (voltage input–light output characteristics) of the single color channels (R, G, and B) will also be dealt with in detail. This issue is often called the "gamma problem" because the exponent of the power function fit to the measured tone curves is often designated by $\gamma$. The value of $\gamma$ depends on the gain/offset setting of the display and on screen position. The use of a value of $\gamma$ determined at another gain/offset setting than that actually used can lead to colorimetric errors of up to $\Delta E^*_{ab} = 21$ and the use of a value of $\gamma$ determined at another screen position than that actually used can lead to errors as high as $\Delta E^*_{ab} = 6$ [4] on typical CRT monitors.

In this section, results of colorimetric measurements at different gain/offset settings of a set of CRT displays will be analyzed and the colorimetric accuracy of the sRGB model will be assessed provided that the user of the display uses sRGB without measuring the actual tone curves in the actual gain/offset settings of the display. The changes of the shape of the tone curves according to the gain/offset changes will also be shown in this section.

The user of the display is usually confronted with the change of the following factors influencing the actual course of display tone curve as well as the phosphor matrix when trying to set up or apply a colorimetrical characterization model:

1) Effect of display technology and graphics card type used to drive the display.
2) Settings of the display such as gain and offset controls, white points, and the controls of the individual color channels.
3) Size of the test color patch measured with a colorimeter or a spectroradiometer.
4) Position of the test color patch on the screen, that is, the effect of spatial nonuniformities.
5) Relative area of bright and dark portions of the screen, that is, the effects of electronic power supply overload and light reflections at the different parts of the screen.
6) Color resolution or bit depth, for example, 6, 8, or 10 bits per color channel.

To elucidate the above considerations, an example of a series of tone curve measurements is shown [5]. In these measurements, single color channel tone curves (red, green, and blue) were measured as a function of the *rgb* value of the display's input voltage. Color differences caused by the differences between actual (measured) tone curves and the sRGB tone curve were computed for 11 CRT monitors ($i = 1–11$).

Two graphics cards were used in these measurements, with a spatial resolution of 640 × 480 pixels and a color resolution of 6 bits per channel. One of them was used for the case of displays $i = 1, \ldots, 8$, and the other one for displays $i = 9, \ldots, 11$. To investigate the effect of changing the graphics card, a supplementary measurement using four different graphics cards was carried out.

The size of the test color patches was either 80 × 80 pixels (*small* patch) or 640 × 480 pixels (*full screen*). For the small patch, the rest of the screen was black. The small patch was either in the center of the screen or in the four corners of the screen (top left, top right, bottom left, and bottom right). The main gain setting and the main offset setting were either at 100% or at 50%.

For each individual color channel ($k$ = red, green, or blue), luminance was measured for the following 14 *rgb* values ($d_j$): 0, 5, 10, 15, …, 50, 55, 60, and 63 ($j = 1, \ldots, 14$). The quantity $d_j$ denotes *r*, *g*, or *b*, while the other two channels were always driven by zero values. Measured luminance values are designated by $L_{k,j}$. This set of luminance values ($L_{k,j}$) was measured for all displays ($i = 1, \ldots, 11$) in nine different gain/offset settings ($s = 1, \ldots, 9$) defined in Table 2.1.

For two selected displays ($i = 5$ and 6), the luminance values corresponding to the settings at zero *rgb* values (display black, $L_{bl}$) and at maximum luminance (display peak white, $L_{pw}$) are shown in Figures 2.1 and 2.2, respectively. A measure of *dynamic range* ($D$) is also shown in these figures computed by dividing peak white luminance by display black luminance in each of the nine settings of Table 2.1.

As can be seen from Figures 2.1 and 2.2, display black and peak white luminance values and dynamic ranges are characteristic for the type of display and its gain and offset settings shown in Table 2.1. For display no. 5 (Figure 2.1), the highest value of peak white luminance and the highest dynamic range have the setting no. 2 (gain = 100% and offset = 100%): $L_{bl} = 0.19$ cd/m$^2$, $L_{pw} = 119.8$ cd/m$^2$, and $D = 631$. For display no. 6, although the setting no. 2 (gain = 100% and offset = 100%) exhibits

**Table 2.1** Settings of the display used in the sample measurement.

| s | Gain (%) | Offset (%) | Patch size | Position | Screen size |
|---|---|---|---|---|---|
| 1 | 100 | 50 | Small | Center | Whole |
| 2 | 100 | 100 | Small | Center | Whole |
| 3 | 100 | 50 | Small | Center | Smaller |
| 4 | 50 | 50 | Small | Center | Whole |
| 5 | 100 | 50 | Small | Top left | Whole |
| 6 | 100 | 50 | Small | Top right | Whole |
| 7 | 100 | 50 | Small | Bottom left | Whole |
| 8 | 100 | 50 | Small | Bottom right | Whole |
| 9 | 100 | 50 | Full screen | — | Whole |

## 2 Colorimetric and Color Appearance-Based Characterization of Displays

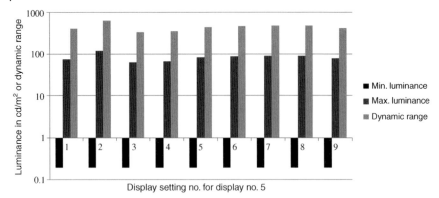

**Figure 2.1** Luminance values corresponding to the settings at zero *rgb* values (minimum luminance or display black, $L_{bl}$) and at maximum luminance (display peak white, $L_{pw}$) as well as display dynamic range (*D*) computed by dividing peak white luminance by display black luminance in each of the nine settings of Table 2.1, for the display no. 5.

the highest peak white luminance ($L_{pw} = 250.1\,cd/m^2$), this is associated with a disadvantageous display black of $L_{bl} = 7.7\,cd/m^2$ (actually dark gray) and the highest dynamic range occurs for the case of setting no. 4 (gain = 50% and offset = 50%, $D = 36.5$).

But the setting no. 4 exhibits a peak white luminance of $L_{pw} = 47.5\,cd/m^2$ only. This makes the visual system work in a suboptimal state because many important colors of the display's color gamut become mesopic, that is, "slide down" to the twilight range where they tend to lose chroma and saturation and hue changes also occur [6]. The dynamic range of $D = 36.5$ is very low leading to a visually unacceptable global contrast (see Section 3.5.5). The setting no. 2 of display no. 5 has a much higher dynamic range at an acceptable peak white luminance, which means that the important memory colors (Section 3.4) tend to appear in the photopic range of vision where the cone photoreceptors work in their optimum luminance range.

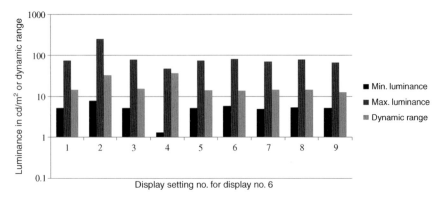

**Figure 2.2** The same as Figure 2.1 but for display no. 6.

To test the accuracy of the sRGB standard characterization model for the example of the above-mentioned monitors ($i = 1–11$) and for the settings of Table 2.1, the following calculation was carried out. For all possible combinations of the integers $j_1$, $j_2$, $j_3$ ($j_1 = 1–14$, $j_2 = 1–14$, $j_3 = 1–14$), the CIE $XYZ$ tristimulus values were calculated from the $rgb$ values $j_1, j_2, j_3$ by the sRGB standard phosphor transformation matrix [1] $\mathbf{P}_{sRGB}$ with a peak white luminance of $L_{pw} = 80\,\text{cd/m}^2$:

$$\mathbf{P}_{sRGB} = \begin{bmatrix} 0.4124 & 0.3576 & 0.1805 \\ 0.2126 & 0.7152 & 0.0722 \\ 0.0193 & 0.1192 & 0.9505 \end{bmatrix} L_{pw} \qquad (2.1)$$

The CIE $XYZ$ tristimulus values were calculated both by using the measured $RGB$ values (denoted by the index "meas" below) and by using the $RGB$ values ($D_{j,sRGB}$, $D = R$, $G$, or $B$) predicted by sRGB (denoted by the index "sRGB" below) from the $rgb$ values ($d_j = r$, $g$, or $b$) in the following way, as required by the standard [1]:

$$D_{j,sRGB} = \left[\frac{(d_j/63) + 0.055}{1.055}\right]^{2.4} \qquad (2.2)$$

Note that for $(d_j/63) \leq 0.04045$, the sRGB standard requires

$$D_{j,sRGB} = \frac{d_j/63}{12.92} \qquad (2.3)$$

CIELAB color differences $\Delta E^*_{ab}$ between the measured and standard sRGB $XYZ$ values were calculated for each one of the $14^3$ ($j_1$, $j_2$, $j_3$) triads, while the CIELAB reference white was the sRGB standard white point (D65) at the luminance level of $80\,\text{cd/m}^2$. As an example, Figures 2.3–2.5 show the difference between the shape of the measured tone curve and the shape of the sRGB standard tone curve for the display no. $i = 5$, for its settings $s = 2$ and $s = 4$ (see Table 2.1), for the red, green, and blue channels, respectively.

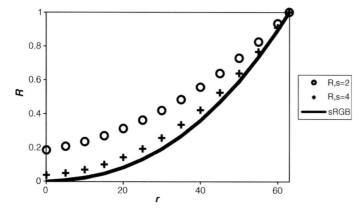

**Figure 2.3** Difference between the shape of the measured tone curves (circle or plus symbols; red color channel) and the shape of the sRGB standard tone curve (black curve) for the display no. $i = 5$, for its settings $s = 2$ and $s = 4$ (see also Table 2.1) [2]. Reproduced with permission from *Displays*.

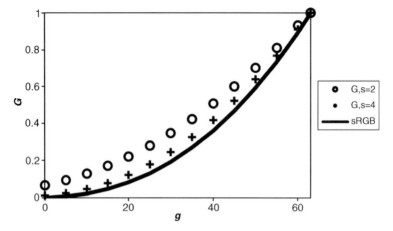

**Figure 2.4** Same as Figure 2.3 but for the green color channel [2]. Reproduced with permission from *Displays*.

As can be seen from Figures 2.3–2.5, setting no. 4 ($s=4$) of 50%–50% gain and offset setting is close to the sRGB tone curve, while the setting no. 2 ($s=2$) of 100%–100% gain and offset setting is further away from it. Table 2.2 shows the mean $\Delta E^*_{ab}$ values for the 11 monitors' nine settings. Mean values were calculated by taking all $14^3$ possible combinations of the measured *rgb* values $j_1, j_2, j_3$ ($j_1 = 1, \ldots, 14, j_2 = 1, \ldots, 14, j_3 = 1, \ldots, 14$) into account.

As can be seen from Table 2.2, for the display no. 5, setting no. 2 (with a tone curve far away from sRGB) produces a mean $\Delta E^*_{ab}$ value of 35 (large color error) and setting no. 4 (with a tone curve closer to sRGB) produces a mean $\Delta E^*_{ab}$ value of 13 (less color error). Note that all mean color errors of Table 2.2 are above $\Delta E^*_{ab} = 6$. This value would correspond to an easily perceptible color difference if a reference sRGB color

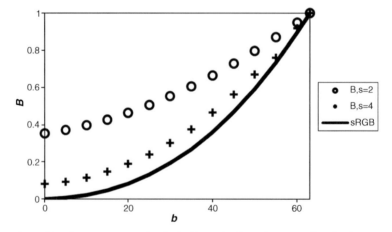

**Figure 2.5** Same as Figure 2.3 but for the blue color channel [2]. Reproduced with permission from *Displays*.

**Table 2.2** Mean $\Delta E^*_{ab}$ values between measured and sRGB XYZ for 11 displays (i) and 9 settings (s) [5].

| s | \ | | | | i | | | | | | | Mean |
|---|---|---|---|---|---|---|---|---|---|---|---|---|
| | 1 | 2 | 3 | 4 | 5 | 6 | 7 | 8 | 9 | 10 | 11 | |
| 1 | 12 | 15 | 33 | 44 | 34 | 39 | 14 | 14 | 20 | 22 | 13 | 24 |
| 2 | 38 | 35 | 25 | 45 | 35 | 34 | 16 | 16 | 29 | 54 | 22 | 32 |
| 3 | 12 | 16 | 36 | 30 | 31 | 41 | 13 | 19 | 20 | 11 | 10 | 22 |
| 4 | 32 | 28 | 12 | 30 | 13 | 15 | 14 | 15 | 40 | 17 | 19 | 21 |
| 5 | 9 | 16 | 35 | 44 | 31 | 42 | 10 | 12 | 20 | 10 | 12 | 22 |
| 6 | 9 | 16 | 37 | 47 | 36 | 43 | 10 | 11 | 17 | 10 | 17 | 23 |
| 7 | 7 | 16 | 36 | 43 | 32 | 43 | 11 | 10 | 28 | 14 | 17 | 23 |
| 8 | 9 | 17 | 37 | 43 | 35 | 55 | 11 | 12 | 30 | 13 | 18 | 25 |
| 9 | 11 | 18 | 36 | 46 | 37 | 43 | 10 | 9 | 21 | 23 | 33 | 26 |
| Mean | 15 | 20 | 32 | 41 | 32 | 39 | 12 | 13 | 25 | 19 | 18 | |

Mean values were calculated by taking all $14^3$ possible combinations of the rgb values $j_1, j_2, j_3$ ($j_1 = 1, \ldots, 14, j_2 = 1, \ldots, 14, j_3 = 1, \ldots, 14$) into account (see text). Reproduced with permission from *Displays*.

appearance were also visible, for example, on a reference display. In the absence of the visible reference (the usual case), the actual colors of the display and their combinations are judged in comparison with long-term memory colors (Section 3.4), preferred colors (Section 3.5), and harmonic color combinations (Section 6.3). In these visual comparisons, a color difference of $\Delta E^*_{ab} = 6$ may cause a significant color distortion.

As can be seen from Table 2.2, color differences as large as $\Delta E^*_{ab} = 46$ also occur. This may correspond to a change of color name (a semantic or categorical color change) leading to a significant reduction of the display's color quality. The magnitude of color differences changes considerably among the measured displays; for example, for display no. 7, the mean color difference is $\Delta E^*_{ab} = 12$, while for display no. 4 it has a significantly higher value of $\Delta E^*_{ab} = 41$.

As seen above, as the sRGB tone curve is fixed, it cannot account for the different display settings. To obtain good colorimetric accuracy with sRGB, it is essential to use a setting where the tone curve of the display comes close to the sRGB tone curve. Alternative display characterization models can also be used (see Section 2.2.2.1). For comparison, Equation 2.4 shows the example of a tone curve derived from electron physics for 6 bits per channel color resolution CRT displays [7, 8].

$$\begin{aligned} &\text{If} \quad Adj + [1-63A] \geq 0 \\ &\text{then} \quad D_j = (Adj + [1-63A])^\gamma \\ &\text{else} \quad D_j = 0 \end{aligned} \quad (2.4)$$

There are two parameters in Equation 2.4, $\gamma$ and $A$. Both parameters depend on the gain and offset settings of the monitor. To compare the electron physics tone curve model and the sRGB model, Equation 2.2 is rewritten in a form that is similar to Equation 2.4:

If $(1/63)d_j + [-0.04045] > 0$
then $D_{j,\text{sRGB}} = ((1/(63 \cdot 1.055))d_j + [0.055/1.055])^{2.4}$  (2.5)
else $D_{j,\text{sRGB}} = (d_j/63)/12.92$

The advantage of Equation 2.4 is that its parameters ($\gamma$ and $A$) can be adjusted to fit a wide range of settings of different CRT monitors, while Equation 2.5 operates with fixed parameter values.

To investigate the effect of changing the graphics card in the computer used to drive the CRT, colors resulting from four different graphics cards were measured using the above-described display color measurement scheme, for the same CRT display and the same setting [5]. Changing the graphics card did not influence the shape of the tone curves significantly. Mean color differences resulting from changing the graphics card were below $\Delta E_{ab}^* = 0.8$.

To characterize the extent of tone curve difference, the following quantity ($\delta$) was introduced [2]:

$$\delta = \sum_{k=R,G,B} \sum_{j=1}^{14} (D_{k,j,\text{measured}} - D_{j,\text{sRGB}})^2 \quad (2.6)$$

The quantity $\delta$ expresses the mean squared difference between the measured RGB values and those predicted from the sRGB tone curve for the above-described $14^3$ combinations of $rgb$ values. The value of the mean color error (for each CRT monitor and each setting) by using sRGB instead of a measured tone curve could be predicted by curve fitting ($r^2 = 0.93$) from the value of $\delta$ by using the following equation [2]:

$$\Delta E_{ab,\text{mean}}^* = 26.426\delta^{0.3418} \quad (2.7)$$

White point differences from the sRGB white point (i.e., D65) cause further colorimetric errors. Table 2.3 shows the total mean color error computed for the

**Table 2.3** Total mean color error ($\Delta E_{ab}^*$) arising from white point difference plus tone curve difference [5].

| | | $\delta$ | | |
|---|---|---|---|---|
| | 0.00 (sRGB) | 0.07 | 1.56 | 4.00 |
| 12 700 K | 11.8 | 16.8 | 35.4 | 45.0 |
| 10 000 K | 5.7 | 12.2 | 33.7 | 44.4 |
| D65 (sRGB) | 0.0 | 10.2 | 32.5 | 43.5 |
| D55 | 3.4 | 10.7 | 32.2 | 43.3 |
| D50 | 5.7 | 11.5 | 32.1 | 43.2 |

Mean $\Delta E_{ab}^*$ values were calculated by using Equation 2.7 and taking all $14^3$ $d_j$ value combinations into account (see text). The chromaticities of the R, G, and B primaries of the monitors were assumed to be equal to the sRGB chromaticities. Display peak white correlated color temperatures are shown in the first column. Reproduced with permission from *Displays*.

above-described $14^3$ rgb value combinations between measured colors and sRGB if both the display's peak white and the tone curves differ from sRGB, for four values of δ, ranging between 0 and 4.00. This is a typical range of δ corresponding to Table 2.2.

### 2.1.3
### Additivity and Independence of the Color Channels

The basic assumptions [9] underlying the simple two-step display characterization method described in Section 2.1.2 are phosphor constancy (i.e., that the chromaticity of a given color channel does not depend on the magnitude of the *r, g,* or *b* value of that channel), independence of the color channels, spatial independence, spatial homogeneity, and temporal stability. This section deals with the independence of the color channels. Color channel independence means for CRT displays "that the output of one electron gun and therefore the excitation of one channel at a given pixel is not dependent on the other channels at that pixel" [10].

Generally (including non-CRT display technologies), color channel independence means that the output of a given color channel depends only on the digital color value of that channel; for example, the output of the red color channel at a pixel depends only on the red digital color value. Color channel independence means that the color of a given pixel can be predicted from the *rgb* values by the additive mixture of independent color channels (additivity) by using the phosphor matrix. The lack of channel independence can be called channel interdependence. Reasons of channel interdependence depend on display technology, including power supply overload (all technologies), deficiencies of the magnetic lenses, and the CRT shadow mask and inter-reflections among the pixels (for all technologies).

In those applications that require displaying color stimuli on different screen backgrounds accurately (e.g., for the implementation of color atlases), it is important to predict the effect of channel interdependence and to provide a method to display colorimetrically accurate colors even in the presence of color interdependence. Reasonably accurate descriptors of color appearance can only be calculated if the color *stimuli* are displayed in a colorimetrically accurate way.

### 2.1.4
### Multidimensional Phosphor Matrices and Other Methods

Brainard stated that improved accuracy of calibration might be achieved by formulating a more sophisticated method of predicting color channel interdependence [9]. Berns *et al.* calculated the CIE *X, Y,* and *Z* tristimulus values of CRT displays in their model inspired by electron physics [7] by using the phosphor matrix determined either by measuring the tristimulus values of the red, green, and blue peak display primary colors (this is the phosphor matrix **P**) or by sampling the color gamut of the display and estimating the values of the $3 \times 3$ matrix by multiple linear regression (this will be called matrix **M**).

Berns *et al.* remark that "characterization accuracy can be improved greatly using the regression method for monitors exhibiting a lack of channel independence" [7].

They suggest that "for some displays, covariance terms are also needed in addition to the linear terms" to account for channel interdependence [7]. This idea was also supported by Motta for CRT monitors [11] who found that an expected tristimulus value $T$ was always greater than the measured one $T'$. Their ratio $T'/T$ depended on the electronic load on the CRT display's color channels.

The above ideas were adopted in another method [12] by introducing a $3 \times 9$ phosphor matrix. This method was originally developed for CRT displays but it can be used for any technology. The $3 \times 9$ matrix ($\chi$) is a result of a multiple linear regression on a measured sample of the display's color gamut similar to matrix **M**. But in addition to the $RGB$ values themselves, matrix $\chi$ takes *products* across the $RGB$ values into account. Thus, the independent variable of the $RGB$–$XYZ$ transformation is a nine-dimensional column vector **S** consisting of $RGB$ values ($T_1 = R$, $T_2 = G$, $T_3 = B$) and all six possible products of them, as shown in Equation 2.8.

$$S_i = T_i, \quad i = 1, 2, 3; \quad S_4 = T_1 T_1, \quad S_5 = T_1 T_2, \quad S_6 = T_1 T_3,$$
$$S_7 = T_2 T_2, \quad S_8 = T_2 T_3, \quad S_9 = T_3 T_3 \tag{2.8}$$

The $RGB$–$XYZ$ transformation can be written in the following form:

$$\begin{pmatrix} X \\ Y \\ Z \end{pmatrix} = \chi \mathbf{S} \tag{2.9}$$

In Equation 2.8, the terms $S_4, \ldots, S_9$ are intended to account for channel interdependence [12].

In the following part of this section, the results of an experimental investigation [12] of the performance of the methods using **P**, **M**, and $\chi$ are summarized. In this series of display measurements, different color patches of different sizes were measured colorimetrically on the display in the presence of different border and background colors on the screen and the extent of violation of channel independence and its reduction by the matrices **M** and $\chi$ was assessed. In the experiments, six CRT displays ($m = 1$–6) were used representing different quality levels. They were switched on 30 min before the measurement with a full-screen peak white to allow for the important issue of temporal stabilization. For the measurements themselves, the *rgb* values of a color patch were varied while the other pixels of the screen were set constant producing a constant border on the screen.

A colorimeter was used for the measurement of the CIE $X$, $Y$, and $Z$ tristimulus values of the displayed colors. Display black luminance ($L_{bl}$) was also measured. If the border is white, then light is scattered from the border to the measured color patch. Display black luminance ($L_{bl}$) was always less than $L_{bl} = 4 \, \text{cd/m}^2$. For an accurate characterization model, it is important to include display black by adding a constant to the model equation.

Every one of the six CRT displays had a color resolution of 6 bit/channel. They had nested gain/offset controls except display $m = 3$ (a high-end display for advanced graphics) that had separate gain/offset controls for the red, green, and blue channels allowing for a free setting of its white point chromaticity. The displays $m = 1$ and

$m = 4$ were inexpensive devices for the office workplace, and displays $m = 2, 5$, and 6 were suitable for graphics applications.

The size of the measured homogeneous color patch of variable color (displayed in the middle of the screen) can be characterized by the display fill factor $f$ representing the number of pixels of the color patch divided by the total number of pixels. The fill factor of the color patch was either $f = 0.02$ ("small", $s = 1$) or $f = 1.00$ ("full screen", $s = 2$). Peak white borders of different constant thickness were also displayed around the color patch in the middle of the screen. Thickness means the distance between the edge of the screen and the square-shaped color patch. Five different borders ($k = 0$–4) were used with fill factors of $f = 0.00$ (no border), 0.15, 0.30, 0.60, and 0.98 (white background, 2% smaller than the whole screen area).

A sample of 125 combinations was chosen from the following set of *rgb* values: 0, 15, 31, 47, and 63. Directly measured colors are designated by the index "m" and calculated colors based on **P**, **M**, or χ by the index "c". CIELAB color differences $\Delta E^*_{ab}$ were computed between measured and calculated colors. Channel interdependence was characterized by the luminance ratio $Q = Y_m/Y_c$. The quantities $\Delta E^*_{ab}$ and $Q$ were computed for each condition ($k$: border; $s$: small bar or full screen situation; $m$: monitor type, $m = 1$ and 2 will be shown) and for each sample color (1–125) using six different transformation matrices including **P**, **M**, and χ determined by using the measured colorimetric data of the given test scheme ($k, s, m$). The three additional matrices (**P₀**, **M₀**, and χ₀) represent the special cases of using the measured colorimetric data of the following condition: $k = 0$ (no border), $s = 1$ (small bar), $m = 1$ or 2. Matrices **P₀**, **M₀**, and χ₀ demonstrate the effect of using a transformation matrix determined from the colorimetric data of a small bar of variable color on a black background instead of the transformation matrices corresponding to the given test condition.

Table 2.4 contains the average CIELAB $\Delta E^*_{ab}$ differences and the average $Q$ values for the different test conditions. Averages were calculated taking all 125 colors into account. Columns 1–3 of Table 2.4 specify the test schemes in terms of monitor type $m$, border size $k$, and small or full screen bar size $s$ values. Columns 4–9 contain the average $\Delta E^*_{ab}$ differences calculated by using matrices **P**, **M**, χ, **P₀**, **M₀**, and χ₀, respectively. Columns 10–15 contain similar average $Q$ values.

For the case of the small bar of variable color ($s = 1$) in Table 2.4, the phosphor matrices determined in the absence of the border (on black display background) and for a small color patch (i.e., using **P₀**, **M₀**, and χ₀) cause considerable color differences as the size of the white border increases (see columns 7–9 of Table 2.4). This finding is not a consequence of the light scattered from the white border to the measured color patch because the latter effect was allowed for by using an additive constant in the characterization model. In the case of scattered light, the measured luminance would be higher than the calculated one.

Just the opposite can be seen from columns 13–15 of Table 2.4 (showing $Q$, that is, the ratio of measured luminance divided by predicted luminance): all $Q$ values in columns 13–15 are less than 1.00. Also, the color differences in columns 7–9 of Table 2.4 increase as the size of the white border increases. The maximum average color difference of about 24 $\Delta E^*_{ab}$ units arises in the case of the white background

**Table 2.4** Average $\Delta E^*_{ab}$ and $Q$ (measured luminance divided by computed luminance) values for the different test schemes [12].

| | | | | $\Delta E^*_{ab}$ | | | | | | $Q$ | | | | |
|---|---|---|---|---|---|---|---|---|---|---|---|---|---|---|
| 1 | 2 | 3 | 4 | 5 | 6 | 7 | 8 | 9 | 10 | 11 | 12 | 13 | 14 | 15 |
| m | k | s | P | M | $\chi$ | $P_0$ | $M_0$ | $\chi_0$ | P | M | $\chi$ | $P_0$ | $M_0$ | $\chi_0$ |
| 1 | 0 | 1 | 0.97 | 0.89 | 0.77 | — | — | — | 1.01 | 1.01 | 1.01 | — | — | — |
| 1 | 1 | 1 | 0.99 | 0.89 | 0.7 | 3.74 | 3.71 | 3.61 | 1.01 | 1.01 | 1 | 0.89 | 0.89 | 0.89 |
| 1 | 2 | 1 | 1.02 | 0.96 | 0.8 | 6.39 | 6.39 | 6.31 | 1.01 | 1.01 | 1.01 | 0.82 | 0.82 | 0.81 |
| 1 | 3 | 1 | 1.05 | 0.92 | 0.84 | 12.51 | 12.46 | 12.36 | 1.02 | 1.02 | 1.01 | 0.69 | 0.69 | 0.69 |
| 1 | 4 | 1 | 1.06 | 0.88 | 0.77 | 24.36 | 24.32 | 24.2 | 1.02 | 1.01 | 1.01 | 0.55 | 0.55 | 0.55 |
| 1 | 0 | 2 | 3.71 | 11.77 | 2.09 | 8.36 | 8.31 | 8.2 | 0.96 | 1.13 | 0.99 | 0.83 | 0.83 | 0.83 |
| 2 | 0 | 1 | 1.21 | 1.29 | 0.94 | — | — | — | 1.01 | 1.01 | 1.01 | — | — | — |
| 2 | 1 | 1 | 0.89 | 0.91 | 0.61 | 1.71 | 1.76 | 1.57 | 1.02 | 1.02 | 1.01 | 0.96 | 0.96 | 0.96 |
| 2 | 2 | 1 | 0.65 | 0.64 | 0.42 | 1.65 | 1.65 | 1.52 | 1.01 | 1.01 | 1.01 | 0.96 | 0.96 | 0.96 |
| 2 | 3 | 1 | 1.05 | 0.93 | 0.73 | 3.87 | 3.89 | 3.76 | 1.02 | 1.02 | 1.01 | 0.91 | 0.9 | 0.9 |
| 2 | 4 | 1 | 0.97 | 0.78 | 0.56 | 25.01 | 25.02 | 24.89 | 1.01 | 1.01 | 1 | 0.58 | 0.58 | 0.58 |
| 2 | 0 | 2 | 3.47 | 10.43 | 3.59 | 6.24 | 6.2 | 5.98 | 1.04 | 1.16 | 0.98 | 0.96 | 0.96 | 0.96 |

Averages were calculated taking all 125 test colors into account (see text). Columns 1–3: test conditions (CRT display type $m$, border size $k$, and small or full-screen color patch $s=1$ or $s=2$). Columns 4–9: average $\Delta E^*_{ab}$ differences calculated by matrices P, M, $\chi$, $P_0$, $M_0$, and $\chi_0$. Columns 10–15: similar average $Q$ values. Reproduced with permission from *Displays*.

($k=4$). These color differences can be reduced to be on the order of one $\Delta E^*_{ab}$ unit for every border size by using a separate transformation matrix for each border size, each determined in the presence of the corresponding border, that is, by using matrices **P**, **M**, or $\chi$ (see columns 4–6 of Table 2.4).

Then, $Q$ values also approach unity (compare columns 10–12 with 13–15). Instead of the decrease of the average $Q$ values down to $Q=0.55$ as the border size $k$ increases, $Q$ values remain approximately constant (and equal to unity) in the columns of matrices **P**, **M**, and $\chi$ (in columns 10–12) indicating the importance of using matrices **P**, **M**, or $\chi$. Above tendencies are similar for the displays $m=1$ and $m=2$ except that a white border of $k=2$ or $k=3$ causes greater color differences (columns 7–9) for $m=1$ than for $m=2$. This is a sign of high spatial interdependence for $m=1$ (see Section 2.1.5).

As can be seen from Table 2.4, best general agreement between the measured and predicted colors (in terms of minimal $\Delta E^*_{ab}$ and $|Q-1|$ values) can be achieved by applying matrix $\chi$ while matrix **M** does not perform well. To understand the role of matrix $\chi$, it is essential to see how the color differences between the colors measured on the display and the colors predicted by the characterization model vary inside the color gamut of the display. To this end, the color gamut was divided into eight sections (so-called octaves) in terms of *rgb* values (see Table 2.5).

In the cube of all *rgb* value combinations, the gamut octaves of Table 2.5 can be thought of as eight small subcubes of equal size. Mean CIELAB $\Delta E^*_{ab}$ color differences in each *rgb* gamut octave are shown in Table 2.6 for the test scheme $k=0$ (no border), $s=2$ (full screen bar), and the display $m=1$, as an example.

**Table 2.5** Eight sections (octaves) of the color gamut of the displays of 6 bits per channel color resolution in terms of *rgb* values [12].

| Octave no. | Red | Green | Blue |
|---|---|---|---|
| 1 | $0 \leq r < 32$ | $0 \leq g < 32$ | $0 \leq b < 32$ |
| 2 | $31 \leq r < 64$ | $0 \leq g < 32$ | $0 \leq b < 32$ |
| 3 | $31 \leq r < 64$ | $31 \leq g < 64$ | $0 \leq b < 32$ |
| 4 | $0 \leq r < 32$ | $31 \leq g < 64$ | $0 \leq b < 32$ |
| 5 | $0 \leq r < 32$ | $0 \leq g < 32$ | $31 \leq b < 64$ |
| 6 | $31 \leq r < 64$ | $0 \leq g < 32$ | $31 \leq b < 64$ |
| 7 | $31 \leq r < 64$ | $31 \leq g < 64$ | $31 \leq b < 64$ |
| 8 | $0 \leq r < 32$ | $31 \leq g < 64$ | $31 \leq b < 64$ |

Reproduced with permission from *Displays*.

**Table 2.6** Mean CIELAB $\Delta E^*_{ab}$ color differences between the color stimuli measured on the CRT display and the color stimuli predicted by the characterization models [12].

| Octave no. | P | M | $\chi$ | $P_0$ | $M_0$ | $\chi_0$ |
|---|---|---|---|---|---|---|
| 1 | 2.33 | 9.99 | 3.06 | 4.82 | 4.82 | 4.69 |
| 2 | 2.63 | 24.9 | 2.34 | 8.18 | 8.18 | 7.98 |
| 3 | 4.05 | 16.11 | 2.38 | 9.65 | 9.54 | 9.6 |
| 4 | 2.22 | 7.51 | 2.03 | 6.21 | 5.99 | 6.02 |
| 5 | 2.39 | 7.52 | 1.71 | 5.64 | 5.62 | 5.40 |
| 6 | 3.93 | 8.38 | 1.75 | 8.70 | 8.85 | 8.71 |
| 7 | 6.56 | 3.13 | 1.71 | 10.97 | 10.99 | 11.04 |
| 8 | 4.39 | 2.81 | 1.55 | 7.05 | 6.95 | 7.03 |
| Average | 3.71 | 11.77 | 2.09 | 8.36 | 8.31 | 8.20 |

Averages result from *rgb* values in the color gamut octaves taken from the sample set of 125 *rgb* values (see text and Table 2.5). Example of the test scheme: $k = 0$ (no border), $s = 2$ (full screen bar), and display $m = 1$. Reproduced with permission from *Displays*.

Comparing the average $\Delta E^*_{ab}$ color differences of the columns of matrices $P_0$, $P$, and $\chi$, in Table 2.6, matrix $\chi$ seems to diminish the color differences of the column of matrix $P_0$ more strongly than matrix $P$ primarily for light and desaturated colors of the color gamut, that is, for the colors of the sixth, seventh, and eighth octaves. This is a consequence of *high channel interdependence* for these octaves; see also Section 2.1.5 containing a more detailed explanation of channel interdependence in association with spatial interdependence for CRT displays.

## 2.1.5
### Spatial Uniformity and Spatial Independence

Spatial independence can be defined in the following way: "When characterizing a portion of the display, it is assumed that the luminance and chromaticity at that portion are stable and invariant to colors displayed in other portions of the display" [10]. The lack of spatial independence is called spatial interdependence. For the case of CRT displays, spatial interdependence is often coupled to channel interdependence because, in general, the extent of channel interdependence depends on the size of the more luminous (e.g., non-black) parts of the display, as seen in Section 2.1.4.

As learned from Section 2.1.4, the measured color of a patch of variable color on a CRT display deviates from the predictions of the calibration model. Such deviations either can arise from the influence of luminous borders displayed around the small patch of variable color (e.g., with a fill factor of $f=0.02$) or can be produced by displaying a large patch of variable color (e.g., a full-screen color patch, $f=1.00$) instead of the small patch used for the characterization measurements.

To explain the above findings, a distinction should be made between two kinds of color patches displayed on the CRT screen, the first one whose color is being varied in the current application (and to be predicted accurately) and a second one whose color is being held constant on the display for a longer time. The first color patch can be called the target patch and the second one the constant patch. Both the target patch and the constant patch are supposed to be homogeneous. For example, in a CRT display-based implementation of a color atlas, the target patch is the color currently being changed by the user while the constant patch is the set of all other constant color patches (e.g., a white frame and a gray background) displayed on the screen by the software driving the display.

By displaying a new non-black color patch on the screen at a place where it was originally black (or dark/dim), the extent of color channel crosstalk (channel interdependence) may increase and the output of the color channels may decrease. The origin of this effect depends on display technology.

The increment of channel interdependence and the decrement of the output of the color channels depend on the display fill factor $f$ and on the *rgb* values of the recently displayed non-black color patch. A large fill factor and large *rgb* values give rise to a large increment of channel crosstalk and a large decrement of the luminous output of the color channels. The results of the experimental test of the characterization model described in Section 2.1.4 (see Tables 2.4 and 2.6) can be explained in the following way.

For the case of the small color patch ($s=1$) with variable white border ($k=0$–4), the target patch is the small bar of variable color and the constant patch is the white border. Turning on a white border, the output of the CRT's color channels (i.e., the electron guns) decreases on the whole screen. Therefore, using a phosphor matrix determined in the absence of the border, the luminance of the target patch will be overestimated by the colorimetric characterization model.

Changing the *rgb* values of the target patch does not further influence the extent of channel crosstalk and the output of the color channels because the fill factor of the target patch is smaller than the fill factor of the white border. That is why the use of phosphor matrices determined in the presence of the white border yielded a better accordance with measured colors.

For the case of the full-screen color patch ($s=2$) with no border ($k=0$), the target patch is the whole screen and there is no constant patch. Changing the *rgb* values of the target patch strongly influences the extent of channel interdependence and the output of the color channels. In this case, the use of matrix **P** instead of matrix $\mathbf{P_0}$ reduces the color differences between measured and predicted colors because matrix **P** was determined by measuring the full-screen primary colors (peak red, green, and blue) of the CRT display.

Matrix $\chi$ reduces the color differences more strongly than matrix **P** primarily for the colors of the sixth, seventh, or eighth octaves (see Table 2.5) as in the case of these colors, all the channels are driven by higher *rgb* values and, as a result, channel interdependence manifests itself more explicitly. It is matrix $\chi$ – thanks to its covariance terms – that takes this channel interdependence into account.

From the above considerations, the coupled nature of channel interdependence and spatial interdependence on a CRT display is obvious. Displaying a color patch of large fill factor causes spatial interdependence. But it also leads to an increased level of channel interdependence. To describe a given setting of a display from the point of view of channel interdependence and spatial interdependence, it is desirable to decouple these two display artifacts.

One way to separate the effect of spatial interdependence from channel interdependence is to measure the luminance of the display's primary colors at their maximum *rgb* values (so-called *peak* red, green, and blue) by displaying patches of these colors of different fill factors. To characterize the extent of spatial interdependence, the following ratio can be introduced:

$$R(f) = Y(f)/Y(f_0) \tag{2.10}$$

In Equation 2.10, the function $Y(f)$ means the luminance of a color patch of a given color of fill factor $f$. The symbol $f_0$ refers to a reference fill factor of the smallest patch of the given color that can be displayed and measured (e.g., $f_0 = 0.02$). The value of $Q$, that is, the ratio of measured and calculated peak white luminance as a function of the fill factor, can be calculated from the $R(f)$ functions of the peak primary colors and from the $R(f)$ function of the peak white. These $Q(f)$ functions can be calculated by the following formula:

$$Q(f) = Y_{\text{peak white}}(f)/[Y_{\text{peak red}}(f) + Y_{\text{peak green}}(f) + Y_{\text{peak blue}}(f)] \tag{2.11}$$

Figure 2.6 shows an example of measured $R(f)$ functions for peak primary colors (peak red, green, and blue) and the peak white for the CRT display $m = 2$.

As can be seen from Figure 2.6, the peak white curve exhibits the lowest values of $R$, as a result of channel interdependence. It can also be seen that there are small differences (up to 3%) among the $R(f)$ functions of the three peak primary colors (red, green, and blue) but their tendencies are the same – in accordance with the above – they are decreasing functions of the fill factor $f$. The $R(f)$ functions of the peak primary colors represent expressive descriptors of spatial interdependence.

Figure 2.7 shows an example of a measured $Q(f)$ function for the CRT display $m = 2$. As can be seen from Figure 2.7, the value of $Q$ remains above 90% until a critical fill factor $f_{cr}$ (its value is $f_{cr} = 0.6$ for $m = 2$) and then it begins to fall rapidly to reach its minimum of 62% for the case of $f = 1$. This function is a representative

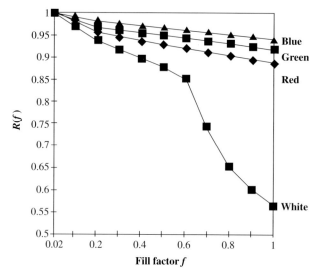

**Figure 2.6** Example of measured R(f) functions (Equation 2.10) for the peak primary colors (peak red, green, and blue) and the peak white of the CRT display $m = 2$ [12]. Reproduced with permission from *Displays*.

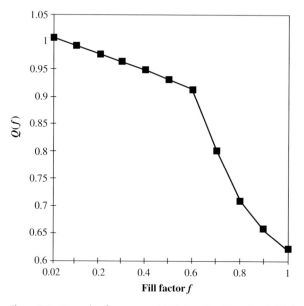

**Figure 2.7** Example of a measured Q(f) function (Equation 2.11) for the CRT display $m = 2$ [12]. Reproduced with permission from *Displays*.

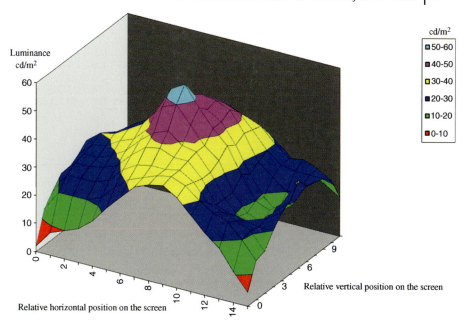

**Figure 2.8** An example of bad spatial nonuniformity for the case of a CRT display. Peak white luminance (cd/m²) is shown as a function of relative position on the screen.

descriptor of channel interdependence. One technological reason for the case of displaying peak white on a CRT display is that – above a fill factor of $f = 0.6$ – the current drawn from the power supply of the display becomes so high that electronic compensation breaks down. This leads to a decrease of luminance.

Concerning spatial uniformity, it is desirable that the same *rgb* values at different pixels of a color patch result in the same color stimulus, that is, the same *XYZ* tristimulus values at all pixels of that color patch across the whole screen. In other words, displayed colors should exhibit the same luminance and chromaticity – within the limits of visually tolerable changes – at every position of the screen. Display measurements show that – in some cases – large spatial nonuniformities may occur. Figure 2.8 shows a bad example of peak white luminance across the screen of a very nonuniform CRT display.

As can be seen from Figure 2.8, peak white luminance decreases rapidly toward the corners of the screen and this causes an annoying visual artifact. Better spatial uniformity was measured for the case of an LCD. Colorimetric measurements of an intentionally homogeneous whole screen peak white (i.e., maximum *rgb* values in all pixels) were carried out in 25 different display positions distributed evenly on this LCD (with position nos. 1–5 in the upper row from left to right, 6–10 in the next row, …, and 21–25 in the bottom row) [13]. The position no. 13 was the center. Position no. 13 was also the reference white to compute the CIELAB differences between the center and any position. Similarly, CIE 1976 UCS chromaticity differences were also computed [13]. Table 2.7 shows the values of peak white

**Table 2.7** Spatial uniformity of an LCD.

| Position | $L_{pw}$ (cd/m²) | $\Delta L^*$ | $\Delta C^*$ | $\Delta u'v'$ | $\Delta E^*$ |
|---|---|---|---|---|---|
| 1 | 170.3 | −3.7 | 2.6 | 0.003 | 4.5 |
| 2 | 171.9 | −3.3 | 2.0 | 0.002 | 3.9 |
| 3 | 172.8 | −3.1 | 2.4 | 0.002 | 3.9 |
| 4 | 171.2 | −3.5 | 3.4 | **0.004** | 4.9 |
| 5 | **163.2** | **−5.3** | 3.1 | 0.003 | **6.1** |
| 6 | 166.6 | −4.5 | 2.0 | 0.003 | 4.9 |
| 7 | 182.8 | −1.0 | 1.3 | 0.002 | 1.6 |
| 8 | 182.3 | −1.1 | 2.6 | 0.003 | 2.8 |
| 9 | 181.2 | −1.3 | 1.8 | 0.002 | 2.3 |
| 10 | 165.8 | −4.7 | 3.0 | 0.003 | 5.6 |
| 11 | 167.4 | −4.3 | 1.8 | 0.003 | 4.7 |
| 12 | 183 | −1.0 | 1.4 | 0.002 | 1.7 |
| 13 | 187.6 | 0.0 | 0.0 | 0.000 | 0.0 |
| 14 | 178.2 | −2.0 | 2.7 | 0.003 | 3.3 |
| 15 | 172.2 | −3.3 | 2.3 | 0.003 | 4.0 |
| 16 | **164.8** | **−4.9** | 2.5 | **0.004** | **5.5** |
| 17 | 176.3 | −2.4 | 1.8 | 0.002 | 3.0 |
| 18 | 177.9 | −2.0 | 1.4 | 0.001 | 2.5 |
| 19 | 178.2 | −2.0 | 2.9 | 0.003 | 3.5 |
| 20 | 167.2 | −4.4 | 1.6 | 0.002 | 4.6 |
| 21 | **164.1** | **−5.1** | 3.1 | **0.004** | **5.9** |
| 22 | 166.5 | −4.5 | 1.3 | 0.002 | 4.7 |
| 23 | 168.5 | −4.1 | 0.7 | 0.001 | 4.1 |
| 24 | 167.2 | −4.4 | 2.4 | 0.002 | 5.0 |
| 25 | 162.5 | −5.4 | 1.5 | 0.002 | 5.6 |

Peak white luminance ($L_{pw}$), CIELAB lightness difference ($\Delta L^*$), chroma difference ($\Delta C^*$), total color difference ($\Delta E^*$), and CIE 1976 UCS chromaticity difference ($\Delta u'v'$) in 25 positions of the display [13]. Bold numbers indicate nonuniform portions of the display. No. 13 was the reference position.

luminance ($L_{pw}$), CIELAB lightness difference ($\Delta L^*$), chroma difference ($\Delta C^*$), total color difference ($\Delta E^*$), and CIE 1976 UCS chromaticity difference ($\Delta u'v'$) as a function of display position.

As can be seen from Table 2.7, going away from the center, screen luminance decreases. The lowest luminance ($L_{pw} = 162.5$ cd/m²) was measured at position no. 25 (bottom-right corner). This value represents a luminance decrease of about 13% corresponding to a CIELAB lightness difference of $\Delta L^* = -5.4$, a disturbing difference. If the luminance and color transitions are continuous on the display, then these changes lose their visual conspicuity. Positions 5, 16, and 21 also indicate nonuniform parts on the display. Chromaticity differences are less than $\Delta u'v' = 0.05$, which means an acceptable uniformity across the screen [14]. Chroma differences (max. 3.4) and total color differences (max. 6.1) are situated in the visible range but as they appear far away from each other in the visual field and as they are connected by continuous color transitions they do not cause a visual artifact.

## 2.1.6
### Viewing Direction Uniformity

The surfaces of certain display types (especially LCDs) do not represent Lambertian light sources. This means that their radiance depends on viewing angle at every pixel. If such a display is viewed by multiple users or a single user views it from different directions consecutively, then considerable luminance and chromaticity distortions can be observed. Even viewing a spatially completely uniform display perpendicularly and from the same point by sitting in front of the display, the color stimulus changes in different parts of the screen because the viewing angle of the different parts varies.

As an example, the result of a measurement on a typical active-matrix LCD (AM-LCD) is shown below [15]. In this measurement, the LCD was measured by a luminance type spectroradiometer kept always at the same position by changing the direction of its optical axis to image each actual part of the LCD. Thus, the user's typical observation task was reproduced. The 17 measuring points can be seen in Figure 2.9. Spatial resolution was $640 \times 480$ pixels and the measured homogeneous color patch had $200 \times 200$ pixels in the middle of the screen. This size corresponded to a viewing angle of $10° \times 10°$ from the position of the spectroradiometer. The center of the image was assigned the coordinates (3; 3).

Seven test colors were measured in each of the 17 measuring points of Figure 2.9, peak white (W) as well as the R, G, and B peak primary colors and their combinations cyan (C), magenta (M), and yellow (Y). Chromaticity differences ($\Delta u'v'$) and color differences ($\Delta E^*_{ab}$) were computed between the center and each measuring point for each test color. Mean and maximum results are shown in Table 2.8.

As can be seen from Table 2.8, even the maximum $\Delta u'v'$ values are less than 0.006 resulting in an acceptable viewing direction uniformity (for chromaticity). The blue peak primary color exhibited the largest viewing direction dependence but it is still in the visually acceptable range, that is, less than $\Delta u'v' = 0.02$. Concerning the total color differences, the maximum value of $\Delta E^*_{ab} = 4.6$ is within the visually acceptable range (i.e., less than 5) for viewing direction uniformity.

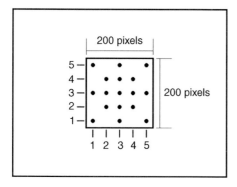

**Figure 2.9** Measuring the angle-of-view dependence of displayed colors on a $200 \times 200$ pixel homogeneous color patch subtending $10° \times 10°$ of viewing angle as viewed by the spectroradiometer. Colors were measured in the indicated points.

**Table 2.8** Mean and maximum chromaticity differences and color differences due to viewing direction nonuniformity measured on an AM-LCD display (example).

|  | Test colors | | | | | | |
|---|---|---|---|---|---|---|---|
|  | R | G | B | C | M | Y | W |
| $\Delta u'v'$ (mean) | 0.0012 | 0.0005 | 0.0024 | 0.0012 | 0.0012 | 0.0009 | 0.0010 |
| $\Delta E^*_{ab}$ (mean) | 1.8 | 1.8 | 2.4 | 1.8 | 2.1 | 2.1 | 2.0 |
| $\Delta u'v'$ (max.) | 0.0033 | 0.0014 | 0.0059 | 0.0025 | 0.0030 | 0.0023 | 0.0020 |
| $\Delta E^*_{ab}$ (max.) | 4.0 | 3.7 | 4.5 | 3.7 | 4.1 | 4.6 | 4.3 |

Seven test colors are shown (see text).

### 2.1.7
### Other Visual Artifacts

In addition to the technological artifacts described in Sections 2.13–2.16 and causing more or less serious colorimetric inaccuracies of the displayed color stimuli, the following effects will be dealt with in this section: the changing luminous output due to display warm-up, effect of nonzero display black, inter-reflections among the different parts of the display, pixel faults, the color fringe error, raster artifacts (aliasing) including the disturbing Moiré pattern, the spatial color quantization artifact, and discomfort glare.

After switch-on, the electro-optical components of the display approach thermal equilibrium continuously. For example, at the moment of turning on a CRT display, the shadow mask may be curved directing a part of the electron beam to the false phosphor causing visible color distortions. To display accurate color stimuli, the warm-up property of each device has to be measured. Accurate color stimuli can be displayed only after they do not change noticeably in time. As an example, Figure 2.10 shows the change of luminance during the warm-up process of an AM-LCD. The luminance of the full-screen peak white was measured as a function of time.

As can be seen from Figure 2.10, there is a rapid increase of luminance within the first 3 min (observe the logarithmic timescale) but there is no visible change after 10 min. Another issue is image stabilization time: if a displayed image is changed, then some time (usually a couple of seconds) is needed to stabilize the color of the new image even after the warm-up process.

Both of the next two artifacts (nonzero black point and internal reflections) cause a desaturation of all displayed colors: if the display's black point ($r=0, g=0$, and $b=0$) exhibits nonzero luminance like for the case of LCD monitors, this (neutral) offset light is added to all colors including the peak primary colors that become desaturated and the display's color gamut decreases. Internal reflections among different parts of the display have in general a similar desaturating effect.

Moreover, ambient light can be reflected from the screen if the display or the light source in the room is incorrectly designed or positioned. Reflected light is added to

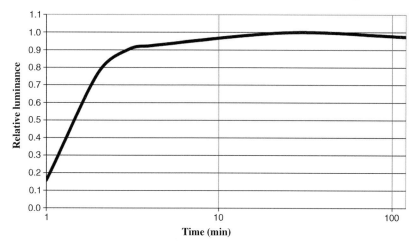

**Figure 2.10** Change of luminance during the warm-up phase. Relative luminance of full-screen peak white as a function of time (logarithmic scale) for an AM-LCD.

the light scattered from other self-luminous parts of the screen. Together with the possible nonzero light of display black (e.g., for LCD monitors), these factors (reflected and scattered light) can be accounted for by using an additive constant in the characterization model. Obviously, reflections should be avoided because they can significantly reduce the saturation of individual colors and all local contrasts on the display threatening its usability [14].

Pixel faults (those pixels that do not work correctly) interrupt the continuous image especially if clusters of pixels break down. Aliasing is a visual artifact due to too low a resolution of the display's pixel pitch compared to a resolution necessary to render a characteristic spatial pattern visible in the original. A typical example is the so-called *Moiré pattern* that is a visually very conspicuous spatial pattern visible on the display but not present in the original image.

The simplest display raster artifact arises from low resolution (the raster becomes visible for tilted lines or edges). A variation of the latter artifact is the so-called color fringe error (see also Section 5.3). This artifact arises on color displays using subpixel resolution. As a subpixel is red, green, or blue, fine edges of black, gray, or white lines at a subpixel exhibit a disturbing color fringe. Spatial raster artifacts can be reduced by dedicated image rendering algorithms (see also Section 5.4).

The spatial color quantization artifact arises because of the low color resolution of the display's color channels. Continuous shadings in the original image where a fine spatial color transition between two similar colors should be observed (e.g., delicate shadings of blue sky) cannot be rendered because the color resolution is not enough, and consequently, disturbing homogeneous areas (surrounded by visible edges) appear.

Discomfort glare occurs if the display has a too high luminance because it is set up in a dim room (e.g., in a badly illuminated conference room). Modern backlit displays represent a typical example of the discomfort glare artifact. It can be reduced by constraining the display's luminous output and increasing the general luminance level of the room.

## 2.1.8
### The Viewing Environment: Viewing Conditions and Modes

As mentioned at the beginning of this chapter, the colorimetric characterization model of the display is necessary but insufficient to obtain a *perceptually* accurate color appearance of the image. The characterization model is used to find the accurate color *stimulus* at every pixel of the display. The color *appearance* (the way observers perceive colors on the display) of a color element of the display (a homogeneous color patch, that is, a contiguous set of pixels subtending a visual angle of, for example, 1–2°) depends not only on the color stimulus of that color element (defined, for example, in terms of its *XYZ* values) but also on the other color stimuli in the total field of view (the so-called adapting field). This is the result of neural (retinal and post-retinal) interactions among the different parts of the retina and its neural projections to the visual cortex of the brain. The total field of view is usually divided into the background (a 10° field immediately surrounding the color element considered) and the surround (outside the background) [16].

Therefore, a mathematical model of color appearance has to be applied to the self-luminous visual display. Such a model goes beyond basic colorimetry (i.e., specification of the stimuli in terms of *XYZ* values) and predicts (among others) the perceived hue, lightness, and chroma of the color element considered by the user of the display. Predicting means mathematical equations of so-called hue, lightness, and chroma *correlates* (descriptor quantities) depending not only on the *XYZ* tristimulus values of the color element but also on a set of parameters characterizing the actual viewing condition of the display [16].

Viewing condition parameters of *non-imaging* color appearance models include (1) reference white chromaticity, (2) surround parameters, (3) adapting field luminance, and (4) degree of chromatic adaptation [17, 18]. This section and the next section concentrate on the currently widely used CIECAM02 color appearance model [17, 18] that balances between complexity and functionality and that is suitable for high-speed color management systems.

The CIECAM02 model consists of two basic parts, a chromatic adaptation transform to a reference viewing condition and equations for the perceptual correlates. In a reverse mode, the model can compute the CIE *XYZ* tristimulus values if the values of the perceptual correlates are given (e.g., hue, chroma, and lightness). In turn, by using a colorimetric characterization model, the *rgb* values can also be computed for any required color appearance on any self-luminous display.

In addition to hue, chroma, and lightness, the CIECAM02 model also provides correlates of brightness, saturation, and colorfulness. The workflow of the model [17, 18] is not described here because this book concentrates on explaining the way of its application to self-luminous displays. A free numerical spreadsheet implementation of the model (both in forward and in reverse mode) can be downloaded from the Internet http://www.cis.rit.edu/fairchild/files/CIECAM02.XLS.

CIECAM02 represents a significant improvement compared to CIELAB and CIELUV due to its perceptually more uniform hue and chroma scales and the possibility of fine-tuning a broad set of viewing condition parameters accounting for numerous effects of color appearance. CIELAB and CIELUV can change *reference white only* and use a simplistic and outdated chromatic adaptation transformation. Effects of the proximal field of the color element, that is, simultaneous color contrast, local adaptation, and spatial color vision (spatial frequency filtering of the different chromatic channels), are not considered in the CIECAM02 model, either. These aspects are covered by the so-called *image color appearance* framework iCAM [19].

The so-called *mode of color appearance* [20] of reflecting surface colors, for example, in a living room lit by a light source, is different from the color stimuli appearing on a self-luminous display. In the case of reflecting surface colors, the mode of color appearance is the so-called "surface" mode in which certain mechanisms of color constancy can be more active than those for self-luminous displays. Namely, in the latter case, the color stimuli often lose their direct connections to other parts of the visual field including the reference white. This effect diminishes the degree of chromatic adaptation.

## 2.1.9
### Application of CIELAB, CIELUV, and CIECAM02 to Self-Luminous Displays

This section describes application principles for the CIECAM02 color appearance model (and also for the two color spaces CIELAB and CIELUV) to typical viewing conditions of self-luminous displays. The starting point is the color stimulus $XYZ$ in any pixel of the display and the question is how to set CIECAM02's viewing condition parameters to obtain the appropriate numerical correlates of color appearance in that pixel (or how to choose the $XYZ$ values of *reference white* in CIELAB and CIELUV).

*Reference white* is an important attribute of the viewing environment because the visual system relates the appearance of single color stimuli to it. For example, the visual system compares the brightness and colorfulness of color stimuli [16] with the brightness of reference white, thus creating two additional internal psychological scales, that is, lightness in addition to brightness and chroma in addition to colorfulness. Therefore, to apply a color appearance model to a self-luminous display correctly, it is essential to define a suitable reference white.

Two kinds of reference white can be distinguished: adopted and adapted. An *adopted* reference white is a computational one: in the absence of any knowledge about what the visual system considers *white* in a given viewing environment of the display, one can speculate and consider, for example, the display's peak white as reference white even if the images actually shown do not contain peak white. Sometimes, if the application requires accurate color appearance, a peak white border is shown on the display to represent the so-called *adapted* reference white, that is, the actual reference white of the human visual system.

In the presence of ambient light, the adapted white point may be between the peak white of the display (e.g., a cold white tone with a correlated color temperature of

9000 K) and the color stimulus of the white walls of the room where the display is situated (e.g., illuminated by a white phosphor LED lamp with a color temperature of 2700 K). This is the so-called mixed adaptation situation [21]. In this case, the adopted reference white can be an intensity-weighted average of display peak white and the white of the walls.

Displays are rarely used in dark rooms because very dark viewing conditions compromise the efficient (i.e., photopic) working of color vision. As mentioned in Section 2.1.7, discomfort glare can also occur. Ambient light is usually switched on around the display providing an average surround luminance and the so-called *surround ratio* can be computed by dividing the average surround luminance by the luminance of the display's peak white. This ratio helps predict incomplete chromatic adaptation for the case of lower adapting field luminance.

The CIECAM02 model uses six viewing condition parameters: $L_A$ (adapting field luminance), $Y_b$ (relative background luminance between 0 and 100), $F$ (degree of adaptation), $D$ (degree of adaptation to the adopted reference white), $N_c$ (chromatic induction factor), and $c$ (impact of the surround). In the absence of a measured value, the value of $L_A$ can be estimated by dividing the value of the adopted white luminance by 5. For typical pictorial images on the display, the value of $Y_b = 20$ can be used or, intentionally, colors can be displayed on an appropriate neutral gray background of a given value of $Y_b$.

The values of $F$, $c$, and $N_c$ depend on the surround ratio: if the surround ratio is equal to 0 then the surround is called *dark*, if it is less than 0.2 then the surround is *dim*, otherwise it can be considered an *average* surround. The values of $F$, $c$, and $N_c$ are equal to 0.8, 0.525, and 0.8 for dark, 0.9, 0.59, and 0.95 for dim, and 1.0, 0.69, and 1.0 for average surrounds, respectively. For intermediate surround ratios, these values can be interpolated. In the CIECAM02 model, the value of $D$ is usually computed from the values of $F$ and $L_A$ by a dedicated equation. But the value of $D$ can also be forced to a specific value instead of using that equation. For example, it can be forced to be 1 to ensure complete adaptation to the adopted reference white.

Concerning the CIELAB and CIELUV color spaces [22], calculations start from the *XYZ* tristimulus values of the color stimulus on the display and provide numerical correlates of lightness, hue, and chroma (CIELAB) or lightness, hue, saturation, and chroma (CIELUV). Before applying these color spaces to color displays, the tristimulus values of the adopted reference white (usually the display's peak white) have to be specified. The main goal of these color spaces is the evaluation of color differences if the adapted reference white is not far away from a phase of daylight in an *average* viewing condition. For small color differences ($\Delta E^*_{ab} < 5$), the CIEDE2000 color difference formula [22] yields an improved prediction.

CIECAM02 has the following advantages compared to CIELAB and CIELUV: (1) viewing condition parameters can be set to represent the display's viewing conditions; (2) reference white can be changed over a wide range reliably; (3) they provide more numerical correlates for all perceived attributes of color (colorfulness, chroma, saturation, hue, brightness, and lightness); (4) numerical scales of these correlates correspond better to color perception than CIELAB and CIELUV correlates; and (5) CIECAM02-based color difference formulas and uniform color spaces have been established for small and large color differences exploiting the advantages of CIECAM02 [23].

## 2.2
## Characterization Models and Visual Artifacts of the Different Display Technologies

After an overview of contemporary display technologies and their applications, this section describes specific characterization models and visual artifacts of different display technologies including CRT, PDP, LCD, head-mounted display (HMD), head-up display (HUD), DMD, LCD projectors, and organic light emitting diode (OLED) displays. Various LCD technologies and their viewing direction uniformity are discussed. The Lambertian emission of OLEDs is compared with the angular emission characteristics of LCDs. Methods are presented to test the *angle-of-view* dependence and ensure viewing direction uniformity for LCDs.

The widespread use of non-CRT display technologies including FPDs (flat panel displays such as LCD and PDP) and different projector technologies requires the well-established CRT colorimetric characterization model to be revisited [24]. First, the well-tried CRT gamma characteristic cannot be used any more particularly in the higher DAC regions of LCD monitors and projectors since this may result in large colorimetric errors [25], while certain types of non-CRT displays mimic CRT tone curves (*gamma curves*) in order to come closer to the sRGB standard. Second, the chromaticity of certain types of LCDs is not constant over the whole range of *rgb* values.

There are also certain displays on the market exhibiting contrast enhancing features that cause the *XYZ* output of given *rgb* values to be dependent on the size of the displayed color patch. For any display, it is especially important to keep in mind that the parameters of the colorimetric characterization model depend on the settings (e.g., white point, gain, offset) of the device. The settings have to be fixed to yield repeatable data; in other words, the display has to be calibrated [25, 26] before its colorimetric characterization.

A good colorimetric characterization model can be implemented by a fast computer algorithm and it can be inverted effectively [27]. In general, according to Thomas *et al.* [26], colorimetric characterization models can be classified into three groups: (1) physical models that model the physical processes generating light inside the display; (2) numerical models that follow a black-box approach by measuring a set of test colors and model *XYZ* values from *rgb* values by fitting continuous functions; (3) look-up table (LUT) models using three one-dimensional LUTs (or one three-dimensional LUT) connecting *rgb* values with *XYZ* values and interpolations in between. The disadvantage of three-dimensional LUTs is that they require much computing power and are not straightforward to invert requiring the interpolation of a sparse 3D data set [27]. Instead of LUTs, neural network methods have also been proposed [28].

In the absence of light measuring instruments, the user can also estimate the value of $\gamma$ visually [29]. However, automatic characterization to be found in some high-end displays is a more accurate approach using a color sensor placed on the screen and software that calibrates the display to, for example, sRGB (see Section 2.1.2).

## 2.2.1
### Modern Applications of the Different Display Technologies

Modern display technologies have a wide range of applications including television, home movie theaters, 2D and 3D cinemas, computer monitors, presentations, digital camera displays, cell phones and PDAs (palmtop computers), automobile dashboards, aircraft instrument panels, in-car and in-flight infotainment systems (also in HD mode), virtual reality domes, and telepresence. For every application, several more or less suitable technologies can be used. According to Silverstein [30], available technologies can be classified in the following way: they are either projection displays or direct-view displays while head-mounted displays and head-up displays (Section 2.2.2.4) as well as holographic displays represent distinct categories.

For projection, CRT, LCD, and DMD can be used (Section 2.2.2.5) where inorganic LEDs represent an important emerging light source technology. In specific applications, such as theaters and planetaria, laser projectors are used. LCoS (liquid crystal on silicon) technology is utilized in small projectors. Direct-view technologies include CRTs (Section 2.2.2.1) and FPDs. FPDs can be grouped into emissive and nonemissive categories. Emissive displays include field emission displays (FEDs), vacuum fluorescent displays (VFDs), electroluminescent displays, inorganic LED displays (both very large LED mosaics and LED backlights, Section 2.3.2), OLED displays (Section 2.2.2.6), PDPs (Section 2.2.2.2), and backlit LCDs (Section 2.2.2.3).

Nonemissive displays include electrochromic and electrophoretic displays (the latter is used in e-book reader devices). For large-area direct-view displays, LED mosaics, PDPs, and LCDs dominate. For LCDs, the use of inorganic LED backlighting with optimized color filters enhances the color gamut and the spatiotemporal resolution of the display. Small displays utilize transmissive and transflective LCDs (in turn, small RGB LED backlights and optimal color filters enhance their color quality), OLEDs, and electrophoretic devices.

Three-dimensional displays represent an emerging application for science and entertainment with several available technologies including direct-view displays and projection displays with special glasses (binocular or two-image displays), autostereoscopic high-resolution TFT LCDs with microlenses creating a separate image for each eye, and virtual reality domes providing total immersion into a computer-generated visual environment. So-called holoform displays are intended to provide smooth motion parallax between the different points of view [31]. Volumetric displays reproduce the object in a volume in space; hence, the observer is able to view it from a wide range of viewpoints and angles similar to holographic displays that reconstruct the wavefront of the light from the original object [31]. Virtual reality (VR) systems are used in a variety of applications within industry, education, public, and domestic settings.

## 2.2.2
### Special Characterization Models of the Different Displays

This section describes and compares the specific colorimetric characterization models of today's most widely used color display technologies.

#### 2.2.2.1 CRT

CRT displays usually exhibit good phosphor constancy (Section 2.1.3) and good viewing direction uniformity (Section 2.1.6) but the extent of channel and spatial interdependence varies over a wide range among low-end and high-end CRT displays. Typical CRT artifacts include flicker sensation if refresh rates are low. For example, a refresh rate of 50 Hz causes a deterioration of visual performance and visual fatigue, while 100 Hz facilitates a reasonable visual performance [32].

Also, CRT displays are bulky and heavy reducing the ability of their users to change their position relative to the screen, for example, in an office application [32]. CRTs are sensitive to magnetic field variations [33]. Moving or turning the CRT display can lead to color distortions. Some of them have an onboard degaussing knob. This knob should be pressed (sometimes repeatedly) before making characterization measurements. Large CRT displays sometimes suffer from luminous inhomogeneity near the edges of the screen. This is sometimes corrected by built-in additional focusing elements (so-called *elliptical correction*).

Warm-up characteristics were shown in Section 2.1.7. Concerning image stabilization time (i.e., if the image is changing after warm-up), a full-screen peak white patch was displayed for 5 min on two CRT monitors in a dedicated experiment. Then, peak red, green, and blue patches of fill factor $f = 0.11$ were switched on in the middle of the screen. On both CRT displays, luminance fluctuations remained within $\pm 0.3\%$ after 10 s.

Concerning CRT characterization models, an equivalent form of Equation 2.4 can be derived by using a physical CRT model [7, 8, 10], in the following way. Each pixel of red, green, or blue phosphor is excited by an electron beam and its radiance depends on beam current. This current, in turn, depends on acceleration voltage controlled by the driving video signal (voltage) depending on the *rgb* values ($d_i$, $i$ = red, green, or blue, $0 < d_i < 2^n - 1$, $n$ representing the color resolution of each color channel) converted into analog voltages by a digital–analog converter (DAC). This is why *rgb* values are also called DAC values. The video signal $v_i$ depends on the *rgb* values in the following way:

$$v_i = (v_{max} - v_{min})\left(\frac{d_i}{2^n - 1}\right) + v_{min} \qquad (2.12)$$

In Equation 2.12, the maximum and minimum video signals are denoted by $v_{max}$ and $v_{min}$, respectively. Video signals are amplified to get the acceleration voltages $w_i$:

$$w_i = a_i v_i + b_i \qquad (2.13)$$

In Equation 2.13, the gain of the video amplifier is denoted by $a_i$ and its offset voltage is denoted by $b_i$. Most CRT displays have nested gain and offset controls (knobs) for all three color channels. High-end CRT displays have separate controls for each channel allowing for the adjustment of white point (peak white) chromaticity. The gain control knob is sometimes incorrectly called the contrast knob but in fact this knob controls maximum luminance (hence the magnitude of all values in the phosphor matrix) and the shape of the tone curve (see, for example, Figure 2.5). The offset knob is sometimes incorrectly called the brightness knob but this one adjusts display black luminance ($L_{bl}$) influencing especially dynamic range ($D$). Black level correction is very important for all display types. This is discussed for LCDs (Section 2.2.2.3) in detail together with the criteria of an optimum display setting.

If the setting of the CRT display is changed, then its color stimuli and the whole color gamut also change. This is why obtaining an optimum display setting should be the first step (calibration). Once this setting has been fixed, accurate colorimetric characterization should be carried out as a second step. Going back to Equation 2.13, the current of the electron beams $j_i$ depends on the acceleration voltages $w_i$ in the following way [7]:

$$j_i = \begin{cases} (w_i - w_{Ci})^{\gamma_i}, & w_i \geq w_{Ci} \\ 0, & w_i < w_{Ci} \end{cases} \quad (2.14)$$

In Equation 2.14, the so-called cutoff voltage is denoted by $w_{Ci}$ and the parameter $\gamma$ introduced in Section 2.1.2 gets a physical meaning. The spectral radiance of the red, green, or blue phosphors $M_{\lambda i}$ is proportional to the electron beam currents $j_i$ with a spectral constant of each phosphor $k_{\lambda i}$:

$$M_{\lambda i} = k_{\lambda i} j_i \quad (2.15)$$

Equation 2.15 expresses phosphor constancy (i.e., *relative* spectral radiance does not depend on electron beam current) and channel independence (the electrons of the red channel are focused correctly onto red phosphor and they do not excite the green and blue phosphors). For some CRT monitors, Equation 2.15 is not true and then matrix $\chi$ shall be used instead of matrix **P** (see Section 2.1.4), especially for colors of higher luminance [26], for example, in the seventh octave of Table 2.5. Using Equations , the maximum spectral radiance of a channel ($M_{\lambda i,\max}$) can be written in the following form:

$$M_{\lambda i,\max} = k_{\lambda i}(a_i v_{\max} + b_i - w_{Ci})^{\gamma_i} \quad (2.16)$$

The gain and offset quantities $k_{gi}$ and $k_{oi}$ are introduced as

$$k_{gi} = \frac{a_i(v_{\max} - v_{\min})}{a_i v_{\max} + b_i - w_{Ci}} \quad (2.17)$$

$$k_{oi} = \frac{a_i v_{\min} + b_i - w_{Ci}}{a_i v_{\max} + b_i - w_{Ci}} \quad (2.18)$$

And then, the relative spectral radiance ($T_i = M_{\lambda i}/M_{\lambda i,\max}$) can be written in the following form:

## 2.2 Characterization Models and Visual Artifacts of the Different Display Technologies

$$D_i = \left[k_{gi}\left(\frac{d_i}{2^n-1}\right) + k_{oi}\right]^{\gamma_i} \quad \text{if} \quad \left[k_{gi}\left(\frac{d_i}{2^n-1}\right) + k_{oi}\right] \geq 0$$

$$D_i = 0 \quad \text{if} \quad \left[k_{gi}\left(\frac{d_i}{2^n-1}\right) + k_{oi}\right] < 0$$

(2.19)

$D_i$ values (so-called *display tristimulus values*) are equivalent to the $RGB$ values ($D_1 = R$, $D_2 = G$, and $D_3 = B$) introduced in Section 2.1. In the literature, Equation 2.19 is called the GOG (gain–offset–gamma) model. As $k_{gi} + k_{oi} = 1$, by introducing $A_i = k_{gi}/(2^n - 1)$, Equation 2.19 can be written in the following form:

$$\begin{aligned} D_i &= [1 + A_i(d_i - (2^n - 1))]^{\gamma_i} \quad &\text{if} \quad A_i(d_i - (2^n - 1)) + 1 \geq 0 \\ D_i &= 0 \quad &\text{if} \quad A_i(d_i - (2^n - 1)) + 1 < 0 \end{aligned}$$

(2.20)

If the display's color resolution equals $n = 6$ and every channel has the same value of $A_i = A$ and $\gamma_i = \gamma$, then Equation 2.20 becomes equivalent to Equation 2.4. Instead of the "physical" tone curve model of Equation 2.19 or 2.20 – in a "black-box" approach – tone curves can be measured directly and piecewise linear interpolation, cubic spline interpolation, or a polynomial approximation can also be used. It is also possible to model $\log(D_i)$ by $d_i$ or by $\log(d_i)$. The second phase of colorimetric characterization consists of summing up the luminous output of the red, green, and blue channels. In this step, matrices **P**, **M**, or $\chi$ shall be used (see Section 2.1.4).

### 2.2.2.2 PDP

Plasma technology is based on gas discharge phenomena [34, 35]. The term plasma refers to ionized gas (mixture of Ne, Xe, or He) excited by high voltage. When a gas ion transforms back to its stable form (ground state), the excess energy is released as visible or UV radiation. While in monochrome type plasma displays emitted light is used directly, in color PDPs, UV radiation is converted to visible light by a phosphor similar to CRT technology. For the sake of color fidelity, the intrinsic tone curves (or in other words, electro-optic transfer functions or optoelectronic transfer functions, OETFs) of the red, green, and blue channels of the PDP are usually *transformed* into CRT-like tone curves by built-in hardware. PDP users are confronted with these tone curves only, although the PDP's technology of phosphor excitation is different from the CRT's. This transformation is important in computer-controlled color image rendering systems that often use sRGB.

PDPs exhibit high peak white luminance and a high dynamic range (HDR) but often suffer from degradation of peak white luminance over time (e.g., 50% by the end of their lifetime) [35]. False contours represent another typical artifact, especially for moving objects. This artifact arises due to the so-called *binary coding* that establishes the PDP's tone curves via temporal nonuniformity of the light emission scheme. Binary coding means that the intensity levels are generated within a time frame (16.7 ms) divided into eight subfields of different durations. A PDP is incapable of producing analog gray levels. Gray levels must be achieved by temporal dithering of the image. Another visible artifact is jitter due to statistical lag time of a discharge related to the number of seed electrons in a PDP cell when the address pulse is applied [35].

In this section, the colorimetric characterization of a 42″ PDP display is described [24]. In the experiment underlying this section [24], tone curves, phosphor matrix, display black radiation, and spectral power distributions were measured by a Photo Research PR-705 spectroradiometer [24]. The subject of the measurement was a 42″ (diagonal) high-definition plasma display panel (HD-PDP). The PDP was used in 1024 × 768 pixel (native) spatial resolution and 8 bits per channel color resolution with a warm-up time of 1 h. The so-called *color mode* of the PDP was set to its 9481 K white point. All spectral measurements were conducted in a dark room. The spectroradiometer was set up on a tripod in front of the PDP. Its optical axis was aligned perpendicular to the plane of the PDP's screen. The measuring distance was 80 cm.

Although the front lens of the spectroradiometer was surrounded by a protection cylinder to prevent reflections from the PDP, an additional black mask was placed in front of the screen allowing light to come out only from a 5 cm side length square hole in the middle of the screen where the spectroradiometer was focused. 60 × 60 pixel color patches were measured in the middle of the PDP. The background was gray ($r=g=b=128$), represented by $RGB$ digital counts of (128, 128, 128).

The general problem of colorimetric characterization is to choose a set of assumptions that allow the characterization to be performed with a practical number of measurements and that accurately predict the device's colorimetric behavior [9]. To characterize the PDP, three versions of the GOG model were used: (1) a simple model; (2) a model including channel interdependence; and (3) a model including black emission.

The simple GOG model (first model) computes $RGB$ values from $rgb$ values by Equation 2.19. In turn, $RGB$ values are transformed into $XYZ$ values by multiplying the $RGB$ vector by matrix $\mathbf{P}$ consisting of the $XYZ$ values of the peak primary colors, that is, the maximum values of the stand-alone red, green, and blue channels (Section 2.1.4):

$$\begin{pmatrix} X \\ Y \\ Z \end{pmatrix} = \begin{pmatrix} X_{r,max} & X_{g,max} & X_{b,max} \\ Y_{r,max} & Y_{g,max} & Y_{b,max} \\ Z_{r,max} & Z_{g,max} & Z_{b,max} \end{pmatrix} \begin{pmatrix} R \\ G \\ B \end{pmatrix} \tag{2.21}$$

To account for channel interdependence (second model), matrix $\mathbf{T}$ of the IEC standard [36] was used. Matrix $\mathbf{T}$ is defined in the following way:

$$\begin{pmatrix} X \\ Y \\ Z \end{pmatrix} = P \cdot T \cdot \begin{pmatrix} 1 \\ R \\ G \\ B \\ R \cdot G \\ G \cdot B \\ B \cdot R \\ R \cdot G \cdot B \end{pmatrix} \tag{2.22}$$

In Equation 2.22, matrix **P** (3 × 3) denotes the dominant relationship between *RGB* and *XYZ*. Matrix **T** (3 × 8) accounts for the interdependence among the red, green, and blue channels. Observe the difference between matrix χ (Equation 2.9) optimized to predict the relationship (R, G, B, R·R, R·G, R·B, G·G, G·B, B·B) → *XYZ* and matrix **T** (Equation 2.22) optimized to predict the relationship (R, G, B, R·G, G·B, B·R, R·G·B) → **P**$^{-1}$(*XYZ*). Matrix **T** was obtained by using the measurement result of the 32 colors (including different gray, red, green, blue, cyan, magenta, and yellow colors) defined in the IEC standard [36].

For certain displays, it is essential to account for black radiation (third model) because the luminance of display black ($r=g=b=0$) is not zero. PDPs always maintain a minimal excitation state. Hence, even when the desired image is black, atoms in the gas mixture are in continuous excitation. This makes the atoms transform into an unstable state. Returning to their original state, UV radiation is emitted and the phosphor will glow although the driving values are zero.

This light emission of the display black state causes chromaticity shifts of the primary colors with changing *rgb* values (and this constitutes the major part of *color tracking*, see below). This effect is significant for lower *rgb* values ($r, g, b < 60$). If the luminance of display black is not negligible, then it is reasonable to measure display black emission and take it into consideration in the characterization model. However, some instruments have insufficiently low sensitivity to measure display black emission.

Berns et al. [37] proposed a method to estimate the black level emission of computer-controlled devices in the absence of a colorimeter or spectroradiometer that is sensitive enough in this low luminance range. This estimation technique was used to estimate the CIE *XYZ* values of the black level of the PDP. It is based on the following assumption. Because the optimal black level results in channel chromaticities that are invariant with luminance level, the target function of the estimation was defined as the sum of chromaticity variances of each channel over a range of measurements. Minimizing this target function resulted in an estimate of display black.

If display black emission ("leakage") is taken into consideration in the colorimetric characterization model, then the *XYZ* values of display black must be subtracted from all measured values before setting up the phosphor matrix **P**. Note that, alternatively, matrices **M**, χ, or **T** can also be used instead of matrix **P**. Display black ($X_k, Y_k, Z_k$) should be added to the right-hand side of Equation 2.21 in the following way:

$$\begin{pmatrix} X \\ Y \\ Z \end{pmatrix} = \begin{pmatrix} X_{r,\max}-X_k & X_{g,\max}-X_k & X_{b,\max}-X_k \\ Y_{r,\max}-Y_k & Y_{g,\max}-Y_k & Y_{b,\max}-Y_k \\ Z_{r,\max}-Z_k & Z_{g,\max}-Z_k & Z_{b,\max}-Z_k \end{pmatrix} \begin{pmatrix} R \\ G \\ B \end{pmatrix} + \begin{pmatrix} X_k \\ Y_k \\ Z_k \end{pmatrix} \quad (2.23)$$

To compute the values of the parameters for the three models (GOG, GOG with channel interdependence, and GOG including display black correction, except for matrix **T** where the 32 colors of the IEC standard were used), the following *rgb* values were measured for every channel: $r, g, b = 10, 20, \ldots, 240, 250, 255$ (i.e., 3 × 26 measurements). These measurements represent the basis for the estimation of the characterization model parameters. Fitting the functions to the measured tone curves

**Table 2.9** Sample coefficients of the three colorimetric characterization models for the PDP of Section 2.2.2.2.

|  | First (GOG)/second (GOG with channel interdependence) | | |
|---|---|---|---|
|  | Red | Green | Blue |
| $k_g$ | 1.09 | 1.15 | 0.96 |
| $k_o$ | −0.10 | −0.16 | 0.04 |
| $\gamma$ | 1.72 | 1.51 | 2.23 |
|  | Third (GOG including display black correction) | | |
|  | Red | Green | Blue |
| $k_g$ | 1.06 | 1.11 | 0.96 |
| $k_o$ | −0.09 | −0.09 | −0.08 |
| $\gamma$ | 1.79 | 1.57 | 2.22 |

Note that the first and the second models have the same coefficients.

was carried out by a least-square estimate method and coefficients are listed in Table 2.9.

Note that the tone curve parameters of the first (GOG) model and the second (GOG model with channel interdependence) are the same since they differ only in matrix **T**. It is interesting to look at the numerical values of matrix **P** (first and second models) found for this PDP:

$$\begin{pmatrix} 88.03 & 55.66 & 57.57 \\ 46.66 & 122.99 & 28.9 \\ 1.7 & 13.83 & 288.59 \end{pmatrix}$$

Compare the above matrix with the one using display black correction where the values of display black are already subtracted:

$$\begin{pmatrix} 87.37 & 55.0 & 56.91 \\ 45.95 & 122.27 & 28.19 \\ 0.85 & 12.97 & 287.74 \end{pmatrix}$$

Figure 2.11 shows the tone curves of the PDP display. As mentioned above, the similarity to CRT tone curves (compare with Figure 2.4) is the result of a built-in hardware conversion in the PDP.

The performances of the above three PDP characterization models and sRGB (Section 2.1.2) were tested by computing CIELAB color difference between 157 different measured and predicted color stimuli on the PDP. The 157 test colors included all permutations of the *rgb* values {0, 63, 127, 191, 255} and the 32 IEC test colors to determine matrix **T** (see above). Mean color differences, their standard deviations (STDs), and maximum values are listed in Table 2.10.

As can be seen from Table 2.10, the GOG model performed best provided that display black radiation was taken into account (see the third column of Table 2.10). The mean CIELAB color difference is equal to 2.09 in this case. Although the maximum error is almost 15 CIELAB units, the STD value (1.48) indicates that this is an exceptional case.

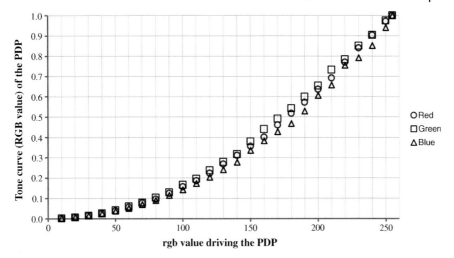

**Figure 2.11** Tone curves of the PDP display [24] of Section 2.2.2.2. Reprinted with permission of IS&T: The Society for Imaging Science and Technology, sole copyright owners of CGIV 2004 – Second European Conference on Color in Graphics, Imaging, and Vision.

Comparing the simple GOG model (first column of Table 2.10) and the GOG model using matrix **T** (second column), the mean difference is higher for the case of the simple model (3.71 against 2.84) and the STD values are almost equal. The maximum color difference is higher for the model with matrix **T**, which overall performed better. It seems that the channel interdependence artifact is not decisive for the accurate characterization of this PDP but display black correction is essential.

The sRGB model performed worst because – although the PDP tone curve characteristics are of the gamma (power function) type – sRGB's value of $\gamma = 2.4$ differs from the values of $\gamma$ in Table 2.9 and also the PDP's correlated color temperature is 9481 K instead of 6500 K. For a better understanding of the mean color differences of Table 2.10, the frequency of the individual color difference values between the 157 different measured and predicted PDP colors is plotted in the histograms of Figure 2.12. As can be seen from Figure 2.12, the third model that takes into account the radiation of display black has the best performance although the mean colorimetric error does not decrease below 2 CIELAB units.

**Table 2.10** Performance of the three PDP colorimetric characterization models and sRGB.

|      | GOG   | GOG using matrix T | GOG black | sRGB  |
| ---- | ----- | ------------------ | --------- | ----- |
| Mean | 3.71  | 2.84               | 2.09      | 32.01 |
| STD  | 2.90  | 2.87               | 1.48      | 7.76  |
| Max  | 19.48 | 23.30              | 14.79     | 50.22 |

CIELAB color differences between 157 different measured and predicted color stimuli on the PDP. Mean color differences, their standard deviations (STDs), and maximum values are shown.

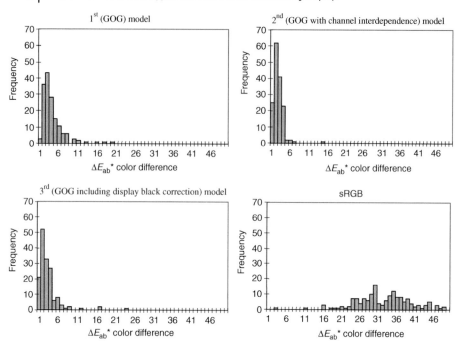

**Figure 2.12** Histograms of the individual color difference values between the 157 different measured and predicted PDP colors for the three colorimetric characterization models [24]. Reprinted with permission of IS&T: The Society for Imaging Science and Technology, sole copyright owners of CGIV 2004 – Second European Conference on Color in Graphics, Imaging, and Vision.

Since the sRGB model (Section 2.1.2) was developed for CRT monitors, it is vital for other display device types to follow the CRT's power function tone curve to display accurate color stimuli. The nature of the basic PDP tone curve (electro-optic transfer function) differs from the power function. Actually, there is a linear native relationship between the number of AC pulses driving the PDP per frame and its luminous output. But this relationship is modified by hardware to mimic a power function [38]. Further errors of the colorimetric characterization model arise from the so-called adaptive brightness intensifier: color patches of small fill factors exhibit a significantly higher luminance (up to five times for a peak white of, for example, $f = 0.04$ compared to the full-screen peak white).

#### 2.2.2.3 Various LCD Technologies and Their Viewing Direction Uniformity

The first active-matrix liquid crystal display [34, 39] was produced in 1972. Since then, three main problems of their colorimetric characterization have been identified. The first artifact is *color tracking*, that is, the dependence of the chromaticity of their RGB primary colors on the $r$, $g$, or $b$ value (this was called phosphor inconstancy for CRT displays). LCD primary colors shift in the CIE $x$, $y$ chromaticity diagram with increasing *rgb* values because the spectral transmittance of the liquid crystal cell changes with applied voltage [40].

The second artifact is color channel interdependence (see also Section 2.1.3). This arises on an LCD because the applied voltages of the neighboring LCD pixels may interfere with each other due to capacitive coupling [41, 42]. Third, pixel defects represent a less serious though noticeable problem. Due to the high manufacturing costs of large LC panels, some manufacturers allow 5–8 pixel defects (unchanging bright or dark spots) on each screen.

The fourth (most serious) artifact is viewing direction nonuniformity. Namely, the surface of an AM-LCD monitor is not a Lambertian radiator (see Section 2.1.6). Viewing angles providing acceptable color appearance and luminance contrast (e.g., 100: 1) are usually specified by the manufacturer, for example, 120° (horizontal) and 45° (vertical).

The visually acceptable viewing angle of an LCD depends strongly on the specific LC modes applied with longitudinal or transversal (or fringing) electric fields. If the directors of the liquid crystal are tilted out of the plane (e.g., in the twisted nematic, vertical alignment, and bend cell LCD), then the viewing angle is narrow and asymmetric. To widen it, so-called *optical phase compensation films* are applied to compensate for the light leakage at oblique angles [43].

Various *wide viewing angle* (WAV) LCD technologies were developed including film-compensated single-domain structures, multidomain structures, and film-compensated multidomain structures. The in-plane switching (IPS), film-compensated multidomain vertical alignment (MVA), film-compensated patterned vertical alignment (PVA), and film-compensated advanced super-view (ASV) LC modes can achieve a luminance contrast of 100: 1 within an 85° viewing cone [43].

In this section, the example of the colorimetric characterization of a desktop AM-LCD monitor (with a color resolution of 6 bit/channel) is shown [44]. The spectral radiance distributions of 84 small color patches were measured on this LCD, in the middle of the screen by the aid of an imaging type spectroradiometer. Five different colorimetric characterization models were used including different look-up tables and a model based on the hyperbolic tangent (tanh) function. Color differences between measured and predicted color stimuli will be shown.

First, the spectral radiance distributions of the individual RGB tone curves were measured for the following *rgb* values: $r, g, b = 0, 5, \ldots, 55, 60, 63 (= 2^6 - 1)$. These measurements constituted the basis for establishing the colorimetric characterization model. Then, the spectra of 42 additional colors were measured to control the accuracy of the model predictions (control measurements). These control colors had the following *rgb* values: $(d, 0, 0), (0, d, 0), (0, 0, d), (d, d, 0), (d, 0, d), (0, d, d)$, and $(d, d, d)$ with $d = 10, 20, 30, 40, 50, 63$. From the measured spectral radiance distributions of the control colors, the XYZ values $X_m$, $Y_m$, and $Z_m$ were calculated, and then the CIELAB $L_m^*$, $a_m^*$, $b_m^*$ values, and also the $u'_m$, $v'_m$ values on the CIE 1976 UCS diagram (index "m" is associated with the control colors).

The above quantities were also calculated from the predictions of five colorimetric characterization models based on the characterization measurements: $X_c$, $Y_c$, $Z_c$, $L_c^*$, $a_c^*$, $b_c^*$, $u'_c$, and $v'_c$ (index "c" is associated with the model predictions). Then, in turn, for each of the five models, 42 chromaticity differences $\Delta u'v'$ and 42 color differences $\Delta E^*_{ab}$ were calculated between the 42 control colors and the colors predicted from their *rgb* values by one of the five models. As mentioned in Section 2.1.6, the visually acceptable range for colorimetric characterization accuracy is less than $\Delta u'v' = 0.02$ or less than $\Delta E^*_{ab} = 5$.

Following models were considered for the characterization of the AM-LCD: (1) $3 \times 3$-LUT; (2) 1-LUT; (3) 1-tanh; (4) $3 \times 3$-tanh; and, for the sake of comparison, (5) sRGB. These models are described below.

1) **The $3 \times 3$-LUT model**: The X, Y, and Z values of the individual color channels (R, G, and B) were measured for the *rgb* values of 0, 5, ..., 55, 60, and 63: $X_r(r)$, $X_g(g)$, $X_b(b)$, $Y_r(r)$, $Y_g(g)$, $Y_b(b)$, $Z_r(r)$, $Z_g(g)$, and $Z_b(b)$. Display black radiation $(X_k, Y_k, Z_k)$ was subtracted from these functions to obtain the following functions: $f_{Xr}(r) = X_r(r) - X_k$; $f_{Xg}(g) = X_g(g) - X_k$; $f_{Xb}(b) = X_b(b) - X_k$; $f_{Yr}(r) = Y_r(r) - Y_k$; $f_{Yg}(g) = Y_g(g) - Y_k$; $f_{Yb}(b) = Y_b(b) - Y_k$; $f_{Zr}(r) = Z_r(r) - Z_k$; $f_{Zg}(g) = Z_g(g) - Z_k$; and $f_{Zb}(b) = Z_b(b) - Z_k$. The measured functions $f_{Ji}(i)$ ($J = X, Y, Z$; $i = r, g, b$) represent a look-up table that can be interpolated linearly between the LUT points.

The functions $f_{Ji}(i)$ can fully account for the color tracking effect because (1) display black radiation is subtracted and (2) different tone curves are used for X, Y, and Z instead of the three tone curves $R(r)$, $G(g)$, and $B(b)$ introduced in Section 2.1.1. The predicted XYZ values $X_c(r, g, b)$, $Y_c(r, g, b)$, and $Z_c(r, g, b)$ of the $3 \times 3$-LUT model are obtained by

$$X_C(r, g, b) = f_{Xr}(r) + f_{Xg}(g) + f_{Xb}(b) + X_k$$
$$Y_C(r, g, b) = f_{Yr}(r) + f_{Yg}(g) + f_{Yb}(b) + Y_k \quad (2.24)$$
$$Z_C(r, g, b) = f_{Zr}(r) + f_{Zg}(g) + f_{Zb}(b) + Z_k$$

In Equation 2.24, display black radiation $(X_k, Y_k, Z_k)$ is included. This is the correct way of tackling it: subtracting it from the measured XYZ values of the single color channels three times and then adding it to the final equation once [35, 45]. Note that Equation 2.24 presumes channel independence (Section 2.1.3) that is usually met in case of AM-LCDs.

2) **The 1-LUT model**: In this model, only one LUT is used based on the average of $f_{Ji}(i)$ ($J = X, Y, Z$; $i = r, g, b$). The value of 1 is set for the maximum *rgb* values (63 in the present example). In this model, every tone curve is represented by the single linear interpolated average function $f(d)$: $R(r) = f(r)$, $G(g) = f(g)$, and $B(b) = f(b)$. In the second step of this model, Equation 2.23 is used. The numerical example of the black-corrected matrix in Equation 2.23 for the AM-LCD of this section is as follows. This matrix reproduces the luminance of the displayed colors in cd/m² units and has a peak white chromaticity of $x = 0.372$, $y = 0.387$ and a correlated color temperature of about 4000 K:

$$\begin{pmatrix} 107.15 & 81.78 & 19.48 \\ 58.22 & 144.93 & 13.64 \\ 6.03 & 31.48 & 96.79 \end{pmatrix}$$

Alternatively, matrices **M**, **χ**, or **T** can also be used (see Section 2.2.2.2).

**Table 2.11** Estimated parameters of the fitted f(x)-type (Equation 2.25) functions in the 3 × 3-tanh model of Equation 2.25.

|   | $f_{xr}$ | $f_{yr}$ | $f_{zr}$ | $f_{xg}$ | $f_{yg}$ | $f_{zg}$ | $f_{xb}$ | $f_{yb}$ | $f_{zb}$ |
|---|---|---|---|---|---|---|---|---|---|
| a | 0.5910 | 0.5895 | 0.5233 | 0.5584 | 0.5544 | 0.5288 | 1.0209 | 1.1585 | 0.9967 |
| b | 0.0602 | 0.0603 | 0.0670 | 0.0627 | 0.0629 | 0.0663 | 0.0398 | 0.0382 | 0.0401 |
| c | 48.9361 | 48.8238 | 40.4169 | 45.9948 | 45.5110 | 41.7009 | 63.7201 | 66.8199 | 63.1251 |

3) **The 1-tanh model**: A hyperbolic tangent function is fitted to the $f(d)$ function in the previous model, for example, by using a least-square estimate method:

$$f(x) = a\{\tanh[b(x-c)] + 1\} \quad (2.25)$$

In Equation 2.25, $x$ represents any *rgb* value and $a$, $b$, and $c$ are parameters to be estimated from the 1-LUT method. For the example of the AM-LCD of this section, following parameter values were obtained: $a = 0.59545$, $b = 0.053886$, and $c = 48.4106$.

4) **The 3 × 3-tanh model**: Here, 3 × 3 tanh functions are fitted to the normalized $f_{ji}(i)$ functions of the 3 × 3-LUT method with 3 × 3 different $a$, $b$, $c$ parameter sets. For the example of the AM-LCD of this section, Table 2.11 contains the numerical values of these parameters.

Figure 2.13 shows the nine tanh functions fitted to the normalized $f_{ji}(d)$ functions with the parameters of Table 2.11 in comparison with the average

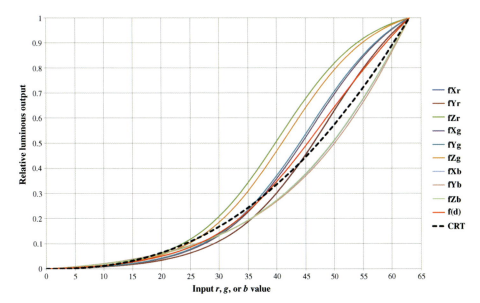

**Figure 2.13** 3 × 3 tanh functions fitted to the normalized $f_{ji}(d)$ functions (J = X, Y, Z; i = r, g, b) with the parameters of Table 2.11 in comparison with the average $f(d)$ function of the 1-LUT model and a typical CRT tone curve of $\gamma = 2.4$ [44].

$f(d)$ function of the 1-LUT model and a typical CRT tone curve of $\gamma = 2.4$. Compared to the CRT tone curve, LCD tone curves exhibit an S-shape modeled by the tanh function. The oddity of the $f_{Yb}(d)$ and $f_{Zb}(d)$ functions is obvious. This behavior causes peculiar color tracking characteristics that cannot be accounted for by black level correction only. Therefore, it is worth combining the $3 \times 3$-tanh model with black level correction. Note that in some LCDs, there is an integrated circuit transforming the native S-shaped tone curve into a *gamma* (i.e., CRT-like) tone curve.

Figure 2.14 shows the result of a prediction of color tracking characteristics for this AM-LCD based on the $3 \times 3$-tanh model with the model parameters of Table 2.11 and the sample black-corrected matrix shown above for the 1-LUT model. Symbols next to the spectral locus are associated with $r$, $g$, or $b = 63$, while the chromaticity of peak white corresponds to display black in this computational

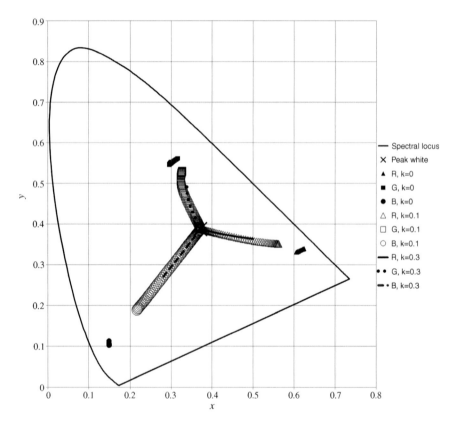

**Figure 2.14** Prediction of color tracking characteristics of an AM-LCD based on the $3 \times 3$-tanh model with the model parameters of Table 2.11 and the sample black-corrected matrix of the 1-LUT model. The *XYZ* values of display black were simulated by multiplying the *XYZ* values of the white point by a factor $k$. $k = 0$: color tracking contribution caused by different tone curves shapes for *X*, *Y*, and *Z*; $k = 0.1$: a lower value of display black; $k = 0.3$: a higher value of display black (desaturation of peak primary colors).

example. The XYZ values of display black were simulated by multiplying the XYZ values of the white point by a factor $k$ ($0.0 < k < 1.0$). Color tracking characteristics of $k = 0$ (i.e., in the absence of any display black radiation) correspond to RGB tone curve shape differences among X, Y, and Z, that is, the changes of the relative spectral radiance distributions of the individual color channels with changing *rgb* values [25].

The value of $k = 0.1$ corresponds to a lower level of display black, while $k = 0.3$ corresponds to a higher level of display black that causes a desaturation of the RGB peak primary colors. Note that a higher level of display black can arise either from the "light leakage" of the LCD or from neighboring pixels or the ambient light reflected from the screen. The sum of these contributions causes the main part of color tracking equivalent to overall desaturation of displayed colors and hence also a reduction of the display's color gamut.

5) **sRGB model**: This model is included for the sake of comparison for the case of a user attempting to predict the monitor colors by this widely available standard designed for CRT monitors.

Table 2.12 lists the mean and standard deviation values of the 42 chromaticity differences $\Delta u'v'$ and 42 color differences $\Delta E^*_{ab}$ calculated between the 42 measured control colors and their predictions by using one of the five models.

As can be seen from Table 2.12, the 3 × 3-LUT and 3 × 3-tanh models performed best. Fitting the tanh functions even enhances model performance slightly. The other models do not yield an acceptable colorimetric performance. This means that it is worth measuring and using 3 × 3 LUTs, that is, separate tone curves for each XYZ value and for each color channel to completely account for color tracking. sRGB could not be used for this AM-LCD monitor because the sRGB tone curve is designed for CRT monitors and not for the S-shaped LCD transmission voltage curves. The white point setting of the AM-LCD of this section was 4000 K instead of 6500 K as required by the sRGB standard.

An alternative method of removing channel crosstalk is the so-called *two-primary crosstalk model* [41] considering that channel interdependence is only due to two-channel crosstalk. The so-called *masking model* [42] takes both effects into consideration: channel interdependence and color tracking. The concept of the masking model is the so-called *under color removal*; that is, for any given color of the display, the

**Table 2.12** Mean and STD values of the chromaticity errors and color errors of the five colorimetric LCD characterization models.

| Model | $\Delta u'v'$ | | $\Delta E^*_{ab}$ | |
| --- | --- | --- | --- | --- |
| | Mean | STD | Mean | STD |
| 1. 3 × 3-LUT | 0.0094 | 0.0127 | 3.5 | 2.6 |
| 2. 1-LUT | 0.0296 | 0.0288 | 11.3 | 4.4 |
| 3. 1-tanh | 0.0263 | 0.0250 | 10.2 | 3.8 |
| 4. 3 × 3-tanh | 0.0107 | 0.0142 | 3.3 | 2.3 |
| 5. sRGB | 0.0454 | 0.0361 | 21.8 | 11.7 |

same amount of each *rgb* value is replaced by gray. If for this given color, for example, the condition $b < g < r$ holds, then the amount of $b$ is replaced by gray.

Then, in the red and green channels, the amount of $(g - b)$ is replaced by yellow. In the red channel, the amount of $(r - g)$ remains red. The corresponding replacements are carried out also for the other cases; for example, if $r < g < b$, then the amount of $(g - r)$ is replaced by cyan, and so on. The *XYZ* values are computed by adding the output values from, for example, gray, yellow, and red tone curves; hence, to implement this model, cyan, magenta, yellow, and gray tone curves shall also be measured in addition to the $R(r)$, $G(g)$, and $B(b)$ tone curves.

For example, a yellow tone curve $Y(r = g, b = 0)$ arises if the values of $r = g$ of a color patch are varied between, for example, 0 and 255 with $b = 0$ and the relative luminous output of the color patch is measured. A limitation of the masking model is that it assumes no color tracking for the combined primaries cyan, magenta, yellow, and gray and this requirement is not satisfied by LCDs [27].

Besides color tracking characteristics, the viewing direction dependence of the color stimuli displayed on an LCD represents another source of colorimetric error, and accordingly, the IEC standard [13] requires the measurement of viewing direction dependence. Figure 2.15 shows the example of another AM-LCD with 8 bit/channel color resolution. Tone curves of the red, green, and blue channels were measured by an imaging type colorimeter that was first aligned perpendicular to the display (0°) and then the display was tilted forward by 45° so that its principal axis was pointing downward at 45°. Observe the loss of dynamic range and color gamut in the 45° observation condition in Figure 2.15. In recent years, wide viewing angle LCDs were developed significantly reducing but not completely eliminating viewing direction dependence [40].

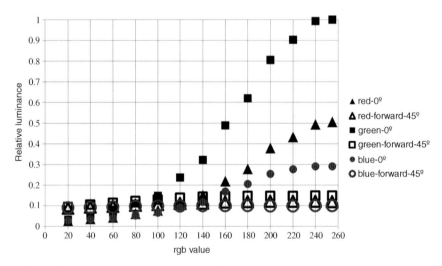

**Figure 2.15** Viewing direction dependence of AM-LCD tone curves (example). Red, green, and blue channels. The imaging type colorimeter was first aligned perpendicular to the display (0°) and then the display was tilted forward by 45°.

### 2.2.2.4 Head-Mounted Displays and Head-Up Displays

For traffic safety, it is essential not to distract the driver's or the pilot's attention from the environment. However, instrument or infotainment displays are often located away from the zone of visual attention required for driving. Combining the light from the display with the dynamic image of the real environment represents a nice way to view the information with "heads up" and looking in a forward direction to the *head-up display*. A head-up display represents a type of *augmented reality* because the information of the HUD is superimposed upon (usually projected onto) a so-called combiner (e.g., the windscreen of a vehicle) usually reflecting a projected image from a projector using different technologies.

A variation of the HUD device is attached to the head of the observer. This is a so-called *see-through* head-mounted display, its important application being *wearable computing*. Another type of HMD is a so-called *closed-view* HMD that displays video camera images of the real environment or computer-generated images (so-called *virtual reality*) on one or two microdisplays and does not allow any direct view of the real world [46]. In a HUD, a monochrome display may be adequate for certain applications but closed-view HMDs usually require full color.

HUD projection technologies include CRT, LED, DMD, or LCD projectors. HMD microdisplays use CRT, OLED, LCoS, MEMS, or AM-LCD technologies. For the colorimetric characterization of a HUD or HMD, specific models of these technologies can be used [47] and the spectrally selective reflection or transmission properties of the optical components (e.g., polarizing beam splitters) must be included in the colorimetric characterization model [48]. In a HMD, it is essential to intercalibrate the different microdisplays and to characterize their viewing direction characteristics [49, 50].

In augmented reality, colorimetric characterization can be used (1) to correct the color of the virtual object so that it matches the colors of the environment or (2) to correct the color of the video camera image so that it matches the color of the virtual object. For example, in a see-through HUD, the user encounters various scenes of various viewing conditions (especially changing brightness) and the visual system adapts to the real world. Hence, especially the brightness level of the virtual objects has to match the user's viewing condition [47, 51].

HMDs are suitable to display stereoscopic images as they often have two separate microdisplays. However, a visual artifact can occur resulting from the mismatch of distance between accommodation and convergence of the human eyes [52]. Such artifacts are common in virtual reality HMDs and 3D theaters. They are called *virtual reality-induced symptoms and effects* (VRISE) and include nausea, oculomotor, and disorientation symptoms [53].

HMDs are single-user devices to display augmented reality or virtual reality. Interactive stereoscopic experience can be realized in so-called virtual reality domes for an entire group of observers wearing spectacles and using specific pointing devices to interact with several objects "flying around" in the dome in 3D. These multi-user virtual environments are usually based on multiple projections on large curved display screens where the colorimetric characterization of each projector together with an intercalibration of all projectors (white point, tone curve, color gamut) is necessary [50].

## 2.2.2.5 Projectors Including DMD and LCD

Tone curve characteristics (also called electro-optic transfer functions) of projectors depend on technology. Figure 2.16 shows the black-corrected red, green, and blue tone curves of an LCD projector, as an example. Both the measured data (measured by an imaging type colorimeter on the screen) and the 1-tanh model of Equation 2.25 are shown in comparison with the average $f(d)$ function of the AM-LCD monitor and the CRT monitor (Section 2.2.2.3).

As can be seen from Figure 2.16, Equation 2.25 with the hyperbolic tangent function works well for this LCD projector, too. Observe the odd tone curve of the blue channel (the blue tone curves are different from the other tone curves in Figure 2.13, too) and the conspicuous difference from the shape of the CRT tone curve. Blue exhibited the largest chromaticity variation for another LCD projector, too [25]. The value of $Q(f=1.0)$ characterizing channel interdependence was equal to 0.95 for this LCD projector (see Equation 2.11). Equation 2.22 can also be used to quantify channel interdependence (sometimes also called interchannel dependence) [54].

For projectors, *calibration* of the device is especially important to find an optimum setting for a large dynamic range, good tone curves, and large color gamut. A criterion for a good tone curve is that if the luminance ($Y$ value) of each channel is transformed into CIECAM02 $J$ or a similar perceptually uniform lightness scale, then the $J(d)$ functions ($d=r, g$, or $b$) should be linear. A linear function means that increasing an $rgb$ value ($\Delta d$) by one unit corresponds to the same increase of perceived lightness independent of the value of $d$. In other words, the tone curves should represent perceptually uniform lightness scales in terms of the $rgb$ values (also called input digital values or $rgb$ counts). Then, the whole dynamic lightness range of the display can be exploited uniformly.

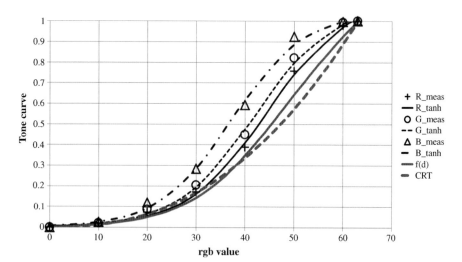

**Figure 2.16** Black-corrected red, green, and blue tone curves of an LCD projector. Measured data and the 1-tanh model of Equation 2.25. For comparison: average $f(d)$ function of the desktop AM-LCD monitor and the CRT tone curve of Figure 2.13.

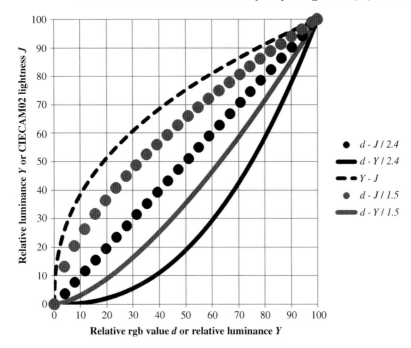

**Figure 2.17** Shape of the CIECAM02 $J(Y)$ function for a grayscale (upper dash curve) viewed in a dark viewing condition with the following values of the viewing condition parameters (Section 2.1.9): $F=0.8$, $c=0.525$, $N_c=0.8$, and $L_A=40\,\text{cd/m}^2$. $Y$: relative luminance of the grayscale values related to the luminance of the peak white of the projector expressed on a 0–100 scale. Two grayscale tone curves $Y(d)$ are also shown with $d=r=g=b$ (black and gray continuous curves) with gamma values of 2.4 and 1.5. The $rgb$ values ($d$) are related to their maximum value. They are expressed on a 0–100 scale.

Figure 2.17 shows the shape of the CIECAM02 $J(Y)$ function for a grayscale (upper dash curve) viewed in a dark viewing condition with the following values of the viewing condition parameters (Section 2.1.9): $F=0.8$, $c=0.525$, $N_c=0.8$, and $L_A=40\,\text{cd/m}^2$. In Figure 2.17, $Y$ denotes the relative luminance of the grayscale values related to the luminance of the peak white of the projector expressed on a 0–100 scale. Note that, in order to predict the color appearance of projectors by CIECAM02, the *dark* viewing condition parameters should be used in contrast to direct-view displays such as desktop LCD monitors corresponding to average or dim parameters [55].

In Figure 2.17, two grayscale tone curves $Y(d)$ are also shown ($d=r=g=b$, black and gray continuous curves) with gamma values of 2.4 and 1.5, respectively. Note that the $rgb$ values $d$ are related to their maximum value and expressed on a 0–100 scale. As can be seen from Figure 2.17, the $J(d)$ functions exhibit different shapes for the two different values of gamma. Using the value of 2.4 (as in sRGB, Section 2.1.2) yields a uniform lightness scale (black dots), while the value of 1.5 results in a perceptually nonuniform lightness scale (gray dots).

A too high contrast setting of a projector leads to the so-called *clipping effect*, that is, losing the dark range of the tone curve [25]. The dynamic range $D$ of an LCD projector changed dramatically, that is, between 35 and 350 by modifying its brightness/contrast setting [25]. Even if the color resolution (bit depth) of the color channels of the display is very high (e.g., 10–12 bits per channel), the first *rgb* values (0, 1, 2, 3, etc.) tend to be buried in the electronic noise of the driving circuitry of the display even in the absence of nonzero black luminance (e.g., for an LCD).

In cinematographic projection, to satisfy the observers' high visual expectations of continuous color gradations, a color resolution of 12 bits per channel is necessary (i.e., *rgb* values between 0 and 4096) [56]. For normal displays and TV applications, 8–10 bits per color channel (i.e., *rgb* values between 0 and 256–1024) is desirable, while 10 bits per channel corresponds to today's high-end computer-controlled consumer displays.

The screen should reflect the light of the projector diffusely (Lambertian surface) and in a spectrally neutral way (ideally reflecting white). To carry out the spectral or colorimetric measurements, an imaging type spectroradiometer or colorimeter can be placed, for example, above the projector and the radiance of the test colors shall be measured on the screen.

An LCD projector exhibited large spatial interdependence especially when the background of the varied color patch had high chroma and high lightness [25] possibly due to *flare* (i.e., internal light scattering from other parts of the screen). Concerning warm-up and temporal stability, the same projector [25] did not reached a stable state within 4 h while an LCD monitor reached stability after about 10 min. LCD projectors tend to exhibit noticeable spatial nonuniformity. In a full-screen peak white colorimetric measurement of 25 different locations, CIELAB color differences of up to 11.6 units (including chroma shifts up to 6 units) were found by taking the brightest part of the screen as reference white [57]. A special artifact of LCD projectors is the so-called *color mottle* [58] arising from the polarized nature of light they emit.

For the S-shaped tone curves of LCD projectors, an alternative hyperbolic tone curve model function [25] was proposed instead of Equation 2.25:

$$R = A_r (r/r_{max})^{\alpha r} / [(r/r_{max})^{\beta r} + C_r] \tag{2.26}$$

In Equation 2.26, only the R channel is shown. $(r/r_{max})$ represents a normalized *rgb* value, where $r_{max}$ is the number of bits per color channel, for example, 63 or 255. Similar equations were proposed for G and B with different parameters ($A_g, C_g, \alpha_g, \beta_g, A_b, C_b, \alpha_b,$ and $\beta_b$). The average CIELAB color difference for a set of measured and predicted test colors was equal to 2.7. To account for the color tracking characteristics of this LCD projector [25], additional terms using the first derivatives of the hyperbolic function in Equation 2.26 were added to the right-hand side of Equation 2.26. These terms represent the dependence of, for example, R on the value of g and b and average CIELAB color differences of 1.2 were achieved [25]. Note that, as a result of built-in image processing hardware, several commercial LCD and DMD projectors exhibit gamma tone curves [58].

In one-chip DMD projectors, the beam of the projector light source is filtered by red, green, and blue filters mounted on a rotating color filter wheel. As the wheel rotates, the light beam reaching the DMD becomes red, green, or blue, in three consecutive time slots. The DMD reflects a red, green, or blue image accordingly and this reflected beam is projected onto the screen by the projector's output lens. Note that DMD projectors are sometimes also called DLP (digital light processing) projectors with DLP being a trademark owned by Texas Instruments.

In three-chip DMD projectors, each color channel (R, G, and B) has its own DMD. This increases temporal resolution and avoids the typical motion artifact of color sequential displays such as the one-chip DMD projector. Such an artifact is caused by sudden head movements or by moving objects projected onto the screen. This artifact is called *rainbow artifact* or *color breakup* (CBU) artifact [59].

A DLP projector (a one-chip DMD) exhibited more spatial nonuniformity than an LCD projector in the above-mentioned study [57], up to 21.7 CIELAB units (including chroma shifts up to 6 units), which is a visible nonuniformity [57]. The luminance of full-screen peak white of a DLP projector was still fluctuating $\pm 5\%$ 8 h after switch-on [60]. The DLP projector in the latter study exhibited the same tone curve shapes for all three color channels unlike LCD projectors (see above). The tone curves of this DLP projector were close to sRGB. This DLP projector exhibited negligible color channel interdependence. For DLPs, channel interaction may arise from the imperfect synchronization of the DMD with the rotating color wheel.

The above-mentioned LCD projectors and DMD projectors with RGB filter wheels can be characterized by the above-described methods for displays with three primary colors (RGB). In some DMD projectors, however, there is also a fourth clear segment in the filter wheel to increase the maximum luminance of the projector to about 145% of the sum of the RGB primaries [58] and the RGB methods cannot be used. The algorithm [61–64] adds white, that is, the light of the projector lamp through the clear segment of the filter wheel in three fixed amounts at three *rgb* transition points. Each time the amount of white is increased, an appropriate amount of the RGB signals is removed. To characterize them, the usual $3 \times 3$ RGB phosphor matrix is extended to a $3 \times 4$ RGBW matrix to describe the additive mixture of four primaries, red, green, blue, and white. The inverse model, however, is not uniquely defined.

### 2.2.2.6 OLEDs

The main advantages of OLEDs are their low power consumption and their thin display structure [65]. A disadvantage of OLEDs is their relatively short lifetime. Depending on the substrate material, OLEDs can be made reflective or transparent and rigid or flexible. According to their size, there are OLED microdisplays, desktop monitors, and large OLED displays consisting of several OLED display modules. However, today's large OLED screens (as of 2011) suffer from several visual artifacts.

The most serious one is the artifact of colored reflections of ambient light creating disturbing color shades and decreasing legibility. Unfortunately, a pixel raster mosaic is often visible with thick lines among the light emitting elements due to the small fill factor of the actual light emitting areas. This artifact is visible even from a large viewing distance. Saccadic eye movements combined with the periodic jitter of the large screen tend to accentuate this raster artifact.

The most important question is how a full color gamut can be achieved by OLED manufacturing. To this end, a white light OLED can be combined with a color filter mosaic similar to LCDs. Alternatively, a red, green, and blue OLED subpixel mosaic can also be built, especially in small screen applications such as mobile phones to extend battery recharge periods. The so-called SOLED (stacked organic light emitting device) display technology uses a stack of transparent organic light emitting devices (so-called TOLEDs). This technology stacks the red, green, and blue subpixels vertically, thus enhancing the spatial resolution of the display [65].

OLED light intensity is proportional to current density and tone curves can be established by changing the current in contrast to LCDs that are voltage driven [65]. Pixel addressing can be achieved by using passive and active matrix addressing. Passive addressing is subject to crosstalk between pixels and is less for low-resolution OLED displays consisting of a small number of lines [65]. Active matrix addressing enhances dynamic range and power efficiency. The RGB tone curves of an active matrix OLED using color filters were measured [65]. These tone curves exhibited a quasilinear shape depending on the gain and offset settings of the OLED display. Spatial homogeneity across the whole OLED display was within ±5%. A further advantage of OLED displays is that their viewing direction uniformity is significantly better than that of LCDs.

## 2.3
### Display Light Source Technologies

In this section, different projector light sources and backlight sources of direct-view displays are compared with relevance to their white points, the use of color filters, local dimming, and HDR imaging. The white light source of the display is filtered by a colored subpixel filter mosaic in a direct-view display or by a rotating filter wheel in a projector. In a direct-view display, white OLEDs, white inorganic LEDs, cold cathode fluorescent lamps (CCFLs), or other types of light sources are mounted behind the filter mosaic and the liquid crystal layer as a backlight. In a projector (including front and rear projection displays), a UHP (ultrahigh pressure) or a xenon discharge lamp or an array of white LEDs illuminate the rotating filter wheel consisting of red, green, and blue filter sectors plus sometimes a clear window as an additional white primary color (see Section 2.2.2.5).

Alternatively, saturated red, green, and blue light sources can also be used as the light sources of the display. Direct-view CRT displays (Section 2.2.2.1) and very large RGB LED mosaic displays correspond to this category because the millions of red, green, and blue phosphor dots or RGB LED emitters can be considered as their light sources. *CRT projectors* also use three separate monochrome CRTs (red, green, and blue) as saturated light sources to project an image. Three-chip DMD projectors (Section 2.2.2.5) use saturated red, green, and blue high-brightness LEDs as light sources. Laser projectors depict the image by rapidly moving beams of red, green, and blue lasers. The use of RGB LEDs as a backlight provides highly saturated primary colors and a wide color gamut with the possibility of local dimming to enhance the dynamic range of the display (see Section 2.3.3).

## 2.3.1
### Projector Light Sources

Today, the most widely used projector light source is the UHP lamp to achieve the highest arc luminance (above 1 Gcd/m$^2$) ideal for projection applications [66]. Other light sources include tungsten halogen lamps, metal-halide lamps, xenon lamps, and LEDs [67]. In principle, the lifetime of UHP lamps can exceed 10 000 h. In practice, intensity is reduced considerably after about 2500–4000 h, usually continuously, without a sudden burnout while certain projectors allow a so-called eco-function for a longer lifetime.

For a usable color gamut, a balanced spectral power distribution of the projector lamp is necessary. For high mercury pressures (above 200 bar), more light is emitted within the continuous parts of the spectrum (especially in the important red part) compared to the emission inside the spectral lines [66]. Balancing the spectrum of the projector lamp is important in projectors with only one image rendering device (e.g., one-chip DMD) to provide enough luminous flux within the transmission bands of the red, green, and blue color filters for the primary colors (see Figure 2.18). This figure shows sample spectral power distributions of five illuminants, UHP (300 W), xenon (300 W), RGB LED, and cold white (6800 K) and warm white (2700 K) phosphor-converted LEDs (pcLEDs). The spectral transmittance functions of the RGB filters are modeled by step functions.

As can be seen from Figure 2.18, the spectral transmission range of the red filter is better covered by the warm white pcLED than by the UHP mercury lamp. The emission spectrum of the cold white pcLED fails to overlap with the red filter. For low-pressure mercury lamps, the red, green, and blue primary colors of the projectors are

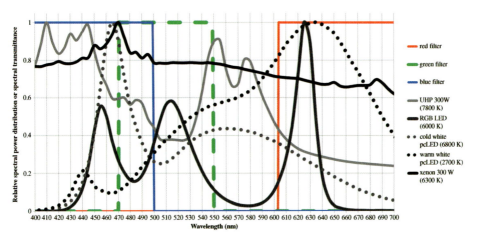

**Figure 2.18** Comparison of the spectral power distribution of an UHP projector lamp (7800 K, 300 W) with a xenon lamp (6300 K, 300 W), an RGB LED (6000 K), and a cold white (6800 K) and a warm white (2700 K) pcLED light source. Modeled spectral transmittance curves of an RGB filter set are also shown. Xenon and UHP emission spectra: courtesy of Osram GmbH, Munich, Germany.

less luminous, less saturated, and spectrally less balanced than for high pressures [66]. With high pressures, it is possible to achieve cold white points such as 7500 K preferred by most display users (see Section 3.5.3).

To create an optimum light source, phosphor-converted LEDs have a high optimization potential by using different types of phosphor materials (YAG or LuAG) and by using additional phosphors (such as a red phosphor). Because of this more flexible spectral optimization, compared to UHP discharge lamps, white LED-based projector light sources can attain an improved color gamut and improved brightness for their primary colors.

LED projector light sources are also more compact and have a longer lifetime. Another possibility is the use of a combination of red, green, and blue LEDs with appropriate emission maxima filling the spectral transmission bands of the red, green, and blue filters, respectively (see the spectrum of the RGB LED light source in Figure 2.18).

The luminous flux of single white LED light sources is significantly lower (a typical value is 100 lm) than that of a UHP lamp (a typical value is 2000 lm) [68, 69]. But this can be significantly improved by constructing complex white illuminator units with a rotating filter wheel. A disadvantage is that rotating filter wheels in a one-chip DMD projector are subject to the rainbow artifact (Section 2.2.2.5) because the subframes of moving objects are not incident on the same position of the human retina while it is scanning the display, so disturbing color patches appear at the edges of moving objects [59].

Conventionally, the color gamut of a display is depicted in the $x$, $y$ chromaticity diagram and the aim of gamut optimization is to cover the set of chromaticities inside the triangle spanned by the NTSC color primaries. Note that in Europe the E.B.U. chromaticities are used. This is the so-called *NTSC triangle* or NTSC gamut with the $x$, $y$ chromaticity coordinates of (0.67, 0.33), (0.21, 0.71), (0.14, 0.08), and (0.310, 0.316) for the red, green, and blue primaries and the white point of illuminant C, respectively (see Figure 2.19).

RGB LED illuminators provide a more extended gamut and high luminous flux (up to 2000 lm as of today). They combine the light of an array of high-power red, green, and blue LEDs with appropriate heat dissipation and project it onto a single DMD or LCoS panel [70]. Other types of RGB LED illuminators are also used in small pocket projectors of low luminous flux. There are also so-called LED–laser projectors that substitute the green and blue LEDs by a laser.

RGB LED illuminators are often applied in a so-called *color sequential* mode. In this mode, the single DMD (or a similar microdisplay) scrolls through the red, green, and blue subframes of the image while the LED illumination changes accordingly in color. This avoids the need of moving parts such as the rotating filter wheel because the RGB LEDs can be switched off when they are not in use. RGB LED illuminators can be extended by cyan and amber LEDs providing a multi-primary color display (see also Chapter 5). If a high-speed microdisplay is used – such as a DMD or a fast LCD or LCoS – then the duty cycle of the colored LEDs can be varied to adjust the white point [71]. In contrast to this, in a rotating color wheel projector, it is the relative filter segment size that determines the white point.

In Figure 2.19, the color gamuts of the light sources of Figure 2.18, that is, the UHP projector lamp (300 W), the xenon lamp (300 W), the RGB LED, and the cold white

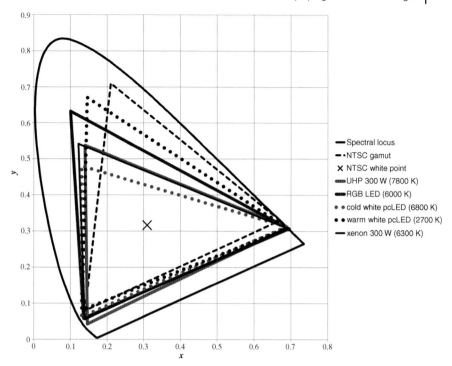

**Figure 2.19** Comparison of different color gamuts in the x, y chromaticity diagram. UHP projector lamp (7800 K, 300 W), xenon lamp (6300 K, 300 W), an RGB LED (6000 K), and a cold white (6800 K) and a warm white (2700 K) pcLED light source. The NTCS color triangle is also shown.

(6800 K) and warm white (2700 K) pcLEDs, are compared. Their emission spectra were filtered by the model step function RGB filter transmissions depicted in Figure 2.18.

As can be seen from Figure 2.19, the RGB LED illuminator (6000 K, covering 104% of the area of the NTSC triangle in the x, y chromaticity diagram) increases the color gamut by 18% compared to UHP (7800 K, covering 86% of the NTSC triangle). The gamut with the cold white phosphor-converted LED (6800 K) covers only 74% of the NTSC gamut, while the warm white pcLED (2700 K) covers 104% (at the price of a not-preferred white point). The performance of the xenon lamp (6300 K, 86%) is comparable to that of the UHP lamp.

### 2.3.2
### Backlight Sources

Backlights can be differentiated according to their structure and their type of light source [72]. The structure of the backlighting device depends on the type of display it is used for; for example, edge backlights are used for notebooks and monitors while direct backlights are used for LCD TVs. In an edge-type backlight, a so-called *light guide plate* is used to direct the side light emitted from the light source to the liquid crystal panel. Direct backlights provide more luminous flux than edge-type backlights [72].

LCD backlight sources include CCFLs, external electrode fluorescent lamps (EEFLs), flat fluorescent lamps (FFLs), white phosphor-converted LEDs, RGB LEDs, hot cathode fluorescent lamps (HCFLs), OLEDs, and electroluminescent (EL) light sources. Today, the CCFL and LED technologies are used most often, the CCFL being the classic light source of AM-LCDs while LEDs and OLEDs are continuously gaining momentum in all applications because of their small size, decreasing costs, increasing efficiency, and large color gamut [72].

Compared to a CCFL backlight, an RGB LED backlight is able to enhance the image quality of LCDs significantly [73]. One of the most serious problems of LCDs is motion blur due to the slow response time of the liquid crystal together with the so-called hold-type driving technique, that is, the driving signal is held for a long time on the LCD. To solve this problem and that of the rainbow artifact, a so-called *mixed color sequential* (MCS) algorithm was proposed for LCD RGB LED backlights controlling RGB LED backlighting blocks simultaneously both in the temporal and in the spatial domain [59].

Besides the reduction of motion artifacts (the rainbow artifact and motion blur [59]) and the achievement of a higher dynamic range, the most important enhancement concerns the color gamut of the display. In combination with an optimized color filter mosaic, RGB LED backlight is able to achieve 110% of the NTSC color gamut [59, 73] enabling natural and preferred color appearance (see also Sections 3.4–3.6). A new method of color gamut optimization based on all perceptual dimensions of color appearance (including lightness) is described in Section 5.1. This method includes a theoretical gamut optimization not only with red, green, and blue primaries but also with additional colors such as cyan, magenta, and yellow.

By the individual control of the different colored LEDs in the backlight of a display operating in color sequential mode, it is the *individual* LEDs' spectral power distributions multiplied by the spectral transmittance of the corresponding color filters that determines the chromaticities of the color primaries and the color gamut of the display unlike the example of Figures 2.18 and 2.19. The backlight of such a color sequential display can be a combination of RGB LEDs. The backlight can also be a combination of two different green LEDs (G1 and G2), one blue LED, and one red LED in a so-called *four-primary display* [74].

In the latter display, the filter mosaic in front of the liquid crystal consisted of RGB color filters and the peak wavelengths of 660 nm (red), 502 nm (G1), 520 nm (G2), and 415 nm (B) were applied. In a sample computation carried out by the present authors, the spectral power distributions of RGB LEDs and RG1G2B LEDs were *optimized* to obtain a *large gamut* in the $x, y$ chromaticity diagram for a simulated color sequential display while maintaining an acceptable white point and reasonable luminance values for all primaries. The spectral power distributions of the LEDs were simulated by Gaussian functions of a fixed FWHM (full width at half maximum) of 25 nm and their peak wavelengths were changed.

In this example of the RGB LED backlight, it was assumed that – in the time sequence when the red LED is on – only the subpixels of the red color

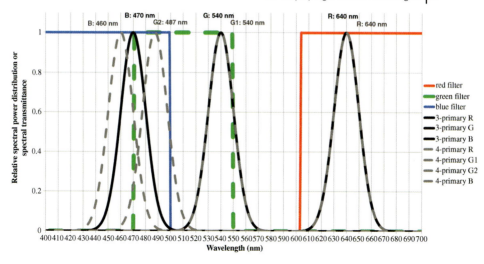

**Figure 2.20** Optimized model spectral power distributions (FWHM = 25 nm) of the RGB LEDs of the three-primary RGB backlight (black curves) and the four-primary RG1G2B backlight (gray dash curves) and the model spectral transmittance functions of the RGB filters. The display is driven by color sequential signals.

filters are transmitting while the subpixels of the green and blue filters are closed, and analogously for the G and B LEDs. For the RG1G2B LED backlight, it was assumed that – in the two time sequences of the G1 and G2 LEDs – only the subpixels of the green color filters are transmitting, and analogously for the R and B LEDs.

The resulting (optimized) backlight spectral power distributions can be seen in Figure 2.20 together with the spectral transmittance curves of the fixed RGB filters from Figure 2.18. In this sample computation, the spectral transmittance of the liquid crystal was ignored. Figure 2.21 shows the resulting color gamuts compared with the NTSC triangle. Note that the purpose of this sample computation is just an illustration of the color gamut issue from the colorimetric point of view while the efficiency of the different LED semiconductor materials was not considered. A comprehensive list of all optimization factors and a more advanced method can be found in Section 5.1.

As can be seen from Figures 2.20 and 2.21, the introduction of a secondary green LED (G2) with a peak wavelength of 487 nm enhances the color gamut by introducing new saturated cyan chromaticities. The blue peak of the four-primary device is shifted to 460 nm instead from the 470 nm of the three-primary display adding further deep blue tones. Thus, the four-primary RG1G2B display of this theoretical example achieved 148% of the NTSC color gamut, while the three-primary RGB display exhibited 123%. It can also be seen from these figures that, although color filters are broadband devices, due to the relatively narrowband colored LED emission spectra (FWHM = 25 nm in this example) the excitation purity of the primary colors increases that in turn increases color gamut [73].

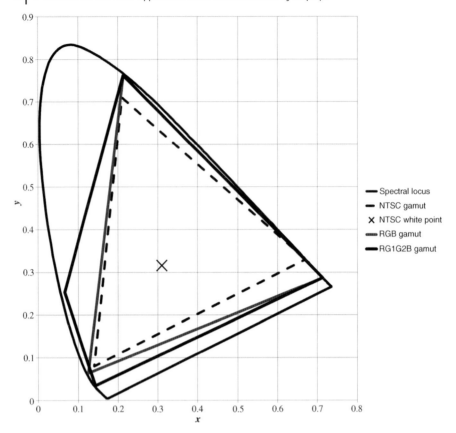

**Figure 2.21** Color gamut of the optimized RGB and RG1G2B model color sequential LED displays for the example of Figure 2.20 compared with the NTSC gamut. In this example, the optimized four-primary color sequential RG1G2B LED display achieves 148% of the NTSC gamut while the optimized three-primary color sequential RGB LED display achieves 123% of the NTSC gamut.

An additional interesting approach is the novel image rendering method of *hybrid spatial–temporal color synthesis* using two types of backlighting spectral power distributions, yellow and blue (e.g., LEDs or fluorescent). This backlighting is combined with a checkerboard subpixel mosaic with only two primary colors consisting of magenta and cyan color filters. The yellow backlight is activated in the first temporal sequence and then red and green light is emitted by the subpixels (red from magenta and green from cyan). Consecutively, blue backlight is emitted from all pixels as both cyan and magenta filters become blue when they are illuminated by blue light. The temporal combination of the two fields yields a full-color display [30].

A further important issue of backlighting is providing uniform chromaticity and luminance across the screen. For LED backlights, spatial uniformity can be achieved by the correct positioning of the LED arrays, optimizing the back reflector, the diverter, and the materials of the optical elements [75]. For RGB LED backlights, spatial uniformity needs to be combined with good color balance between red, green, and blue LEDs. The advantage of RGB LEDs is that the white point is tunable in contrast to white phosphor-converted LED backlighting [76]. The white point is sometimes controlled by feedback sensors to maintain color consistency between the different devices [77, 78]. Most display users prefer a white point of about 9000 K instead of the standard D65 white point.

Concerning viewing direction uniformity, optical simulation results of a wide viewing angle LCD showed that, by using an RGB LED backlight instead of a CCFL backlight, maximum chromaticity shifts arising from tilting the display between $-80°$ and $80°$ can be reduced by a factor of 2–4 [79]. Concerning initial switch-on characteristics, white LEDs in an edge-lit LCD TV exhibit less (9%) warm-up luminance change than CCFL backlights (70%) [80].

### 2.3.3
### Color Filters, Local Dimming, and High Dynamic Range Imaging

The filters of the color wheel of a DMD projector are composed of sectors of different spectral transmissions. When the wheel rotates, the light beam is modulated sequentially by the DMD according to the primary color of the illuminated sector [81]. The *XYZ* tristimulus values of each primary color can be computed by integrating the product of the spectral transmittance of the filters weighted by their relative areas, the spectral power distribution of the light source, and the standard color matching functions (CMFs). The relative areas of the color filters should be chosen to yield a balanced white point. The spectral transmittance curves of the color filters and their relative areas can be optimized for a given light source spectral power distribution to obtain a large gamut covering the most important colors (see also Section 5.1.1).

In an RGB LED backlight with a color filter mosaic, the color filters and the RGB LEDs can be co-optimized so that the filter transmission maxima overlap with the RGB LEDs' emission peaks [77]. Theoretically, a spatially homogeneous RGB backlight can also provide color sequential illumination without the need of a color filter mosaic. This increases the brightness of the display by maintaining a wide color gamut because no light intensity is lost due to the color filters. To obtain a large color gamut, *wide-gamut* CCFL backlights with improved phosphors can also be used. By optimizing the spectral transmittance of the filters of a color filter mosaic consisting of red, green, blue, cyan, and yellow filters and changing the relative areas of the colored subpixels, 110% of the NTSC gamut was achieved by using this type of backlight [82].

The conventional aim of backlighting design has been to provide a completely uniform light distribution behind the liquid crystal layer. But such a design has the disadvantage of LCD light leakage, that is, the nonzero black point of LCDs (see Section 2.2.2.3), reducing the dynamic range and the color gamut of the display. To diminish light leakage, the backlight itself is dimmed behind dark image regions *locally* to make them even darker there [75]. In this *local dimming* (or *area control*) procedure, the backlight itself contributes to the image formation of the display in addition to the liquid crystal structure and the color filter mosaic.

In the so-called *adaptive brightness intensifier method* [83] of local dimming, the luminous intensity of the backlight is dimmed uniformly by a factor $k$ across the whole display. However, if the value of $k$ is too large, it cannot be compensated for by the image signals to keep the original shading of the display. To overcome this difficulty, several groups of backlighting light sources can be isolated vertically and using a different value of $k$ for each group [84]. To avoid blocking artifacts near the boundary of two neighboring groups due to the optical isolators, the backlight RGB LEDs can be grouped into blocks both horizontally and vertically. The intensities of these blocks are dimmed in the intentionally dark image regions. Every RGB channel is dimmed simultaneously by the same ratio.

Chen *et al.* [75] give a good overview about the possible visual artifacts of local dimming. If the luminous intensity of an LED group is computed by the *maximum* gray level of the image block, then the image becomes sensitive to noise. If the *average* gray level is used instead of the maximum gray level, then the brightness of certain pixels decreases. Pixels of decreased brightness are sometimes called *clipped* pixels. Images with several clipped pixels may exhibit a false contour. This is the so-called *clipping artifact*. Several *backlight luminance compensation* (BLC) algorithms have been proposed to reduce this clipping artifact [85, 86].

Spatial uniformity can be increased by applying a low-pass filter to the backlight LED intensities. Local dimming can also cause a *flicker artifact* due to an incorrect algorithm with too frequent backlight intensity changes with respect to the video sequence. To overcome the above difficulties and to establish a dynamic range of up to 10 000: 1 within the same video frame, a new local dimming algorithm including spatial and temporal video filters was developed [75].

Contrast ratios of 100 000: 1 can also be achieved by spatially very fine local dimming. In this case, the backlight is composed of very small white LEDs with a pitch of 5–25 mm. This means that the LED backlight is used as a low-resolution (coarse) primary display and the liquid crystal structure as a secondary (fine) image modulator [87]. Such displays are called *high dynamic range* displays. The HDR image is especially sensitive to short-term temporal light variations or long-term degradations of the backlighting LEDs so that the use of optical feedback sensors is essential [77].

To present a HDR image for the visual system in a perceptually correct manner, both on HDR and on non-HDR displays, special HDR image rendering algorithms are necessary [88]. For non-HDR displays, so-called *tone mapping operators* are needed to compress the dynamic range of the HDR image to the conventional value of about 300: 1 in order to make it somewhat similar to its HDR appearance. Spatially varying tone mapping operators reduce the global contrast of the image without affecting its local contrast to remain in the preferred contrast range of the viewers (see also Sections 3.5.4 and 3.5.5). Tone mapping also represents an important application of the image color appearance model (iCAM) [19].

Although tone mapping operators are able to yield a HDR illusion, they cannot replace the appearance of a real HDR display [88]. Extremely high local contrast appears blurry due to intraocular scatter of light in the vicinity of bright objects. The same effect (so-called veiling luminance) is also responsible for disability glare [89].

However, the interaction between glare and perceived local contrast is very complex because retinal mechanisms compensate for the loss of local contrast due to veiling luminance. This compensation makes a high global dynamic range perception in a HDR image possible [90].

## 2.4
### Color Appearance of Large Viewing Angle Displays

As learned from the previous sections, to obtain a desired color appearance on a self-luminous display, the procedures of colorimetric characterization and the subsequent application of a color appearance model are used to adjust the spectral composition and intensity of radiation of the display at every pixel. The input of the CIECAM02 color appearance model includes not only local spectral composition and intensity (at every pixel individually) but also viewing condition parameters such as adaptation luminance, reference white chromaticity, and degree of chromatic adaptation. Aspects of spatial color vision such as chromatic contrast sensitivity to different spatial frequencies are considered in the so-called image color appearance framework [91, 92].

In addition to the previously mentioned factors, there is a very important parameter influencing the color appearance on today's large self-luminous displays, that is, the *size* of the color stimulus. If the color stimulus subtends a visual angle of more than about 20° (e.g., 50° on a PDP or 220° in a virtual reality dome), then perceived lightness (or brightness, if there is no observable reference white) and chroma (or colorfulness) tend to be greater than expected from CIECAM02 predictions. Nonsystematic hue changes were also observed. This so-called color size effect is the subject of this section. Methods and results of psychophysical experiments will be presented together with two mathematical models for an immersive color display of very large viewing angle and for a 50° display.

### 2.4.1
#### Color Appearance Differences between Small and Large Color Stimuli

Physiological reasons of the color size effect originate from cone density and cone distribution of the human retina. Depending on the size of the visual field, small-field and large-field phenomena can be distinguished, starting from "minutes of arc" domains, and up to the complete immersion of the observer in the color stimulus, yielding a different color perception from the "standard" size (e.g., 2°) color stimulus of the same spectral composition [93]. The perceived colors of stimulus sizes of 22° and 77° were matched with that of a 2° stimulus visually. As a result of modifying stimulus size, a change in excitation purity was found [94, 95].

Recently, with the advance and the widespread use of new experimental techniques, such as color matching on computer-controlled color monitors or the magnitude estimation technique, it became possible to begin a more extensive research on the color size effect in terms of more sophisticated and powerful tools in color science, such as CIELAB and CIECAM02 and the new imaging devices such as well-calibrated and -characterized color displays capable of providing very large self-luminous stimuli of accurate color.

The CIE 1931 standard colorimetric observer [96] contains the color matching functions of an average trichromatic observer, collected from color matching experiments where the subtended visual angle of the matching fields was 2°. The different pigment content of the fovea, that is, the presence of the yellow spot (macula lutea), made the introduction of a supplementary set of CMFs necessary. They are known as the CIE 1964 standard colorimetric observer [96]. Color appearance models usually assume only a 2° color stimulus size. From the experimental results of this section, it will become apparent that CIECAM02 does not account for the color appearance of large size stimuli.

Practical applications such as large self-luminous displays, head-mounted displays, or virtual reality environments require a mathematical model of the color size effect to provide a perceptually accurate color appearance characterization for advanced color design. Also, out of scope of this book, there is an industrial demand to be able to predict the color appearance of the same dye if applied on an indoor surface (walls of a real room) [97] or on a facade surface of large size [98–100].

On a computer-controlled self-luminous display, the viewing situation and the displayed colors are often totally immersive and subject to (sometimes rapid) temporal changes. This needs a special psychophysical experimental design to explore the color size effect. In this section, two experiments will be presented. In the first experiment (Section 2.4.1.1) [93], a totally immersive color stimulus was observed on a large PDP extended by a viewing chamber of mirrors in visual comparison with a smaller CRT monitor to match the large color stimulus with a color stimulus of either 2° or 10° size. In the second experiment (Section 2.4.1.2) [101], the color stimulus subtended either 8° or 50° on an LCD monitor. Both the 8° and the 50° stimuli were matched to a 10° color stimulus on a CRT monitor and the color size effect was modeled between the 10° matching colors corresponding to those of the 8° and 50° stimuli.

### 2.4.1.1 Color Appearance of an Immersive Color Stimulus on a PDP

In this psychophysical experiment [93], two computer-controlled color monitors were used: a 54 cm diagonal CRT monitor and a 106 cm diagonal PDP. The two devices were set up next to each other on a table (see Figure 2.22).

As can be seen from Figure 2.22, the PDP served to show the large (immersive) color stimulus. To achieve the immersion of the observer into the viewing field of the stimulus, that is, to fill the whole visual field of the observer, a viewing booth was set up in front of the PDP. The inner wall of the booth was covered by mirrors. The visual field of the PDP itself subtended 85° (horizontal) and 55° (vertical) and the mirrors provided total immersion. Standard size stimuli on the CRT had to be matched to the large stimuli on the PDP in a so-called "asymmetric" color matching procedure (asymmetric because the two stimuli to be matched visually were presented under different viewing conditions). The CRT monitor had an achromatic background subtending $39° \times 30°$. A black matte separator plate was mounted between the two displays (see Figure 2.22). The separator helped observers maintain their viewing distance to view the matching stimuli under the desired 2° or 10° on the CRT.

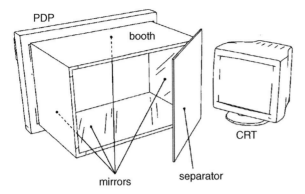

**Figure 2.22** Experimental setup to quantify the color appearance of an immersive color stimulus on the PDP in a mirror booth [93]. Observers had to compare the immersive color appearance with the small (standard size) color stimulus (2° or 10°) on the CRT monitor. Reproduced with permission from *Color Research and Application*.

The 16 large test color stimuli (PDP) were preset and the matching stimuli (CRT) were adjustable. The chromaticities of the test color stimuli were not close to the color gamut boundaries of the CRT monitor to allow for the color size effect. CIELAB $a^*$, $b^*$ color coordinates of the 16 test colors are plotted in Figure 2.23. $L^*$ values are listed in

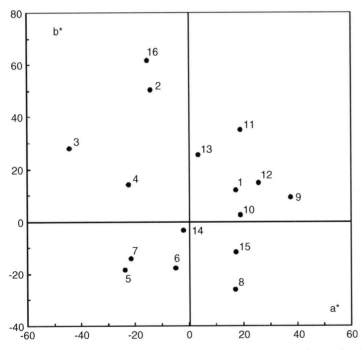

**Figure 2.23** The 16 test color stimuli used in the experiment [93]. $L^*$ values are listed in Table 2.13. Reproduced with permission from *Color Research and Application*.

**Table 2.13** CIELAB $L^*$, $a^*$, and $b^*$ values of the 16 test color stimuli [93] (see also Figure 2.23).

| No. | $L^*$ | $a^*$ | $b^*$ | Color name |
|---|---|---|---|---|
| 1 | 73.92 | 17.59 | 11.78 | Light skin |
| 2 | 84.77 | −14.15 | 50.14 | Pale yellow |
| 3 | 74.74 | −43.99 | 27.82 | Light green |
| 4 | 39.81 | −22.18 | 14.03 | Dark green |
| 5 | 59.88 | −23.70 | −18.52 | Turquoise |
| 6 | 41.68 | −4.83 | −17.90 | Dark blue |
| 7 | 64.88 | −21.27 | −14.20 | Light blue |
| 8 | 31.29 | 17.28 | −26.15 | Dark purple |
| 9 | 65.84 | 37.63 | 9.06 | Pink |
| 10 | 28.25 | 19.32 | 2.31 | Dark mauve |
| 11 | 50.19 | 19.25 | 34.96 | Dark orange |
| 12 | 46.03 | 25.98 | 14.71 | Flesh |
| 13 | 69.02 | 3.67 | 25.42 | Light drab |
| 14 | 77.17 | −1.76 | −3.59 | Light gray |
| 15 | 76.08 | 17.39 | −11.72 | Light purple |
| 16 | 85.25 | −15.08 | 61.46 | Yellow |

Reproduced with permission from *Color Research and Application*.

Table 2.13. CIELAB values were determined from the measured spectral data by the CIE 1931 standard observer and the measured tristimulus values of the maximum achromatic output of the CRT monitor (i.e., its white point).

The luminance of the CRT background was set equal to that of the actual test color stimulus being observed on the PDP. Observers were requested to adjust the color of the matching stimuli on the CRT monitor by the aid of three sliders (lightness, hue, and saturation) until the color perception of the large stimulus on the PDP matched that of the matching color on the CRT. Natural binocular viewing was applied. Each matching sequence started with the observation of the large field presented to the observer on the PDP for 2 s. Then a gray background appeared on the PDP, and subjects moved their head and looked at the CRT monitor to adjust the matching color that was initially mid-gray. The large stimulus appeared repeatedly at fixed time intervals (8 s) automatically but always for 2 s only.

This short viewing time corresponds to real large viewing angle displays where the instantaneous effect of the immersive scene produces a perceptual mismatch between the small and the large versions of the same radiation. The time course of chromatic adaptation [102] showed that a 2 s experience of a color stimulus allowed for only 20% of complete chromatic adaptation.

Another interesting issue related to the immersive color stimulus concerns the prediction of chroma. From the definition of chroma, it is clear that it has no meaning for static immersive colors since there is no achromatic anchor to which to relate colorfulness. Therefore, both colorfulness itself and saturation may be relevant. But, with the above-mentioned exposure time of 2 s, when the color stimulus is followed by the mid-gray background on the CRT, observers do have an anchor. Hence, a chroma perception may evolve in the observer for the immersive stimulus.

**Table 2.14** Interobserver variability in terms of MCDM [103] and color difference between the visually matching standard size colors (CRT; 2° or 10°) and the large color stimuli (PDP; immersive) expressed in CIELAB $\Delta E^*_{ab}$ color difference units averaged for the 16 test colors [93].

|  | Matching stimulus size | |
| --- | --- | --- |
|  | 2° | 10° |
| Interobserver variability (MCDM) | 5.8 | 5.6 |
| Color size effect: mean color difference for 16 test colors × 3 observers × 5 repetitions | 13.0 | 11.3 |
| Range of the color differences | 7.8–23.2 | 5.5–17.3 |
| STD of the mean color difference | 4.1 | 3.5 |

Reproduced with permission from *Color Research and Application*.

Observers had unlimited occasions to look back and forth between the two situations until a perceptual color match was found. Each of three subjects of normal color vision made five repetitions for each of the 16 test colors for both matching sample sizes (2° or 10°) and the mean CIELAB $L^*$, $a^*$, $b^*$ values of the matching color stimuli were calculated. To characterize the variability among observers, individual mean CIELAB $L^*$, $a^*$, $b^*$ values were also calculated for each condition (2° and 10°) and color difference was computed between the individual mean values and the overall mean values of all observers in terms of CIELAB $\Delta E^*_{ab}$ values. Individual deviations from the overall mean values were averaged for the 16 test colors. This average represents the so-called MCDM (mean color difference from the mean) [103]. MCDM values are listed in Table 2.14 together with the average measured color difference values between the visually matching standard size colors (CRT; 2° or 10°) and the large stimuli (PDP).

As can be seen from Table 2.14, average color differences due to the color size effect significantly exceeded the values of interobserver variability (MCDM). In an analysis of variance, the effect of matching stimulus size was found to be significant with less average difference for the 10° field size than for 2°. Significant mean color differences in Table 2.14 indicate the significance of the color size effect. The appearance of each one of the 16 large test colors is compared with their standard size appearance in Figure 2.24.

As can be seen from Figure 2.24, observers perceive the large stimuli to exhibit more lightness than the same color stimulus of standard size (2° or 10°). The mean $L^*$ difference between the large test colors and the matching standard size colors (mean of 16 test colors × 3 observers × 5 repetitions) is $\Delta L^* = 9.9$ (STD = 4.5, max. = 17.6) for the 2° condition and $\Delta L^* = 10.1$ (STD = 4.0, max. = 17.5) for the 10° condition. Chroma and hue changes can also be seen from the $a^*$–$b^*$ graphs of Figure 2.24. Chroma shifts were not systematic, sometimes increasing and sometimes decreasing.

Hue shifts can also be observed, typically for the case of reddish tones (nos. 1, 9, and 12, see Figure 2.24) where the perception of the large stimuli seemed to be more pinkish. Also, the chroma of these stimuli decreased or remained unchanged. The

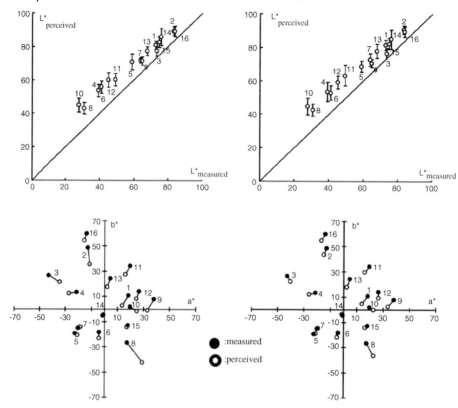

**Figure 2.24** Large PDP color stimuli: comparison of *perceived* colors (in the following sense: observed on the PDP, matched by the observers on the CRT monitor, and then measured on the CRT) and *measured* colors (directly measured on the PDP) [93]. L* graphs: mean *perceived* L* value of the large stimulus in the above sense (ordinate, 3 observers × 5 repetitions) versus measured L* value for the 16 test colors (abscissa). Error bars represent STDs. a*–b*: solid dots represent mean measured a*–b* values connected to the corresponding open circles representing mean perceived values. Left: 2° matching color; right: 10° matching color. Reproduced with permission from Color Research and Application.

darkest reddish tone (dark mauve – no. 10), however, showed only a negligible hue shift and a slight chroma increase. The largest color size effect was observed for this test color, in terms of its lightness increase shown in the upper diagrams of Figure 2.24. The chroma of test color nos. 13 (light drab) and 11 (dark orange) decreased with the increase of stimulus size and no hue change was observed. The chroma of the two yellowish test colors (nos. 2 and 16) decreased and a minor hue shift toward green was observed.

Test color nos. 3 and 4 (light green and dark green) exhibited more chroma and immersive light green exhibited less chroma. Subjects did not perceive any significant chromatic difference for the case of immersed bluish colors similar to the achromatic test color (no. 14) and for light purple (no. 15). A remarkable chroma increase occurred for test color no. 8 (dark purple), the amount of chroma increase

**Table 2.15** Mean matching colors (for 2° and 10°) [93] of the 16 test colors (Table 2.13).

| | | | | | | | | Test color | | | | | | | | |
|---|---|---|---|---|---|---|---|---|---|---|---|---|---|---|---|---|
| | 1 | 2 | 3 | 4 | 5 | 6 | 7 | 8 | 9 | 10 | 11 | 12 | 13 | 14 | 15 | 16 |
| 2° $L^*$ | 82.1 | 90.6 | 76.1 | 54.6 | 72.1 | 56.7 | 72.8 | 43.9 | 72.5 | 45.8 | 61.4 | 61.0 | 78.6 | 87.3 | 81.4 | 90.6 |
| $a^*$ | 12.9 | −12.7 | −36.6 | −27.9 | −22.2 | −4.9 | −19.7 | 29.1 | 32.8 | 24.3 | 15.4 | 23.6 | 1.4 | −2.4 | 17.7 | −17.0 |
| $b^*$ | 3.9 | 36.6 | 21.2 | 13.2 | −20.0 | −22.7 | −13.3 | −42.7 | −0.1 | −1.1 | 27.9 | 9.2 | 18.7 | −3.8 | −15.5 | 55.7 |
| 10° $L^*$ | 82.6 | 91.9 | 77.6 | 54.7 | 69.8 | 53.8 | 73.9 | 43.4 | 71.7 | 45.8 | 64.6 | 60.2 | 79.2 | 86.5 | 83.5 | 90.7 |
| $a^*$ | 13.2 | −16.4 | −41.6 | −27.6 | −22.3 | −5.7 | −20.6 | 23.0 | 33.1 | 22.4 | 15.2 | 26.0 | 1.1 | −2.4 | 15.6 | −18.3 |
| $b^*$ | 6.1 | 44.4 | 23.5 | 13.0 | −18.5 | −21.8 | −14.0 | −36.8 | 3.6 | 0.6 | 29.3 | 10.0 | 18.7 | −2.3 | −13.1 | 55.8 |

Reproduced with permission from *Color Research and Application*.

being less for the 10° than for the 2° matching color. Mean observed data (corresponding colors of the large field for 2° and 10°) are listed in Table 2.15.

#### 2.4.1.2 Xiao et al.'s Experiment on the Appearance of a Self-Luminous 50° Color Stimulus on an LCD

In this experiment [101], 12 self-luminous color stimuli subtending either 8° or 50° on an LCD were displayed. Both the 8° and the 50° stimuli were matched to a 10° color stimulus on a CRT monitor and the color size effect was modeled between the 10° matching colors corresponding to either 8° or 50°. In the experiment, Xiao et al. [101] also studied the color appearance change for reflection paint samples in the range between 2° and 50° but this is out of the scope of this book. The matching color stimulus was displayed in the center of a CRT monitor on black background. The experimental setup was similar to the one shown in Figure 2.22 except that no mirrors and no separating plate were used. Ten observers of normal color vision adjusted the matching color stimuli until the asymmetric match with the test color subtending 8° or 50° on the LCD was achieved. Both the LCD and the CRT had D93 white points. Results showed that the chroma and the lightness of the 50° stimulus increased compared to the 8° stimulus for all test colors but no hue change was found. Matching colors were expressed in CIECAM02 $J$, $C$, and $H$ values and the results were fitted with a mathematical model described in the next section.

The tendency of lightness increase was similar to the immersive condition described in Section 2.4.1.1: the darker the stimulus, the lighter it was perceived. However, in the immersive condition, instead of the general chroma increase it was found that, on the red–green axis ($a^*$), colors exhibited slightly more red. Also, along the yellow–blue axis ($b^*$), blue content increased.

### 2.4.2 Mathematical Modeling of the Color Size Effect

The aim of modeling is to predict how the same color stimulus appears if it subtends a large viewing angle compared to the color appearance of a standard size stimulus

**Table 2.16** Coefficients of the color size effect model (Equation 2.27) for self-luminous immersive stimuli [93] described in Section 2.4.1.1, for two sizes of the matching color stimuli, 2° and 10°.

|  | $c_L$ | $c_{1a}$ | $c_{2a}$ | $c_{1b}$ | $c_{2b}$ | $c_{3b}$ |
|---|---|---|---|---|---|---|
| 2° | 0.757 | 0.953 | 0.191 | −0.0019 | 1.017 | −4.600 |
| 10° | 0.758 | 0.982 | −1.015 | −0.0011 | 0.997 | −3.026 |

Reproduced with permission from *Color Research and Application*.

(e.g., 2° or 10°). Predictors of color appearance attributes (CIELAB $L^*$, $a^*$, $b^*$ values or CIECAM02 $J$, $C$, $H$ values) of the large stimulus are expressed as functions of similar predictors of the standard size stimulus.

Concerning the color appearance of the immersive color stimulus on the PDP (Section 2.4.1.1), the CIELAB $L^*$, $a^*$, and $b^*$ values of the matching colors of the immersive stimuli were expressed as functions of the CIELAB $L^*$, $a^*$, and $b^*$ values of the immersive colors [93] in the following way:

$$L^*_{immersive} = 100 + c_L(L^*_{matching} - 100)$$
$$a^*_{immersive} = c_{1a}a^*_{matching} + c_{2a} \quad (2.27)$$
$$b^*_{immersive} = c_{1b}(b^*_{matching})^2 + c_{2b}b^*_{matching} + c_{3b}$$

Table 2.16 shows the coefficients in Equation 2.27. The mean error between the model predictions of Equation 2.27 and the result of the visual observations for the experimental data described in Section 2.4.1.1 is about $\Delta E^*_{ab} = 5$. This value represents an improvement compared to the mean color differences of about $\Delta E^*_{ab} = 12$ (see Table 2.14) without color size effect modeling.

Concerning the color appearance of the self-luminous 50° color stimulus on the LCD in Xiao et al.'s experiment [101] described in Section 2.4.1.2, the CIECAM02 $J$, $C$, and $H$ values of the matching 10° colors of 50° stimuli were expressed as functions of the CIECAM02 $J$, $C$, and $H$ values of the matching 10° colors of the 8° stimuli in the following way [101]:

$$J_{50°} = 100 + K_J(J_{8°} - 100)$$
$$C_{50°} = K_C C_{8°} \quad (2.28)$$
$$H_{50°} = H_{8°}$$

with $K_J = 0.87$ and $K_C = 1.12$.

As can be seen from Equation 2.28, lightness increase and chroma increase of the 50° stimulus are predicted compared to the 8° stimulus. To compare the two models of Equations 2.27 and 2.28, Equation 2.28 was approximated in terms of CIELAB $L^*$, $a^*$, $b^*$ values by using $L_A = 100 \text{ cd/m}^2$, $Y_b = 20$, D93 illuminant, and average viewing condition parameters in CIECAM02. Correlation coefficients ($r^2$) between original and approximated colorimetric data ranged between 0.998 and 0.999. The following equations were obtained instead of Equation 2.28:

## 2.4 Color Appearance of Large Viewing Angle Displays

$$L^*_{50°} = 0.8158 L^*_{8°} + 16.896$$
$$a^*_{50°} = 1.1342 a^*_{8°} + 0.2337 \quad (2.29)$$
$$b^*_{50°} = 0.000(b^*_{8°})2 + 1.1247 b^*_{8°} - 0.122$$

For comparison, the color size effect model of self-luminous immersive stimuli (Equation 2.27) is rewritten below for the case of the 10° matching color:

$$L^*_{immersive} = 0.758 L^*_{10°} + 24.200$$
$$a^*_{immersive} = 0.982 a^*_{10°} - 1.015 \quad (2.30)$$
$$b^*_{immersive} = -0.0011(b^*_{10°})^2 + 0.997 b^*_{10°} - 3.026$$

Equations 2.29 and 2.30 are compared in Figures 2.25–2.27 for CIELAB $L^*$, $a^*$, and $b^*$, respectively.

Figures 2.25–2.27 show that, compared to the small size situation, the value of CIELAB lightness of immersive self-luminous stimuli increases more than the CIELAB lightness of the 50° self-luminous stimuli. The increase of the absolute value of CIELAB $a^*$ ($b^*$) is higher for the 50° self-luminous condition. There is a decrease for the immersive self-luminous condition for $a^*_{small} > 0$ ($b^*_{small} > 0$). There is no systematic chroma increase in case of the immersive self-luminous stimuli unlike the 50° self-luminous condition. The hue of 50° self-luminous stimuli is unchanged while there are hue changes for the immersive self-luminous stimuli.

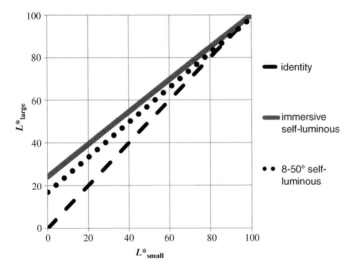

**Figure 2.25** Comparison of CIELAB $L^*$ in the color size effect equations of 50° self-luminous colors and immersive colors, Equations 2.29 and 2.30, respectively: "8–50° self-luminous" (gray dots, Equation 2.29) and "immersive self-luminous compared with 10°" (gray dash lines, Equation 2.30). Identity (dash black line) means no size effect in which case the ordinate would equal the abscissa. Large: either immersive or 50°; small: either 8° or 10°.

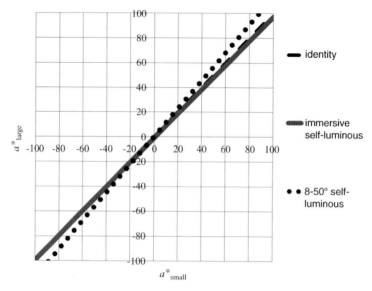

**Figure 2.26** Comparison of CIELAB $a^*$ in the color size effect equations of 50° self-luminous colors and immersive colors, Equations 2.29 and 2.30, respectively: "8–50° self-luminous" (gray dots, Equation 2.29) and "immersive self-luminous compared with 10°" (gray dash lines, Equation 2.30). Identity (dash black line) means no size effect in which case the ordinate would equal the abscissa. Large: either immersive or 50°; small: either 8° or 10°.

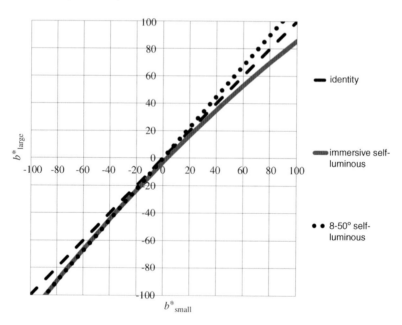

**Figure 2.27** Comparison of CIELAB $b^*$ in the color size effect equations of 50° self-luminous colors and immersive colors, Equations 2.29 and 2.30, respectively: "8–50° self-luminous" (gray dots, Equation 2.29) and "immersive self-luminous compared with 10°" (gray dash lines, Equation 2.30). Identity (dash black line) means no size effect in which case the ordinate would equal the abscissa. Large: either immersive or 50°; small: either 8° or 10°.

# References

1. IEC 61966-2-1:1999 (1999) *Color Measurement and Management in Multimedia Systems and Equipment. Part 2.1. Default RGB Color Space – sRGB*, International Electrotechnical Commission.
2. Bodrogi, P., Sinka, B., Borbély, Á., Geiger, N., and Schanda, J. (2002) On the use of the sRGB color space: the "Gamma" problem. *Displays*, **23** (4), 165–170.
3. Beretta, G. and Moroney, N. (2011) Validating large-scale lexical color resources. Proceedings of AIC 2011 Conference, Interaction of Color & Light in the Arts and Sciences, June 7–10, 2011, Zürich.
4. Bodrogi, P. and Schanda, J. (1994) Farbmetrische kalibrierung von Farbmonitoren (Colorimetric characterization of color monitors). 4th International Symposium on Color and Colorimetry, Bled, Slovenia.
5. Bodrogi, P., Sinka, B., Borbély, Á., Geiger, N., and Schanda, J. (2002) On the use of the sRGB color space: the "Gamma" problem. *Displays*, **23** (4), 165–170.
6. Shin, J.Ch., Yaguchi, H., and Shioiri, S. (2004) Change of color appearance in photopic, mesopic and scotopic vision. *Opt. Rev.*, **11** (4), 265–271.
7. Berns, R.S., Motta, R.J., and Gorzynski, M.E. (1993) CRT colorimetry. Part I. Theory and practice. *Color Res. Appl.*, **18**, 299–314.
8. Berns, R.S., Gorzynski, M.E., and Motta, R.J. (1993) CRT colorimetry. Part II. Metrology. *Color Res. Appl.*, **18**, 315–325.
9. Brainard, D.H. (1989) Calibration of a computer controlled color monitor. *Color Res. Appl.*, **14**, 23–34.
10. CIE 122-1996 (1996) *The Relationship Between Digital and Colorimetric Data for Computer-Controlled CRT Displays*, Commission Internationale de l'Éclairage.
11. Motta, R.J. (1991) All analytical model for the colorimetric characterization of color CRTs. M.S. thesis, Rochester Institute of Technology.
12. Bodrogi, P. and Schanda, J. (1995) Testing a calibration method for color CRT monitors. A method to characterize the extent of spatial interdependence and channel interdependence. *Displays*, **16**, 123–133.
13. IEC 61966-4 (2000) *Multimedia Systems and Equipment – Color Measurement and Management. Part 4. Equipment Using Liquid Crystal Display Panels*, International Electrotechnical Commission.
14. ISO 9241-303:2008 (2008) *Ergonomics of Human–System Interaction. Part 303. Requirements for Electronic Visual Displays*, International Organization for Standardization.
15. Bodrogi, P., Sinka, B., and Ondró, T. (2001) Mathematical models for the colorimetric characterization of AM LCD flat panel monitors. Proceedings of Lux junior 2001, Dörnfeld/Ilm, Germany, pp. 223–230.
16. Hunt, R.W.G. and Pointer, M.R. (2011) *Measuring Color (Wiley-IS&T Series in Imaging Science and Technology)*, 4th edn, John Wiley & Sons, Ltd., p. 504.
17. Moroney, N., Fairchild, M.D., Hunt, R.W.G., Li, Ch. Luo, M.R., and Newman, T. (2002.) The CIECAM02 color appearance model. IS&T/SID 10th Color Imaging Conference.
18. CIE 159-2004 (2004) *A Color Appearance Model for Color Management Systems: CIECAM02*, Commission Internationale de l'Éclairage.
19. Fairchild, M.D. and Johnson, G.M. (2002) Meet iCAM: a next-generation color appearance model. IS&T/SID 10th Color Imaging Conference, Scottsdale, pp. 33–38.
20. Johnson, G.M. and Fairchild, M.D. (2003) Visual psychophysics and color appearance, in *CRC Digital Color Imaging Handbook*, CRC Press, Boca Raton, FL, pp. 115–171.
21. CIE 162-2010 (2010) *Chromatic Adaptation Under Mixed Illumination Condition When Comparing Softcopy and Hardcopy Images*, Commission Internationale de l'Éclairage.
22. CIE 15-2004 (2004) *Colorimetry*, 3rd edn, Commission Internationale de l'Éclairage.

23 Luo, M.R., Cui, G., and Li, Ch. (2006) Uniform color spaces based on CIECAM02 color appearance model. *Color Res. Appl.*, **31**, 320–330.

24 Kutas, G. and Bodrogi, P. (2004) Colorimetric characterisation of a HD-PDP device. Proceedings of CGIV 2004 – Second European Conference on Color in Graphics, Imaging and Vision, Aachen, Germany, pp. 65–69.

25 Kwak, Y. and MacDonald, L. (2000) Characterisation of a desktop LCD projector. *Displays*, **21**, 179–194.

26 Thomas, J.B., Hardeberg, J.Y., Foucherot, I., and Gouton, P. (2008) The PLVC display characterization model revisited. *Color Res. Appl.*, **33** (6), 449–460.

27 Bastani, B., Cressman, B., and Funt, B. (2005) Calibrated color mapping between LCD and CRT displays: a case study. *Color Res. Appl.*, **30**, 438–447.

28 Tominaga, Sh. (1993) Color notation conversion by neural networks. *Color Res. Appl.*, **18** (4), 253–259.

29 Cowan, W.B. (1983) An inexpensive scheme for calibration of colour monitor in terms of CIE standard coordinates. *SIGGRAPH Comput. Graph.*, **17**, 315–321.

30 Silverstein, L.D. (2006) Color display technology: from pixels to perception. *IS&T Rep.*, **21** (1), 1–5.

31 Benzie, P., Watson, J., Surman, P., Rakkolainen, I., Hopf, K., Urey, H.., Sainov, V., and von Kopylow, C. (2007) A survey of 3DTV displays: techniques and technologies. *IEEE Trans. Circuits Syst. Video Technol.*, **17** (11), 1647–1658.

32 Oetjen, S. and Ziefle, M. (2009) A visual ergonomic evaluation of different screen types and screen technologies with respect to discrimination performance. *Appl. Ergon.*, **40** (1), 69–81.

33 Gibson, J.E. and Fairchild, M.D. (2000) Colorimetric characterization of three computer displays (LCD and CRT). Munsell Color Science Laboratory Technical Report.

34 Fedrow, B.T. (ed.) (1999) *Flat Panel Display Handbook – Technology Trends and Fundamentals*, Stanford Resources, Inc., San Jose, CA.

35 Day, E.A., Taplin, L., and Berns, R.S. (2004) Colorimetric characterization of a computer-controlled liquid crystal display. *Color Res. Appl.*, **29**, 365–373.

36 IEC 61966-3 (2000) *Colour Measurement and Management in Multimedia Systems and Equipment. Part 3. Equipment Using Cathode Ray Tubes*, International Electrotechnical Commission.

37 Berns, R.S., Fernandez, S.R., and Taplin, L. (2003) Estimating black-level emissions of computer-controlled displays. *Color Res. Appl.*, **28** (5), 379–383.

38 Choi, S.Y., Luo, M.R., Rhodes, P.A., Heo, E.G., and Choi, I.S. (2007) Colorimetric characterization model for plasma display panel. *J. Imaging Sci. Technol.*, **51** (4), 337–347.

39 Chigrinov, V.G. (1999) *Liquid Crystal Devices: Physics and Applications*, Artech House, Boston, MA/London.

40 Yeh, P. and Gu, C. (2009) *Optics of Liquid Crystal Displays* (*Wiley Series in Pure and Applied Optics*), John Wiley & Sons, Inc.

41 Wen, S. and Wu, R. (2006) Two-primary crosstalk model for characterizing liquid crystal displays. *Color Res. Appl.*, **31**, 102–108.

42 Tamura, N., Tsumura, N., and Miyake, Y. (2003) Masking model for accurate colorimetric characterization of LCD. *J. Soc. Inform. Display*, **11**, 333–339.

43 Lu, R., Zhu, X., Wu, Sh.T., Hong, Q., and Wu, T.X. (2005) Ultrawide-view liquid crystal displays. *IEEE/OSA J. Display Technol.*, **1** (1), 3–14.

44 Bodrogi, P., Sinka, B., and Ondró, T. (2001) Mathematical models for the colorimetric characterisation of AM LCD flat panel monitors. Proceedings of Lux junior 2001, Ilmenau, Germany, pp. 223–230.

45 Jimenez del Barco, L., Daz, J.A., Jimenez, J.R., and Rubino, M. (1995) Considerations on the calibration of color displays assuming constant channel chromaticity. *Color Res. Appl.*, **20**, 377–387.

46 Azuma, R.T. (1997) A survey of augmented reality. *Presence Teleop. Virt.*, **6** (4), 355–385.

47 L'Hostis, D. (2001) Real time computer graphics in augmented reality. Final Report, The University of Hull.

48 Zhang, R. and Hua, H. (2008) Characterizing polarization management in a p-HMPD system. *Appl. Opt.*, **47** (4), 512–522.

49 Moreau, O., Curt, J.N., and Leroux, Th. (2001) *Contrast and Colorimetry Measurements Versus Viewing Angle for Microdisplays*, Eldim Publications.

50 Gadia, D., Bonanomi, C., Rossi, M., Rizzi, A., and Marini, D. (2008) Color management and color perception issues in a virtual reality theater, in *Stereoscopic Displays and Applications XIX* (eds A.J. Woods, N.S. Holliman, and J.O. Merritt), *Proc. SPIE*, **6803**, 68030S-1–68030S-12.

51 Weiland, Ch., Braun, A.K., and Heiden, W. (2009) Colorimetric and photometric compensation for optical see-through displays. Proceedings of UAHCI '09, 5th International Conference on Universal Access in Human–Computer Interaction. Part II. Intelligent and Ubiquitous Interaction Environments, Springer, Berlin, pp. 603–612.

52 Shibata, T. (2002) Head mounted display. *Displays*, **23** (1–2), 57–64.

53 Sharples, S., Cobb, S., Moody, A., and Wilson, J.R. (2008) Virtual reality induced symptoms and effects (VRISE): comparison of head mounted display (HMD), desktop and projection display systems. *Displays*, **29** (2), 58–69.

54 IEC 61966-3 (2006) *Colour Measurement and Management in Multimedia Systems and Equipment. Part 6. Front Projection Displays*, International Electrotechnical Commission.

55 Kwak, Y., MacDonald, L.W., and Luo, M.R. (2001) Colour appearance comparison between LCD projector and LCD monitor colours. Proceedings of AIC Color 2001.

56 Khanh, T.Q. (2004) Physiologische und psychophysische Aspekte in der Photometrie, Colorimetrie und in der Farbbildverarbeitung (Physiological and psychophysical aspects in photometry, colorimetry and in color image processing). Habilitationsschrift (Lecture qualification thesis), Technische Universitaet Ilmenau, Ilmenau, Germany.

57 Thomas, J.B. and Bakke, A.M. (2009) A colorimetric study of spatial uniformity in projection displays, in *Proceedings of CCIW 2009 (Lecture Notes in Computer Science 5646)* (eds A. Tremeau, R. Schettini, and S. Tominaga), Springer, Berlin, pp. 160–169.

58 Stone, M.C. (2001) Color balancing experimental projection displays. Proceedings of 9th IS&T/SID Color Imaging Conference, pp. 342–347.

59 Chen, Y.F., Chen, C.C., and Chen, K.H. (2007) Mixed color sequential technique for reducing color breakup and motion blur effects. *J. Display Technol.*, **3** (4), 377–385.

60 Bakke, A.M., Thomas, J.B., and Gerhardt, J. (2009) Common assumptions in color characterization of projectors. Proceedings of Gjøvik Color Imaging Symposium 2009, Gjøvik, Norway, no. 4, pp. 45–53.

61 Humphreys, G., Buck, I., Eldridge, M., and Hanrahan, P. (2000) Distributed rendering for scalable displays. Proceedings of 2000 ACM/IEEE Conference on Supercomputing.

62 Kunzman, W. and Pettitt, G. (1998) White enhancement for color sequential DLP. SID'98 Digest.

63 Wallace, G., Chen, H., and Li, K. (2003) Color gamut matching for tiled display walls. Proceedings of 7th International Immersive Projection Technologies Workshop and 9th Eurographics Workshop on Virtual Environments (eds J. Deisinger and A. Kunz).

64 Wyble, D.R. and Rosen, M.R. (2006) Color management of four-primary digital light processing projectors. *J. Imaging Sci. Technol.*, **50** (1), 17–24.

65 Stark, P. and Westling, D. (2002) OLED – evaluation and clarification of the new organic light emitting display technology. Master thesis, University of Linköping, SAAB Avionics, Kista, Sweden.

66 Derra, G., Moench, H., Fischer, E., Giese, H., Hechtfischer, U., Heusler, G., Koerber, A., Niemann, U., Noertemann, F.Ch., Pekarski, P., Pollmann-Retsch, J., Ritz, A., and Weichmann, U. (2005) UHP lamp systems for projection applications. *J. Phys. D*, **38**, 2995–3010.

67 Yu, X.J., Ho, Y.L., Tan, L., Huang, H.C., and Kwok, H.S. (2007) LED-based projection systems. *J. Display Technol.*, **3** (3), 295–303.

68 Murat, H., De Smet, H., and Cuypers, D. (2006) Compact LED projector with tapered light pipes for moderate light output applications. *Displays*, **27** (3), 117–123.

69 Cassarly, W.J. (2008) High-brightness LEDs. *Opt. Photon. News*, 19–23.

70 Keuper, M.H., Harbers, G., and Paolini, S. (2004) RGB LED illuminator for pocket-sized projectors. SID'04 Digest, pp. 943–945.

71 Harbers, G., Keuper, M.H., and Paolini, S. (2004) Performance of high power LED illuminators in color sequential projection displays. White Paper, Lumileds Lighting.

72 Lim, S.K. (2006) LCD backlights and light sources. Proceedings of ASID'06, October 8–12, 2006, New Delhi, pp. 160–163.

73 Anandan, M. (2006) LED backlight: enhancement of picture quality on LCD screen. Proceedings of ASID'06, October 8–12, 2006, New Delhi, pp. 130–134.

74 Hiyama, I. *et al.* (2002) 122% NTSC color gamut TFT LCD using 4 primary color LED backlight and field sequential driving. Proceedings of the International Display Workshop IDW'02, AMD1/FMC2–4, pp. 215–218.

75 Chen, H., Sung, J., Ha, T., Park, Y., and Hong, Ch. (2006) Backlight local dimming algorithm for high contrast LCD-TV. Proceedings of ASID'06, October 8–12, 2006, New Delhi, pp. 168–171.

76 Harbers, G. and Hoelen, Ch. (2001) High performance LCD backlighting using high intensity red, green and blue light emitting diodes. SID'01 Digest.

77 Harbers, G., Bierhuizen, S.J., and Krames, M.R. (2007) Performance of high power light emitting diodes in display illumination applications. *J. Display Technol.*, **3** (2), 98–109.

78 Lee, T.W., Lee, J.H., Kim, C.G., and Kang, S.H. (2009) An optical feedback system for local dimming backlight with RGB LEDs. *IEEE Trans. Consum. Electron.*, **55** (4), 2178–2183.

79 Lu, R., Hong, Q., Ge, Z., and Wu, Sh.T. (2006) Color shift reduction of a multi-domain IPSLCD using RGB-LED backlight. *Opt. Express*, **14** (13), 6243–6252.

80 Ko, J.H., Ryu, J.S., Yu, M.Y., Park, S.M., and Kim, S.J. (2010) Initial photometric and spectroscopic characteristics of 55-inch CCFL and LED backlights for LCD-TV applications. *J. Korean Inst. Illuminating Electr. Installation Eng.*, **24** (3), 8–13.

81 Zhao, X., Fang, Z.L., and Mu, G.G. (2007) Analysis and design of the color wheel in digital light processing system. *Optik*, **118**, 561–564.

82 Roth, S., Weiss, N., Chorin, M.B., David, I.B., and Chen, C.H. (2007) Multi-primary LCD for TV applications. SID International Symposium – Digest of Technical Papers, vol. 38, no. 1, pp. 34–37.

83 Funamoto, T., Kobayash, T., and Murao, T. (2001) High-picture-quality technique for LCD television: LCD-AI. Proceedings of IDW2000, pp. 1157–1158.

84 Shiga, T., Kuwahara, S., Takeo, N., and Mikoshiba, S. (2005) Adaptive dimming technique with optically isolated lamp groups. SID'05 Digest, pp. 992–995.

85 Hong, J.J., Kim, S.E., and Song, W.J. (2010) Clipping reduction algorithm using backlight luminance compensation for local dimming in liquid crystal displays. Proceedings of the IEEE International Conference on Consumer Electronics (ICCE) – Digest of Technical Papers, pp. 55–56.

86 Cho, H. and Kwon, O.K. (2009) A backlight dimming algorithm for low power and high image quality LCD applications. *IEEE Trans. Consum. Electron.*, **55** (2), 839–844.

87 Seetzen, H. and Whitehead, L. (2003) A high dynamic range display using low and high resolution modulators. SID'03 Digest of Technical Papers, pp. 1450–1453.

88 Seetzen, H., Heidrich, W., Stuerzlinger, W., Ward, G., Whitehead, L., Trentacoste, M., Ghosh, A., and Vorozcovs, A. (2004)

High dynamic range display systems. *ACM Trans. Graph.*, **23** (3), 760–768.
89 Moon, P. and Spencer, D. (1945) The visual effect of nonuniform surrounds. *J. Opt. Soc. Am.*, **35** (3), 233–248.
90 McCann, J.J. and Rizzi, A. (2007) Camera and visual veiling glare in HDR images. *J. SID*, **15** (9), 721–730.
91 Johnson, G.M. and Fairchild, M.D. (2003) A top down description of S-CIELAB and CIEDE2000. *Color Res. Appl.*, **28** (6), 425–435.
92 Fairchild, M.D. and Johnson, G.M. (2010) Meet iCAM: a next-generation color appearance model. Proceedings of IS&T/SID 10th Color Imaging Conference (CIC10), pp. 33–38.
93 Kutas, G. and Bodrogi, P. (2008) Color appearance of a large homogenous visual field. *Color Res. Appl.*, **33** (1), 45–54.
94 Burnham, R.W. (1952) Comparative effects of area and luminance on color. *Am. J. Psychol.*, **65**, 27–38.
95 Burnham, R.W. (1951) The dependence of color upon area. *Am. J. Psychol.*, **64**, 521–533.
96 CIE 2004:15 (2004) *Colorimetry*, 3rd edn, Commission Internationale de l'Éclairage.
97 Xiao, K., Luo, M.R., Li, Ch., and Hong, G. (2010) Color appearance of room colors. *Color Res. Appl.*, **35**, 284–293.
98 Fridell Anter, K. (2000) What color is the red house? Perceived color of painted facades. Dissertation, Department of Architectural Forms, Royal Institute of Technology, Stockholm, 338 pp.
99 Billger, M. (1999) Color in enclosed space. Department of Building Design, Chalmers University of Technology, Gothenburg.
100 Hårleman, M. (2007) *Daylight Influence on Color Design. Empirical Study on Perceived Color and Color Experience Indoors*, Axl Books, Stockholm.
101 Xiao, K., Luo, M.R., Li, C., Cui, G., and Park, D.S. (2011) Investigation of color size effect for color appearance assessment. *Color Res. Appl.*, **36**, 201–209.
102 Fairchild, M.D. and Reniff, L. (1995) Time course of chromatic adaptation for color-appearance judgments. *J. Opt. Soc. Am. A*, **12**, 824–833.
103 Berns, R.S. (2000) *Billmeyer and Saltzman's Principles of Color Technology*, 3rd edn, John Wiley & Sons, Inc., New York, pp. 97–99.

# 3
# Ergonomic, Memory-Based, and Preference-Based Enhancement of Color Displays

The aim of this chapter is to describe the principles and methods to improve the rendering of color images and user interfaces on color displays from visual ergonomic and esthetic points of view. Visual ergonomics (Sections 3.1 and 3.3) establishes the basic requirements of providing perceivable and comprehensible information for a human observer on a usable and comfortable (interactive) visual display. Esthetic principles go one step beyond comfort and usability and concentrate on a *pleasing* appearance delighting and amusing the user visually. In this sense, the *color quality* of the image is one of the most important aspects.

One way of enhancing color quality is shifting the actual image colors toward so-called *long-term memory colors* stored in human memory in connection with certain important objects often seen in the past (e.g., skin, sky, grass, leaves, orange, and banana) and appearing on the display (Section 3.4). To introduce the above aspects, Section 3.2 summarizes the *objectives* of color image reproduction, either the *exact* reproduction of the spatial and color appearance of the original image or applying some color image transformations according to *esthetic* considerations.

The other way of color quality enhancement is changing the *overall* color and spatial frequency histograms of the image according to the users' color image *preference* independent of specific objects. This method (including image *contrast* preference) is described in Section 3.5 together with an algorithmic framework of *preference-based* color image enhancement in Section 3.6.

## 3.1
### Ergonomic Guidelines for Displays

The visual information presented on the display should be easy to process by the human visual system (HVS) [1]. The background of the display should be bright enough so that the HVS works in an optimum state for optimum visual acuity. Characters (letters and numbers) should be large enough and exhibit high contrast to achieve good legibility and avoid visual overload or eyestrain on a long-term basis. Color should be used carefully to prevent the numerous color artifacts and the distraction of visual attention by overcoloring the display, for example, a

*Illumination, Color and Imaging: Evaluation and Optimization of Visual Displays*, First Edition.
Peter Bodrogi and Tran Quoc Khanh.
© 2012 Wiley-VCH Verlag GmbH & Co. KGaA. Published 2012 by Wiley-VCH Verlag GmbH & Co. KGaA.

computer-based graphic user interface. Principles of *color ergonomics* of self-luminous displays are reviewed separately in Section 3.3.1.

User interfaces represent one important type of displays transmitting huge amounts of information toward the user with a high level of interaction with the visual objects on the display and consequently with the computer controlling the display. To increase work efficiency at the user interface, visual information should be presented in a concise, organized, and understandable manner according to the principles of human visual cognition. For example, the correct segmentation of the display (e.g., the different methods of grouping words or symbols including the use of hue categories) facilitates orientation, search, pointing, and editing.

In this section, the ergonomic principles of visual displays are described in accordance with the ISO international ergonomic standard about the requirements for electronic visual displays [2]. If these ergonomic guidelines are applied to the display hardware (e.g., appropriate spatial resolution, contrast, and color) and to the software generating the image (e.g., coloring the components of a user interface consciously), then work performance, comfort, and satisfaction of the human display user can be increased significantly.

Display ergonomics has two general factors, a "human" factor and a "display" factor. The human factor has anthropometric, visual, cognitive, psychological, intellectual, and social aspects including conventions and standards. The display factor includes the display hardware and the aspects of the software driving the display, the interaction of the display and its environment, workplace illumination, input/output devices of interactive displays, display colors, adaptation luminance, refresh rate, software usability, methods of presenting the information, dialog principles, helping the user of the interface, and other aspects of user interface design [2–7].

Ergonomic guidelines for displays help avoid complaints arising in association with a long-term interaction with them, for example, hurting muscles or bones, visual fatigue, eyestrain, or psychological problems such as losing work motivation. Ergonomic design has to consider the heterogeneity of display users according to their education level, age, gender, nationality, physical ability or inability, visual acuity, and normal or deficient color vision. Ergonomic design is especially important for interactive display applications such as Internet, computer user interfaces, interactive television, and mobile phones. Ergonomic design usually implies trade-offs between increasing human performance and work safety as well as reliability and cost efficiency [2–7].

Ergonomic guidelines for displays were derived from the findings of visual psychophysics and work psychology and also from specific ergonomic studies recording user behavior, task completion time, and error rates or asking display users to fill questionnaires or talk about their visibility or usability problems related to the display. Typical findings of visual ergonomics include the Hick–Hyman law [8, 9] of choice reaction time being proportional to the logarithm of alternatives and implying that a simple interactive display is better than a series of consecutive displays with only a few alternatives to choose from.

Furthermore, Fitts' law [10, 11] for motor performance claims that the time needed to point to an object depends on the size and position of that object. The latter rule helps develop pointing devices and setting their velocity and sensitivity. Users of interactive displays build internal mental models about the user interface they are using. Typical users tend to assume more flexibility and intelligence to the user interface than it actually has. Enquiring and then using the user's mental models (e.g., their expectations about the options of the interaction) to develop the interface is an important ergonomic design principle. Such an interaction model ensures a usable interface easy to control and to understand.

Ergonomic drawbacks of self-luminous displays compared to paper-based technologies include a small visual angle of the working area, less brightness and contrast, and the viewing angle dependence of color and contrast (on LCDs). But these drawbacks are compensated for by the excellent properties of modern interactive graphic user interfaces on modern high-resolution displays with dynamic mixed representations of text and graphics, iconic symbols, rapid pointing, and overlapping windows.

Realistic video sequences, moving (rotating) geometric objects (graphs, molecule models, or mock-ups), and 3D graphics represent a tremendous amount of data easy to capture by the human visual brain that cannot be imagined at a conventional paper-based workplace or by simply displaying raw numerical data in tables. Users, however, may experience difficulties due to the limitations of their working memory and visual attention, for example, when interpreting complex diagrams or large tables. These difficulties can be mitigated by ergonomically co-optimizing the self-luminous interactive display with its input devices, for example, keyboard, mouse, handwriting, speech, touchscreen, eye tracking, specific pointing, and control devices.

Sitting in front of the display, the user's head and eye movements scan the limited range of viewing angles corresponding to the display. In case of interactive displays or computer user interfaces, the user's hands are attached to the input device. The user and the display constitute a closed system restricting the space available to reposition the human body. In contrast to a natural environment, muscle activity patterns become limited and *constrained body postures* can appear.

Reasons of constrained postures include unsuitable font size, bad luminance contrast or bad viewing distances, inappropriate computer chair or desk, improper positioning of the display in relation to the windows, and a bad local or general illumination of the room. Constrained postures can be avoided by ensuring that the eye level is about at the top of the screen (to be able to look slightly downward), the viewing distance is correct, the desk is at a suitable height, elbows are flexed, the posture is erect (the chair is adjustable), the feet lie flat on the floor, and the keyboard can be inclined to be just below the elbow level with enough room for wrist support.

To design the *range* of sizes for ergonomic furniture correctly, typical sizes of human users can be obtained from anthropometric results. These sizes include eye height when sitting, shoulder height, elbow rest height, thigh clearance height, popliteal height when sitting, trunk thickness, elbow to elbow and hip breadth, and hand and foot length. After adjusting to the user's body, modern computer chairs still behave in a *flexible* manner; that is, they allow *dynamic* muscle activity by slight movements of the body (e.g., left and right or back and forth) relaxing the stress of a static posture. Users can be advised to relax actively by standing up, moving around, and looking away from the display on a regular basis to release accommodation workload of the eye lens by focusing on objects at different distances.

For the visual ergonomics of displays, it is essential to consider the basic characteristics of the human visual system (see also Section 1.1) including visual acuity, near and far accommodation of the eye lens, different achromatic and chromatic contrast sensitivities in the different parts of the entire visual field (>180° with different rod and cone densities in the foveal and the peripheral areas of the retina), large dynamic range of luminance adaptation (scotopic, mesopic, and photopic range, 0.001–10 000 cd/m$^2$), and color vision by the aid of three types of cone photoreceptors (L, M, and S, see Section 1.1).

For the visual ergonomics of self-luminous displays, the following principles – derived from the knowledge of the human visual system – are essential. The visual details of small (static) objects (e.g., strokes of letters, numbers, and icons) can only be seen if these objects exhibit a minimal angular size, a minimal (average) luminance, a minimal luminance contrast between the strokes and their background (luminance contrast ratios of at least 3: 1), and a minimal duration of appearance.

Visual acuity (VA, the ability to resolve two separated points of a minimum viewing angle) depends on average background luminance level, retinal position (central or peripheral), age, eye condition, and the chromaticities of the visual object and its background. Visual acuity can be expressed in inverse minute of arc units (min$^{-1}$). The absolute limit of foveal visual acuity is about 2 min$^{-1}$. But in practice it is more reasonable to assume about 0.3 min$^{-1}$ according to the large scatter of viewing conditions and interpersonal differences. Visual acuity is reduced dramatically when the object to be recognized appears further away from the fovea toward the periphery of the retina. Visual acuity increases with increasing background luminance level (see Figure 3.1).

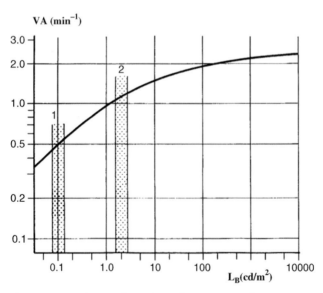

**Figure 3.1** Visual acuity of normal spatial vision expressed in inverse minute of arc units (min$^{-1}$) as a function of background luminance. Range 1: background at about 0.1 cd/m$^2$; range 2: background at about 2 cd/m$^2$. *Source:* Technische Universität Darmstadt, Lecture on Lighting Engineering.

As can be seen from Figure 3.1, in range 1 (i.e., for background luminance levels around 0.1 cd/m$^2$), visual acuity of normal spatial vision is about 0.5 min$^{-1}$ while in range 2 (i.e., for background luminance levels around 2 cd/m$^2$), it is slightly above 1 min$^{-1}$. Values above 2 min$^{-1}$ can be achieved if background luminance is increased above 100 cd/m$^2$.

For a user interface, the height of characters on a display should be at least 16 min. For an ergonomic design, 20–22 min is recommended as viewed from the intended viewing distance in a typical application of the display. The viewing distance should not be less than 40 cm. Character stroke width should be 1/6–1/12 of character height. The size of the same characters should be constant across the display.

For moving objects, acuity remains approximately constant until an angular velocity of 20°/s. For higher velocities, acuity decreases rapidly being somewhat worse for vertically moving objects than for horizontally moving objects. The detection of moving objects (and flicker perception) is better in the periphery than in the fovea because peripheral visual detection signals new objects in the field of view like an alarm indicator. The latter effect has an important application for interactive displays, that is, the high conspicuity of blinking or moving signs appearing at the corners of the display.

For flicker detection, the criterion of critical flicker frequency (CFF) is very important for the flicker-free design of display hardware. If the refresh frequency of the display is higher than the value of the CFF, then no flicker is perceived [2]. According to the Ferry–Potter law [12], CFF is proportional to the logarithm of display overall luminance level between 3 and 300 cd/m$^2$ ranging between 20 Hz (1 cd/m$^2$) and 70 Hz (1000 cd/m$^2$). This means that the higher the average luminance of the display, the higher refresh rate should be used.

The nearest focusable viewing distance increases with increasing age, being about 10 cm for a 20-year-old observer and 100 cm for a 70-year-old observer on average implying the necessity of correcting glasses in case of focusing deficiencies. The range of distances for sharp vision is rather constrained at lower luminance levels. For example, at 0.3 cd/m$^2$, focusing is only possible in the range of 0.3 and 2 m for a young observer (see the arrow in Figure 3.2).

As can be seen from Figure 3.2, to exploit the whole focusing range, background luminance levels above about 100 cd/m$^2$ are necessary. Extensive watching of a display at a fixed viewing distance (e.g., 50 cm) tends to "freeze" the state of accommodation resulting in an extended refocusing time if a different viewing distance is needed (e.g., for car driving), especially for elderly observers and for the case of longer durations of display viewing. A typical value of refocusing time is 8 s for elderly observers (60) and 4 s for young observers (20). This indicates the necessity of pauses and a longer refocusing period before moving to a visual task of different nature, for example, driving.

Adaptation of the human visual system to different luminance levels (from starlight through twilight up to intense sunlight) covers a luminance range of 1: $10^{14}$. Adaptation from light to dark is much slower than the reverse process that takes a couple of seconds. The course of dark adaptation can be determined in a psychophysical absolute detection threshold measurement after bleaching the rods

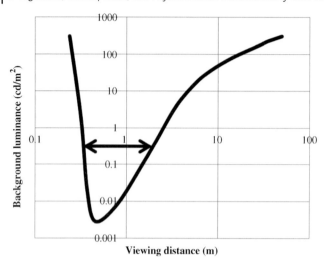

**Figure 3.2** Range of viewing distances for sharp vision (abscissa) as a function of luminance level (ordinate) for young observers. For example, at 0.3 cd/m², focusing is only possible in the range of 0.3 and 2 m (see the arrow). *Source*: Technische Universität Darmstadt, Lecture on Lighting Engineering.

and the cones by a very bright bleaching field (e.g., about 3500 cd/m² [13]). The luminance of the object at the detection threshold decreases with elapsing time after bleaching as the cones and then the rods become active (see Figure 3.3). In this experiment, two kinds of objects were used, a green one with a dominant wavelength of 566 nm stimulating both the cones and the rods and a red one with a dominant wavelength of 629 nm favoring cone responses [13].

As can be seen from Figure 3.3, for this particular subject, the cone threshold (red objects) was reached roughly 10 min after bleaching and the rod threshold was attained about 30 min after bleaching (green objects).

If *rapid* adaptation is required, however, adaptation mechanisms cannot cover more than 2 logarithmic units. Rapid adaptation is required when looking at the different parts of the display or its direct environment. Therefore, the concept of *luminance balance* is an important ergonomic principle; that is, for an optimal visual performance, average luminance should not change more than 1: 10 between the display and its surround (e.g., the office) and not more than 1.7: 1 within the display.

This implies important consequences for the positioning of the display within the room to avoid too bright areas or glare sources such as windows. Both direct glare and glare reflected from the surface of the display should be avoided. Nevertheless, if the display is used for entertainment (e.g., cinema) and not for work purposes, then bright highlights on the display, for example, in high dynamic range imaging, can be desirable (see Section 2.3.3).

Another important feature of visual ergonomics is the issue of the *regions of visual attention*. The arrangement of visual information on the display should be adjusted to the visual search and attention characteristics of the human observer. Object

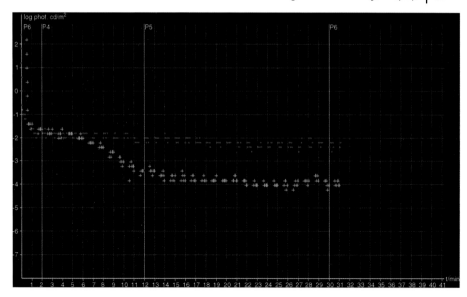

**Figure 3.3** Time course of dark adaptation in a psychophysical absolute detection threshold measurement after bleaching the rods and the cones by a very bright bleaching field (about 3500 cd/m² [13]). Two kinds of objects were used, a green one (green dots, dominant wavelength: 566 nm, stimulating both the cones and the rods) and a red one (red dots, dominant wavelength: 629 nm, favoring cone responses). Abscissa: time elapsed after bleaching in minutes; ordinate: threshold luminance of the object to be detected in log cd/m² units. Sample result of one subject from a joint study of the University Eye Hospital, Tübingen, Germany and the Technische Universität Darmstadt.

conspicuity (Sections 3.3.2 and 3.3.3) depends not only on color but also on the location within the visual field. The optimum visual field of displays subtends about 30° below the horizontal line of sight. Visual acuity is highest within ±1° implying the necessity of scanning the display by so-called saccadic eye movements: within 20–50 ms, the eye can be positioned to the next target (*saccade*).

The next fixation point (e.g., a conspicuous object) shall be visible before the onset of the saccade. For static images, 99% of all saccades are shorter than 15°. *Fixation dwell time* ranges between 0.2 s and several seconds. Visual attention cannot be focused onto the whole display at the same time. Simultaneously observed areas (*fields of attention*) extend up to ±15° with fixed eyes and up to ±50° with eye movements. To bridge large distances, head movements are necessary. These facts should be kept in mind when arranging information on the display, for example, in a set of windows.

The size of the fields of attention depends on the visual complexity of the object the observer is looking for and especially on the complexity of the background (e.g., the number of the objects distracting visual attention). Search time is proportional to the search area and the luminance contrast between the search object and its background. *Cognitive hints* reduce search time, for example, searching for red objects *or* circular objects *only* (see Section 3.3.3).

For visual ergonomics, a clear distinction should be made between *detectability* (or *visibility*) and *legibility*. Visibility relates to the simple fact whether an object as an entity can be seen at all, independent of its details. For example, a very blurred letter "N" can be visible but it is not legible. Legibility is related to the ability of the observer to distinguish the fine visual details inside the object (e.g., the strokes of the letter "N") and hence recognize the object. Legibility is not only related to the text reading task but, more generally, to the ability of identifying a fine spatial structure, that is, visual details containing higher spatial frequencies, for example, the hair structure of a person. Readability is another concept, in turn, related to whether strings of letters (e.g., "the cat") are written in a known language.

The distinction among visibility, legibility, and readability is based on the different steps of processing visual information in the visual brain. For visibility, the presence of low spatial frequencies triggers a detection signal just indicating for the visual brain that "something new" is in the field of view. For legibility, foveal ganglion cells amplify the spatial difference signals of the edges in the image (e.g., the strokes of a letter "N") enhancing high spatial frequencies. The set of edges constitutes the so-called primal sketch of the image. From the primal sketch, objects (such as a letter "N") are formed governed by the so-called Gestalt laws of organization. A set of simple visual components (points, edges, shapes, angles, slopes, and colors) are considered to belong to the same object if they are next to each other, if they are similar, continuous, build a closed shape, or have a common faith of analogously moving components.

Gestalt laws, fields of attention, visibility, legibility, and conspicuity are important characteristics to arrange the information on the display in a usable manner and to design ergonomic diagrams and graphic user interfaces. Diagrams help process electronically stored information by a human user because they exploit the multitask nature of information processing in the visual brain.

The components of a diagram are compared in the user's working memory with diagram types (schemes) stored in long-term memory. Known diagram schemes enable the user to extract the information encoded in the diagram rapidly. But to avoid working memory overload, diagrams should not be overcomplicated. For example, the number of identifying items to be remembered in working memory simultaneously should not exceed 7–11.

Recently, so-called tablet computers (tablets or mobile information terminals) gained extensive use worldwide. These mobile computers possess small displays (often with LED backlights and diagonals ranging between about 8 and 24 cm) usually operated by a touchscreen and a digital pen. These mobile displays exhibit large information content (so-called information-intensive mobile displays [14]), for example, spatial resolutions of $1024 \times 768$ pixels at 132 pixels per inch (ppi) or $1600 \times 1200$ pixels at 667 ppi.

Information is viewed within a small visual angle and under constantly changing illumination, a short viewing distance, and often in shaky environments [14]. Mechanical vibrations tend to blur the edges displayed on the mobile screen. Short viewing distances and holding the display toward one direction (e.g., always on the right side) can cause ophthalmological problems such as decreased accommodative power or the abnormality in convergence of both eyes [15].

Users expect high image quality for their efficient interaction with generally complicated graphical user interfaces or for video viewing but power consumption limitations require a trade-off for white point luminance and the choice of the primary colors. Therefore, for this type of display, specific ergonomic guidelines apply.

Due to their short viewing distances, tablet computers are so-called near-eye displays (NEDs) [14] and they tend to cause nausea and eye fatigue, especially when the user is constantly moving while using it. If pixel density is not sufficient to resolve the high information content of the device, then a *scalable* user interface is an essential requirement. Scaling is sometimes implemented by the so-called pinch-out (to enlarge character size) and pinch-in (to reduce character size) touch operations to be able to adjust the size of the characters to achieve an optimum (*most legible*) size [16].

For black characters on white background on a tablet computer, this optimum size was found to be about $35 \pm 20$ arcmin for alphanumeric characters and about $30 \pm 12$ arcmin for Japanese characters, while the ergonomic standard [2] recommends 16 arcmin for minimum character height [16]. In the latter study, viewing distance increased with the age of the subjects and it was pointed out that elderly tablet users require larger character size. Aged observers may have difficulty in adjusting the viewing distance to become optimal in the range of adjustment allowed by the length of their arm with the tablet in their hand [16].

Due to light reflections from a bright environment or rapid transitions from high luminance adaptation levels to lower luminance levels, contrast sensitivity can be reduced considerably – especially for the case of high spatial frequency content, that is, for small icons or letters [14]. Very bright reflections cause glare rendering the tablet illegible. Reflections of colored illumination distort color appearance and color contrast on the tablet. To overcome the problems of bright (e.g., sunlit) environments, an illumination level adaptive color reproduction method and a flare compensation method were proposed [17].

## 3.2
### Objectives of Color Image Reproduction

In this section, the objectives of color image reproduction [18] on a self-luminous color display are summarized. A realistic scene observed by a human observer is three-dimensional. In reality, the observer scans the spatial, angular, and spectral radiance distribution of the scene by head movements and eye movements scrutinizing every important point of the objects from several different viewing angles consecutively. The observer assesses the perceived attributes of total appearance [19]: color (which in itself consists of three psychological dimensions, hue, chroma, and lightness), gloss, translucency, and texture.

Total appearance cannot be reproduced by reproducing color alone. The other attributes of the objects related to the spatial structure of their light emitting surface and to the angular dependence of light emission (gloss, translucency, and texture) also contribute to their overall appearance. The challenging objective of *reproducing total appearance* on a self-luminous display is difficult. A full-color three-dimensional holographic display would be needed especially to render the attributes of gloss and translucency.

The objective of *multispectral imaging* means that the spectral reflectance function of the object is captured at every point and then it can be reproduced by simulating its appearance under *every* illuminant on a color display. A further aim is the reproduction of *highlights* (e.g., highly translucent colored stain glass windows inside a dark cathedral, sunset or sunrise landscapes, or glaring car headlights on a road) in the image. This can be called the objective of *high dynamic range* (HDR) imaging (Section 2.3.3).

Simple *colorimetric reproduction* of measured colorimetric data of an original image (i.e., reproducing absolute or relative *XYZ* values) captured by, for example, an imaging colorimeter generally results in *false* color appearance because the viewing conditions of capturing and reproducing are different (Sections 2.1.8 and 2.1.9). The image is captured very often in an average viewing condition, while the projection display reproduces it in a dark viewing condition. In this example, *colorimetric reproduction* results in a significant loss of perceived contrast.

*White point changes* between the original and the reproduced image result in significant color appearance distortions if colorimetric reproduction is used. The reference color appearance is the color appearance of an original scene with real colored objects illuminated by a light source. In every point, *XYZ* values are measured and specified in terms of *XYZ* values of object colors. The value of Y yields the luminance factor of the object color in %. The original scene is then reproduced on a self-luminous display by using the same relative *XYZ* values (the display's peak white luminance equaling 100%). If the white chromaticity of the display differs significantly from the chromaticity of the light source illuminating the real colored objects, then a significant change of color appearance can be observed [20].

To avoid the above difficulties, the objective of *color appearance reproduction* can be followed, for example, by using the CIECAM02 color appearance model (Section 2.1.9) in a color management workflow (see Sections 4.1.1–4.1.3) transforming the digital values of the original image into numerical correlates of color appearance attributes (e.g., hue, chroma, and lightness), changing the viewing conditions to those of the output device (the self-luminance display on which the original image needs to be rendered), and finally, applying the color appearance model in reverse mode, thus obtaining the digital color values of each pixel for the rendered image. Note that the image color appearance (iCAM) framework represents a more advanced solution to consider the spatial color features and tone compression for HDR appearance.

The above objectives are related to the *exact* reproduction of the appearance perceived in the original image. As mentioned at the beginning of this chapter, another aim of self-luminous displays is producing an *esthetic image* to delight or entertain the user of the self-luminous display. Esthetic image appearance differs from the original appearance; hence, color image transformations shall be applied according to certain esthetic considerations. These transformations can shift the original colors toward long-term memory colors (this is the objective of *memory color reproduction*, see Section 3.4) or change the overall appearance of the image to get a *preferred* image (this is the objective of *preferred color image reproduction*, see Sections 3.5 and 3.6).

## 3.3
## Ergonomic Design of Color Displays: Optimal Use of Chromaticity Contrast

This section describes the principles of ergonomic color design for color displays (*color ergonomics*) together with the relationship among legibility, conspicuity, and visual search to derive a method of optimal use of chromaticity contrast. This method can be used to optimize visual search performance on the user interface. The important ergonomic issue of chromaticity contrast preference and luminance contrast preference of young and elderly display users is presented.

### 3.3.1
### Principles of Ergonomic Color Design

This section is devoted solely to a brief illustration of color ergonomics because color appearance and color cognition are complex phenomena requiring a *very careful use of color* on the user interface appearing on a self-luminous display. For color graphics and color diagrams, the most important artifact to avoid is *chromostereopsis* related to the spectral changes of refractive power of the eye lens. This implies that blue rays tend to intersect in front of the retina while red rays tend to intersect behind the retina. Consequently, saturated backgrounds should be avoided on self-luminous displays, especially saturated red symbols on saturated blue backgrounds (and vice versa) otherwise a disturbing colored quasi-stereoscopic perception occurs.

Accordingly, it is advisable to use a neutral (white or light gray) background. A light background helps cone photoreceptors operate in their optimal (photopic) state also fostering a wide range of accommodation (see Section 3.1). For good legibility, it is essential to ensure enough luminance contrast between a colored object and its background; hence, colors of low to medium lightness should be used for the graphic objects themselves. Using saturated colors enhances both the conspicuity and the efficiency of categorical identification of the colored objects, for example, the colored lines in a line graphics diagram. Figure 3.4 reproduces Figure 5.3 without considering the principles of ergonomic design.

As can be seen from Figure 3.4, due to the incorrect use of color, the diagram becomes overcrowded, chromostereopsis occurs, and the luminance contrast between the background and the curves disappears. Disturbing simultaneous color contrasts can also be observed. The background of Figure 3.4 is so saturated that – by looking away from it – chromatic afterimages can be perceived. This is very annoying

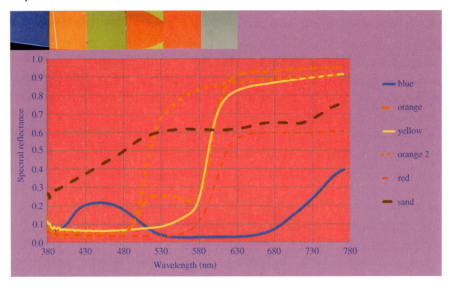

**Figure 3.4** A recolored version of Figure 5.3. The aim of recoloring was to point out the adverse visual effects of ignoring the principles of ergonomic color design.

especially in a long-term interaction with the user interface. Also, thin colored lines of low luminance contrast to the background cannot be seen because the contrast sensitivity of the chromatic channels of the human visual system is lower than the contrast sensitivity of the luminance channel.

In addition to the above, it cannot be overemphasized that the use of color should have a clear motivation and any new color added to the diagrams should have a meaning. Otherwise, colors tend to act as distractors of attention reducing search performance as will be pointed out in the next sections.

### 3.3.2
#### Legibility, Conspicuity, and Visual Search

Legibility (as introduced above) is one of the most salient factors of ergonomic display design. It was also mentioned above that, for good legibility, luminance contrast must be high enough (i.e., at least 3: 1). Visual performance on the self-luminous display can be further improved by adding a certain amount of chromaticity contrast (also called color contrast) to the luminance contrast of the objects (contrast is always related to the background) [21, 22]. This can be used to emphasize, group, and identify the objects [23]. Emphasizing single objects on the display can easily cause overcoloring. To use color contrast in an ergonomic way, so-called *equiluminance legibility* (ELL) rules can be applied [24].

In the literature, it is well known that "color difference between the target and the surround may enhance symbol conspicuousness: ... a bigger color difference improves search performance" [25]. But overcrowding the display by uncontrolled use of color confuses the observer and reduces visual comfort [26]. The correct addition of color contrast to luminance contrast can be controlled by a predictor formula derived from a psychophysical ELL scale derived from experiments with colored text [24]. The result of this ELL experiment was compared with the so-called Ware and Cowan conversion factor (WCCF) formula [27] constructed to predict chromatic brightness perception considering the Helmholtz–Kohlrausch effect (more chromatic objects are perceived to be brighter/lighter than less chromatic objects of the same luminance, see Section 6.5.2)

In this ELL experiment [24], the word "Legible" was displayed with colored 28pt Times New Roman characters on a gray ($x = 0.279$; $y = 0.282$) rectangle background (see Figure 3.5). The height of the characters was 1° of visual angle. The size of the background was 9° (horizontal) × 3° (vertical). The characters and the background had the same luminance, that is, zero luminance contrast. Eight to twelve rectangles were displayed at the same time.

The color of the characters was the same in a given rectangle but it was varied from rectangle to rectangle. The rest of the display was white (the reference white with $Y = 111.0 \, \text{cd/m}^2$; $x = 0.278$; $y = 0.281$). The display was in a dark room where it was the only light source. The distance between the display and the observer's eye was 60 cm. Observers had to rank the rectangles as follows. The rectangle containing the most legible word had to be assigned the number 1, and so on. Based on the rank orders of 13 color normal observers, each of the rectangles was assigned a scale value called ELL between 0 and 1.

In Experiment I, the CIELAB chroma $C_{ab}^*$ of the characters was fixed and their hue angle $h_{ab}$ was varied from rectangle to rectangle from 0° to 330°, in 30° steps. Five such images were tested, with the following fixed chroma values: $C_{ab}^* = 9.5, 19, 28.5, 38$, and $47.5$. An example of ELL images is shown in Figure 3.5.

In Experiment II, the hue angle $h_{ab}$ of the characters was fixed and the chroma $C_{ab}^*$ of the characters was varied from rectangle to rectangle, from a minimal to a maximal value, in equal steps. The minimal and maximal chroma values and the chroma step

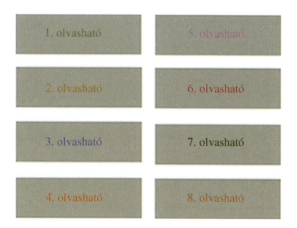

**Figure 3.5** Sample images for scaling ELL visually ("olvasható" = "legible" in Hungarian).

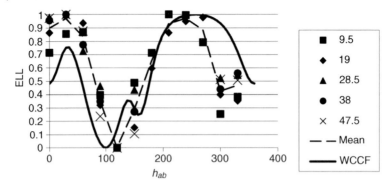

**Figure 3.6** Visual scale values from 13 observers for ELL as a function of the hue angle $h_{ab}$ of the text (see Figure 3.5) for five different chroma levels indicated next to the symbols. Mean curve of the ELL scale values and a curve predicted by WCCF [24]. Reproduced with permission from the International Commission on Illumination (CIE).

depended on the fixed value of the hue angle. Eight such images were investigated, with the following fixed hue angle values: $h_{ab} = 0°$, 45°, 90°, 135°, 180°, 225°, 270°, and 315°.

Figure 3.6 shows the ELL scale values resulting from Experiment I as a function of hue angle $h_{ab}$ for every fixed chroma separately. A mean curve of all five chroma values is also shown together with the WCCF curve [27] (rescaled between 0 and 1) predicting the chromatic contribution to brightness perception.

As can be seen from Figure 3.6, the dependence of ELL on hue angle was very similar for the different fixed chroma values. Also, there was a good correlation between the mean ELL curve and the WCCF curve. As a consequence, WCCF seems to be a good predictor of ELL when chroma is fixed and hue angle is varied. The chroma dependence of ELL turned out to be a linear increasing trend. Table 3.1 shows the correlation coefficients between the ELL scale values resulting from Experiment II and the varied chroma $C_{ab}^*$ for the eight fixed hue angle values together with the correlation coefficients between ELL and WCCF.

As can be seen from Table 3.1, ELL correlates well with $C_{ab}^*$ for very fixed value of hue angle. As a consequence, a linear function of $C_{ab}^*$ can be considered as a good predictor of ELL when hue angle is fixed and chroma is varied. The above results can be used for an ergonomic color design of color displays, especially for interactive color displays such as computer user interfaces. The dependence of ELL on hue angle and on chroma can be used to optimize visual search performance. Colored display items

**Table 3.1** CIELAB hue angle ($h_{ab}$), correlation coefficients ($r^2$) between the ELL scale values and the varied chroma $C_{ab}^*$, and $r^2$ between ELL and WCCF [27].

| $h_{ab}$ (°) | 0 | 45 | 90 | 135 | 180 | 225 | 270 | 315 |
|---|---|---|---|---|---|---|---|---|
| $r^2(C_{ab}^*$–ELL) | 0.88 | 0.91 | 0.89 | 0.94 | 0.87 | 0.97 | 0.92 | 0.95 |
| $r^2$(WCCF–ELL) | 0.83 | 0.54 | 0.91 | 0.94 | 0.85 | 0.97 | 0.92 | 0.92 |

(e.g., symbols, icons, or characters) can be arranged on several "conspicuity levels" depending on the designer's intent or emphasis [28], as will be shown below.

### 3.3.3
### Chromaticity Contrast for Optimal Search Performance

As described above, the concept of luminance contrast (described by, for example, the quantity of luminance ratio of the visual object and its background) has to be applied to describe *legibility*, that is, the user's ability to extract information from the display coded by the fine spatial details containing high spatial frequencies of the visual objects. This, in turn, establishes the possibility of easy recognition of the visual objects on the display, for example, symbols, icons, buttons, cursors, and alphanumeric characters. Actually, as mentioned above, according to the international ergonomics standard [2], the minimum luminance contrast ratio (CR) for acceptable legibility is equal to 3.

In a long and intensive interaction with the display, the user's satisfaction depends not only on legibility but also on a more general feature called *visual comfort* [21, 22]. Visual comfort depends on the ability of the user to perform all kinds of perceptual tasks (not only legibility) including the important task of *searching* for a single specific information on the whole screen. For good visual comfort, the ergonomic standard [2] calls for several general requirements such as clarity, discriminability, conciseness, consistency, detectability, legibility, and comprehensibility. Legibility and visibility (or, in other words, detectability) belong entirely to the field of the *perceptual* aspects whereas the others belong partially to *software ergonomics*, that is, the ergonomic organization of the dataflow between the human and the computer by considering the human's *cognitive* characteristics.

The aim of this section is to investigate the question of how to *direct* the display user's attention toward *relevant* information to improve search performance by adding chromaticity contrast to the luminance contrast of the visual objects on the display [29–31]. For *legibility*, chromaticity contrast and luminance contrast are additive only for low luminance contrasts and for saturated colors [31]. Although pure chromaticity contrast is able to ensure legibility (Figure 3.5), the appearance is quite blurred. Chromaticity contrast and luminance contrast are not interchangeable as chromaticity contrast alone cannot produce acceptable legibility [31].

The effect of chromaticity contrast on reading rate (words/min) was found to depend on the level of luminance contrast [29]. Effects of chromaticity contrast on legibility became only evident when luminance contrast was lowered, especially for small (0.2°) characters. In this case, reading performance with near-equiluminous text was as high as with the normal luminance contrast. From the above findings, the general conclusion is that chromaticity contrast is not a suitable means to guarantee legibility on a display.

For good legibility, luminance contrast must be ensured in the first step of ergonomic color design [23, 26]. Then, adding chromaticity contrast (i.e., by using *colored* symbols instead of black, gray, or white symbols) generally does not further improve legibility. But chromaticity contrast may improve the conspicuity of important (emphasized) visual objects on the display, hence improving visual search performance [25]. But the use of chromaticity contrast must be controlled very carefully to avoid confusing the user by overcoloring the display [25, 26, 32–40].

Visual search is parallel when the visual target (e.g., a colored icon) and its distractor objects (e.g., the other colored icons) differ on a single perceptual dimension only (e.g., only on hue). The condition of *parallel search* means that the observer is able to consider the whole display at the same time. But if there are several perceptual features combined to identify the target, then observers are forced to search serially through the display to find the target [36].

Visual search for color stimuli is mediated by higher order mechanisms of the human visual system combining color-opponent signals and achromatic signals [38]. Search time increases linearly with the number of distractors when the color difference between the target and the distractor is small and it is constant when the color difference is large [39]. It was found that there is a minimum color difference between the target and its distractors to achieve the *condition of parallel search* and this minimum color difference is about 13–38 times the foveal just-noticeable color difference [39].

The color difference between the target and the distractors (both the targets and the identical distracters being on black background) was shown to be an important variable to predict the conspicuity of the target [35]. But in this study [35], the observer's task was not similar to the user's typical search task on a multicolor user interface on which visual search is usually far away from the parallel search condition. In another visual search experiment [28], the situation of a crowded multicolor user interface display was simulated containing targets and distractors in many different colors.

The aim of the latter experiment [28] was to find a mathematical formula from average search time values to predict the conspicuity of a specific colored visual target (so-called search target). In the first experiment [28], all visual targets had a constant luminance contrast to their gray background but they exhibited different chromaticity contrasts. In the second (control) experiment [28], all symbols were set to the same white.

In this experiment, all stimuli were displayed on a colorimetrically characterized color display with good channel independence and good spatial uniformity (Section 2.1). Observers had to search for and find one of the 15 different meaningful computer commands consisting of letters of constant size and font, for example, "Window". A particular command had a constant chromaticity during the experiment but all commands had different chromaticities. Each observer had to find each of the 15 different search commands after each other, on 15 consecutive search screens. This was called a series.

There were 10 series altogether. In a particular search screen, all 15 commands were displayed at pseudorandom positions of the screen and the observer had to find one search command that was displayed before each search in black and white, in a stand-alone dialog box in the middle of the screen. The observer memorized the name of the search command, fixated to the middle of the screen, and pressed the "space" button. At that moment, the search screen was displayed. The observer pressed the "space" button again when the search command was found and the computer program recorded search time ($t_C$) as a function of the chromaticity of the command (CIELAB chroma and hue angle), its length $l$ (number of letters in the command), and its position (its distance $d$ from the center of the screen).

There were also many randomly chosen gray words randomly distributed among the commands (see Figure 3.7).

The experiment in Figure 3.7 was called the "color experiment." All 15 colored search commands were displayed at the same time corresponding to the user's usual search task on a multicolor user interface. To isolate the effect of chromaticity on search, a control experiment was also carried out. All parameters were the same as in the color experiment but the color of the search commands. All 15 search commands had the display's peak white. This experiment was called the "white experiment." In each of the 10 series, the pseudorandom positions of the 15 search commands on the screen were the same as in the color experiment. Search time is designated by $t_W$ in this case. A sample screen is shown in Figure 3.8.

Eleven observers with normal color vision carried out both the "color" experiment and the "white" experiment. Observers had following search cues to find the search command: command name, command length, and, in the color experiment, the

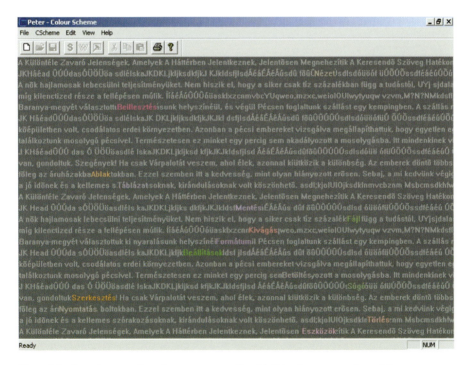

**Figure 3.7** Sample screen from the experiment to explore the role of chromaticity contrast in visual search on a display containing a user interface. "Color" experiment. Search command (e.g., "Ablak" = "Window"), the other 14 distractor commands, and the meaningless gray words. All chromatic words were equiluminous [28]. Reproduced with permission from *Displays*.

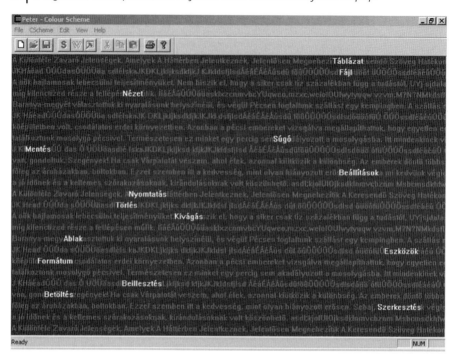

**Figure 3.8** Sample screen from the experiment to explore the role of chromaticity contrast in visual search on a display containing a user interface. "White" experiment. Search command (e.g., "Ablak" = "Window") and the other 14 distractor commands were the same white (the display's peak white) [28]. Reproduced with permission from *Displays*.

chromaticity of the search command. Both experiments can be described by the following variables: observer number $n$ ($n = 1, 2, \ldots, 10, 11$); series number $k$ ($k = 1, 2, \ldots, 9, 10$); and search command number $m$ ($m = 11, 15, 18, 22, 23, 24, 31, 34, 36, 42, 44, 46, 51, 54,$ or $57$).

In the color experiment, the high digit of $m$ indicates the hue angle of the search command and the low digit of $m$ indicates the chroma of the search command. The meaning of the high digit of $m$ is the following: 1: $h_{ab} = 55°$, orange; 2: $h_{ab} = 129°$, green; 3: $h_{ab} = 16°$, red; 4: $h_{ab} = 316°$, blue–purple; and 5: $h_{ab} = 330°$, purple. For each hue angle, the chroma of the search command was proportional to the low digit of $m$. For example, $m = 36$ (high digit 3, low digit 6) represents a saturated red search command. The colors of the 15 search commands are shown in the CIELAB $a^*$–$b^*$ diagram of Figure 3.9.

In both the color experiment and the white experiment, the value of $m$ uniquely determined the command length $l$ (e.g., $l = 6$ for "Window"). According to the pseudorandom positions of the 15 search commands on the screen in the 10 series, the values of $k$ and $m$ uniquely determined the so-called "*search zone*" $z$. The latter was defined as the integer value of $d/100$, where $d$ was the distance of the search

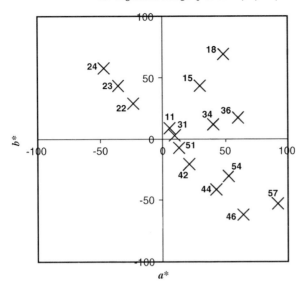

**Figure 3.9** The 15 search commands in the CIELAB $a^*$–$b^*$ diagram (crosses). Search command number $m$ is indicated by bold numbers next to the crosses. All colored commands were always visible in all "color experiments" [28]. Reproduced with permission from *Displays*.

command from the center of the screen in pixels. The value of $z$ was 0, 1, 2, 3, or 4; that is, there were five search zones.

First, the dependence of mean search time from the white experiment on the observer ($n$), series or repetition number ($k$), search zone ($z$), and command length ($l$) was analyzed. Then these effects were taken away from the result of the color experiment to be able to examine the separate effect of chromaticity on visual search. In the white experiment, the overall mean search time was equal to 2.38 s. The effect of $k$ on search time was not significant.

The "search zone" dependence of mean search time was significant and the tendency was similar for all observers, that is, the mean search time increased by going away from the center ($z = 0, 1, 2, 3$) and it decreased slightly at the screen border ($z = 4$). This tendency can be seen in Figure 3.10. The overall effect of the search zone on white command search can be described by the maximum search time difference $\Delta t_W^{(z)} = 0.7$ s on the ordinate of Figure 3.10.

Observers were asked about their search strategy after the experiment. From this survey, the command length $l$ turned out to be a relevant variable. Its effect on white search time was significant. If the observers had to search for a short command, then they have not even inspected the longer commands and vice versa. This common opinion of all observers is well indicated by the analysis of the mean search time by the variable $l$ plotted in Figure 3.11.

As can be seen from Figure 3.11, the maximum of the mean search time versus command length curve is at the medium command length. In the latter case, observers have to inspect both the short and the long commands to find the search command. The overall effect of $l$ was $\Delta t_W^{(l)} = 1.1$ s. To appreciate the individual effect

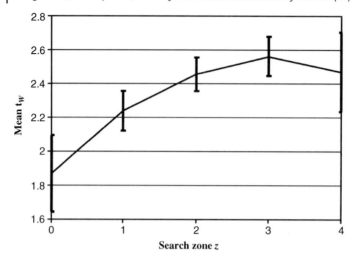

**Figure 3.10** Mean search time in the white experiment in the different search zones ($z$) defined as the integer value of $d/100$, where $d$ is the distance of the search command from the center of the screen in pixels. Ninety-five percent confidence intervals are also shown [28]. Reproduced with permission from *Displays*.

of chromaticity contrast on command search time, the two most relevant additional effects, that is, the search zone and command length effects, were eliminated both from the white experiment and from the color experiment.

Each measured search time value ($t_W$ and $t_C$) was divided by mean white search time in the corresponding search zone. The latter value was divided by the mean

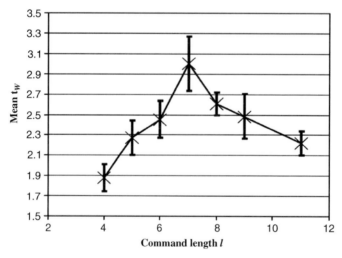

**Figure 3.11** Effect of command length in the white experiment. Mean search time as a function of the command length $l$, the number of characters in the search string. Ninety-five percent confidence intervals are also shown [28]. Reproduced with permission from *Displays*.

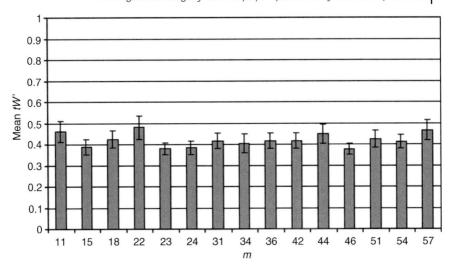

**Figure 3.12** Mean values and 95% confidence intervals for $t'_W$ (see text) as a function of $m$, after eliminating the search zone and command length effects [28]. Reproduced with permission from *Displays*.

white search time of the corresponding command length $l$. The resulting quantities are designated by $t'_W$ and $t'_C$. Their mean values and 95% confidence intervals are depicted in Figure 3.12 (white experiment) and in Figure 3.13 (color experiment) as a function of $m$. Mean values were calculated by taking all answers of all observers for a given value of $m$ into account.

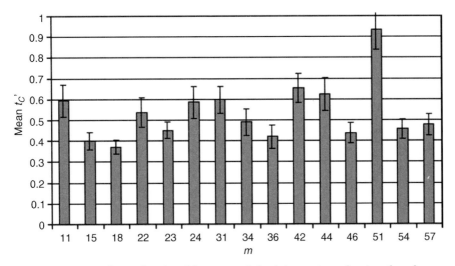

**Figure 3.13** Mean values and 95% confidence intervals for $t'_C$ (see text) as a function of $m$, after eliminating the search zone and command length effects [28]. Reproduced with permission from *Displays*.

As can be seen from Figures 3.12 and 3.13, mean $t'_C$ values were usually higher than mean $t'_W$ values. Overall, mean search time equaled 3.0 s in the color experiment and 2.4 s in the white experiment. This finding implies that the 14 colored command names other than the current search command have played the role of *distractor objects*. Also, the range of search time values was extended by a factor of 2 when different types of chromaticity contrast were added to the letters of the search commands.

In the color experiment, the luminance, font, and letter size were the same for all search commands and the effects of search zone and command length were excluded. Therefore, the only reason for the substantial search time differences is chromaticity contrast. The wide range of search time values in Figure 3.13 implies that the chromaticity of a particular display element (i.e., a visual search target) should be chosen to reflect its relevance on the display, for example, on the user interface of the human–computer dialog.

The designer should assign an appropriate *conspicuity level* to each display object, for example, by the appropriate use of chromaticity contrast; this is discussed in this section in detail. Other ways to improve search target conspicuity include luminance coding, blink coding, and changing font type and font size. Concerning the modeling of the search time characteristics of colored visual objects, several theories were developed [37, 38, 40–42]. In this section, two different approaches of modeling are compared. The first one is based on the color difference between the symbol and its background and the second one is based on the concept of ELL as introduced in Section 3.3.2.

The *measured* conspicuity $C_m$ of a colored target (search command) among many colored distractors can be defined by its inverse mean search time value in the "color experiment" (see Equation 3.1).

$$C_m = 1/t'_C \tag{3.1}$$

Then the value of $C_m$ can be predicted by a theoretical conspicuity value $C$ calculated from the spectral characteristics of the search command. In this section, two such formulas are compared, the CIELAB color difference formula and the *ELL formula* [28]. Concerning CIELAB color difference, mean search time was found to decrease rapidly with increasing CIELAB color difference between the search target and the distractor [35], but in the latter experiment [35], only one type of color distractor was used and both the targets and the distractors were displayed on a black background.

To apply the CIELAB color difference formula in a similar way for the case of the present study, the CIELAB color difference between the search command and its gray background was considered as the first conspicuity model of the colored search command, that is, $C = \Delta E^*_{ab}$. Figure 3.14 shows $C_m$ as a function of $C = \Delta E^*_{ab}$.

As can be seen from Figure 3.14, the correlation is not sufficient ($r^2 = 0.37$). Concerning the ELL formula [28], Figure 3.13 implies that a greater chroma value yields less search time value and hence a greater $C_m$ value (consider the *low* digit of $m$ in Figure 3.13, except for $m = 22, 23$, and 24, that is, with $h_{ab} = 55°$, orange).

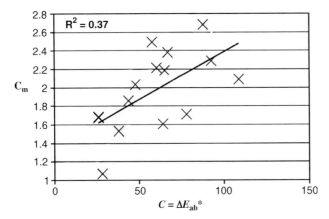

**Figure 3.14** Measured conspicuity $C_m$ as a function of $C = \Delta E_{ab}^*$ [28]. Reproduced with permission from *Displays*.

In the ELL experiment (Section 3.3.2), visually scaled ELL correlated well with the CIELAB chroma value $C_{ab}^*$ of the colored strings and the dependence of ELL on the CIELAB hue angle of the strings was very similar for different fixed chroma values (see Figure 3.6). Based on these findings, $C_m$ can be modeled by the predictor quantities of ELL, $C_{ab}^*$, and the mean $f(h_{ab})$ function of Figure 3.6. A modified version of the latter function, that is, $f(h_{ab}/360°)$, was approximated by a 10th-order polynomial (see Figure 3.15).

As can be seen from Figure 3.15, $f(h_{ab}/360°)$ has two minima and two maxima, similar to the hue angle dependence of the Helmholtz–Kohlrausch effect. The absolute minimum is at $h_{ab} = 120°$ (greenish yellow). The second minimum is at $h_{ab} = 322°$ (bluish purple). The absolute maximum is at $h_{ab} = 21°$ (red). The second

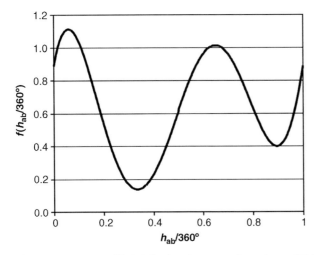

**Figure 3.15** Function $f(h_{ab})$ defined as the average dependence of ELL on CIELAB hue angle $h_{ab}$ (measured in degrees) [28] (see also Figure 3.6). Reproduced with permission from *Displays*.

**Table 3.2** Coefficients of the 10th-order polynomial to approximate $f(h_{ab}/360°)$ [28].

| Order | Coefficient |
|---|---|
| 0 | 0.8893 |
| 1 | 8.3508 |
| 2 | −90.4847 |
| 3 | 227.426 |
| 4 | −117.674 |
| 5 | −115.231 |
| 6 | −62.462 |
| 7 | 277.103 |
| 8 | −126.334 |
| 9 | 0.6890 |
| 10 | −1.3837 |

Reproduced with permission from *Displays*.

maximum is at $h_{ab} = 234°$ (greenish blue). Polynomial coefficients of the function $f(h_{ab}/360°)$ are listed in Table 3.2.

Using the function $f(h_{ab}/360°)$, the following conspicuity function yields a reasonable prediction of $C_m$:

$$C = 1.3973 + 0.0046956 C_{ab}^* + 0.048907 f(h_{ab}/360°) + 0.010548 C_{ab}^* f(h_{ab}/360°) \quad (3.2)$$

With Equation 3.2, the correlation between $C_m$ and $C$ can be improved ($r^2 = 0.61$, see Figure 3.16) and a procedure can be formulated to use chromaticity contrast for optimum visual search performance by selecting an ergonomic color set on the display.

Each colored display object should be assigned a *conspicuity level* according to its importance. First, the colorimetric characterization of the computer-controlled display should be carried out to be able to calculate the background luminance $L_B$ and the symbol luminance $L_S$ as well as $C_{ab}^*$ and $f_{ab}$ for each display element from their *rgb* values (see Chapter 2). Then a (gray) background luminance $L_B$ and a symbol luminance (or a set of symbol luminances) $L_S$ should be chosen for the symbols that will be colored. $L_B$ and the set of $L_S$ values constitute a *luminance outline* of the display with $L_S/L_B \geq 3$ or $L_B/L_S \geq 3$.

It is believed that it is the perceived chromatic brightness (see Section 6.5.2) of the object that attracts the visual attention of the observer so that the object becomes the next fixation point. So the concept of chromatic brightness can be associated with measured conspicuity as defined by the inverse mean search time. The perception of saturated equiluminous (and self-luminous) character–background combinations is something like "glimmering."

It is easy to see this effect on a desktop computer monitor by using graphics software. This fact in turn emphasizes the importance of distinguishing the roles of

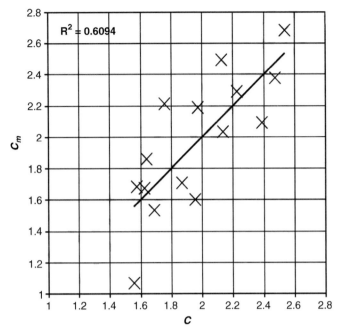

**Figure 3.16** Measured conspicuity $C_m$ as a function of a numerical correlate of conspicuity (C) as defined by Equation 3.2 [28]. Reproduced with permission from *Displays*.

luminance contrast and chromaticity contrast. By the aid of chromaticity contrast, visual search can be controlled without changing the luminance contrast outline of the screen. The latter ensures the perception of fine spatial resolution of the visual details of the display elements.

The last step of the color ergonomics procedure is the selection of the color set to be used on the display. The color set can be predefined [43]. Alternatively, the user may also choose his/her own color set [44]. Whatever method is used, the value of $C(C^*_{ab}, h_{ab})$ from Equation 3.2 should be considered for each color in the color selection algorithm. The goal is to establish several distinct C intervals (i.e., *conspicuity levels*) within the color set chosen.

Search commands or other objects of high priority, that is, those with the expected lowest mean search time values, should be colored to get into a high C interval. This interval can be, for example, $C = [2.4; 2.6]$ (see the abscissa of Figure 3.16). Objects of medium priority should be placed into the interval $C = [2.0; 2.2]$. Objects of low priority should be placed into the interval $C = [1.6; 1.8]$. An example can be seen in Figure 3.17. Its application is demonstrated by a sample user interface in Figure 3.18.

Note that the use of different hues (from left to right in Figure 3.17) can be used for grouping or segmentation. This is a further important ergonomic feature of applying chromaticity contrast.

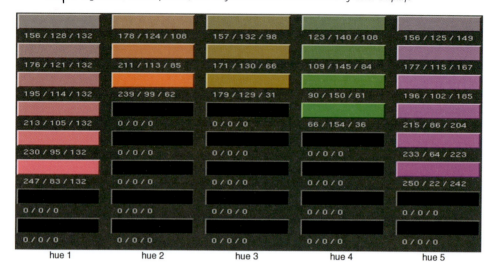

**Figure 3.17** Example of a scheme of colors at different conspicuity levels (increasing from the top to the bottom) for an ergonomic design. Note that the use of different hues (from left to right) can be used for grouping (a further important ergonomic feature of chromaticity contrast).

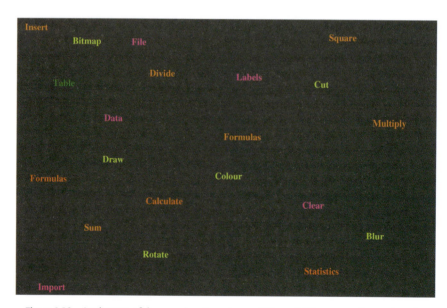

**Figure 3.18** Application of the sample color scheme of Figure 3.17 to a sample user interface.

## 3.3.4
### Chromaticity and Luminance Contrast Preference

One of the most important degradations of the aging human visual system concerns the decrease of spatial contrast sensitivity [45, 46]. The aim of this section is to summarize the characteristics of these changes for both luminance contrast and chromaticity contrast. Another aim is to point out the difference between *threshold* contrast and *preferred* contrast of young and elderly observers on a computer-controlled color display.

Spatial contrast sensitivity decreases rapidly with age, especially at higher spatial frequencies [47]. For color displays, spatial contrast sensitivity plays a major role while – as the refresh rate of displays is today well above the critical flicker frequency – temporal contrast sensitivity is of minor importance. One reason for the decrease of contrast sensitivity with age is that while the overall absorption of the crystalline lens increases, this deficit arises most remarkably in the blue spectral region [48]. This effect also causes color vision losses [49] although the color appearance of elderly observers is remarkably stable throughout their life [50] possibly due to a long-term chromatic rebalancing effect.

Neural changes during aging – as a second factor in the spatial sensitivity decline of the elderly – also occur [51]. At low luminance levels, these neural changes can cause significant sensitivity losses for lower spatial frequencies [52]. Optical reasons also contribute to the loss of contrast sensitivity causing intraocular light scatter and reduced retinal illuminance.

Threshold contrasts for legibility are well investigated in the literature [52–57]. But for display ergonomics – for a long-term interaction with the display – *preferred* luminance contrast is also important. Legibility is known to increase significantly by increasing the contrast from the threshold level to the preferred level.

As seen in Section 3.3.3, the ergonomically correct use of chromaticity contrast decreases search time and it is essential to know the preferred amount of chromaticity contrast and its spatial frequency characteristics. From the visual ergonomics point of view, chromaticity contrast on an achromatic background is of importance because of the serious artifacts on chromatic backgrounds. Therefore, chromatic–chromatic (e.g., red–green) spatial gradations are out of interest in the present discussion.

Preferred contrasts represent a special type of suprathreshold contrast perception [58]. Both young and elderly observers are able to assess preferred contrast values visually [45]. The contrast sensitivity of the human visual system is its capability to discriminate fine details in a visual scene based on the detection of slight variations of the color image. In the early stage of postreceptoral visual processing, these color variations are mediated by the luminance channel and by additional two cone-opponent mechanisms (the so-called chromatic or spectrally opponent channels [59] (see also Section 1.1).

In visual studies of chromaticity contrast perception [55, 56, 60, 61], test images of isoluminant chromatic gratings are generally used. They are usually modulated along either the $L + M - S$ axis (to investigate the blue–yellow opponent channel sensitivity) or the L − M axis (for the red–green opponent channel). As mentioned above, the principles of color ergonomics exclude the usage of isoluminant opponent color pairs (e.g., red text on green background) due to chromostereopsis.

Chromatic gratings of the same hue on a neutral background are more relevant from the ergonomics point of view [45]. Therefore, to test the threshold and preferred chromaticity contrasts, chromatic test patterns of sinusoid gratings of the same hue consisting of different spatial frequencies were displayed with upward decreasing contrast [45]. In addition to these test patterns, achromatic test images were also observed. Figure 3.19 shows the test images of the experiment [45].

As can be seen from Figure 3.19, in the bottom of these test images, the amplitude of the sine wave was maximal and in the top of the image it was zero. In the achromatic contrast (AC) experiment, test images were achromatic gray (see Figure 3.19a). The luminance depended on the phase of the sine wave (horizontal

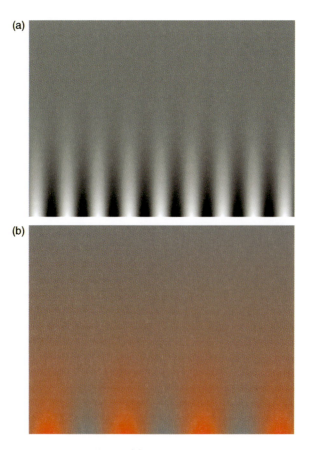

**Figure 3.19** Test images of the visual experiment on threshold and preferred contrasts [45]. The achromatic test image (a) has a greater spatial frequency in this example (2 c/deg) than the chromatic test image (0.4 c/deg) (b). The spatial frequency of the sinusoid grating was constant within each individual test image. Reproduced with permission from *Displays*.

position) and on the amplitude of the sine wave (vertical position). The maximum luminance of the grating was 62.0 cd/m² (peak white) and the minimum was less than 0.1 cd/m² (black). Besides this peak white luminance level, determined by the peak luminance of the display, three further luminance levels were displayed by employing different neutral gray filters in front of the same test images. The average transmittance values of these filters were 13, 26, and 49% to achieve a maximum luminance of 8, 16, and 30 cd/m², respectively, for the peak white.

In the chromatic contrast (CC) experiment, the test images (see Figure 3.19b) were similar but at a constant luminance level. The CIELAB chroma $C_{ab}^*$ of the picture varied with the sine wave periodically in the horizontal direction between achromatic gray and a chromatic shade of a color of constant hue angle in CIELAB. Chroma reached its maximum at the bottom sine wave in each test image. Spectral data were transformed to CIELAB by using the CIE 1931 standard colorimetric observer and the white point of the CRT monitor as reference white ($x_0 = 0.298$; $y_0 = 0.310$). Eight different CIELAB hue angles were observed, that is, the multiples of 45°, while the constant lightness of the test images was $L^* = 50$ corresponding to a luminance of 11.0 cd/m².

In the CC experiment, only this latter luminance level was used. The spatial frequency of the sinusoid grating of a single test image was always constant for both the AC and the CC experiments. Ten spatial frequencies were investigated: 0.1, 0.2, 0.4, 1, 2, 4, 8, 10, 12, and 14 cycles per degree (c/deg). The display, a color CRT monitor exhibiting excellent channel independence, was calibrated and characterized in a dark room. The surround of the test images was black – its luminance level was less than 0.1 cd/m². The spatial resolution of the CRT monitor was 1600 × 1200 pixels at 85 Hz. The color resolution equaled 8 bits per channel. This color resolution corresponded to small enough luminance and chroma steps; that is, these steps were below detectable contrast thresholds for the spatial frequencies investigated.

Observers sat in front of the screen facing it perpendicularly. Test images appeared in the center of the screen. The first task was to mark the region in the test image where the vertical stripes "faded into the background," that is, where the observers could not distinguish any line. The contrast of this level was saved as a "threshold contrast" value. After that, observers had to mark another contrast level in the test image in addition to the threshold where the contrast of the altering stripes was perceived to be *preferred* to comfortably perceive the image pattern.

The preference task turned out to be easy to understand and easy to carry out for all observers. This second contrast level found by the observer was saved as a "preferred contrast" value. After setting the "preference mark" in the test image, the test image disappeared automatically and, after a relaxing stage of 12 s with a homogeneous mid-gray, the next test image of a different spatial frequency was shown. Observers were asked to use normal binocular viewing to come close to the viewing situation of real displays.

Both in the AC and in the CC experiments, the task was the same. Only contrast type was varied: luminance contrast (AC) or chromaticity contrast. If an observer was not able to detect any stripe for a particular image, then he or she was asked to mark the test image at the very bottom, as close as possible. Observers were requested a

5 min adaptation period to a typical test image viewed in that experimental series on the monitor in the dark room before starting the experiments.

Ten healthy adult observers participated in the experiments, five young observers (mean age: 25.8 years; range: 24–29 years) and five elderly subjects (mean age: 66.6 years; range: 61–70 years). All of them had normal or normal-corrected vision. Their color vision was tested by the FM 100 hue test and was found to be normal. In the experiments, all observers were allowed to wear their spectacles if they had any. The 10 different spatial frequency test images appeared five times each. These 50 test images per luminance level were randomly ordered and then separated into two series to be observed. In the CC experiment, each of the 10 spatial frequencies appeared five times again for each of the eight hues. The order of the test images was randomized and the test images were presented in 10 separate series to the observers.

The *rgb* values of the test images were saved both at the threshold contrast levels and at the preferred contrast levels for every observer. CIE XYZ tristimulus values were computed, and then, for the chromatic case, also the values of CIELAB chroma ($C^*$). In the achromatic (AC) case, the contrast value (*Michelson contrast*) was calculated by Equation 3.3.

$$AC = 100\% \times (L_{max} - L_{min})/(L_{max} + L_{min}) \tag{3.3}$$

In Equation 3.3, in the sinusoid achromatic test images, $L_{max}$ ($L_{min}$) represents the maximum (minimum) luminance at the threshold or preference level found by the observer. In the CC experiment, threshold chromaticity contrasts and preferred chromaticity contrasts were determined by Equation 3.4.

$$CC = 100\% \times (C^*_{max} - C^*_{min})/(C^*_{max} + C^*_{min}) \tag{3.4}$$

In Equation 3.4, in the sinusoid chromatic test images, $C_{max}$ ($C_{min}$) represents the maximum (minimum) CIELAB chroma along the horizontal line crossing the point marked by the observers. All analyses (for both AC and CC) were performed on these resulting contrast values given in %. A paired-sample *t*-test was performed to investigate the significance of the overall mean difference between the threshold contrast and the preferred contrast levels set by the observers. Results are listed in Table 3.3.

As can be seen from Table 3.3, regardless of luminance level, age group, or spatial frequency, the overall achromatic threshold contrast ($AC_{thr}$) level is significantly

**Table 3.3** Paired-sample comparisons between the mean threshold ($\mu_{thr}$) and the mean preferred ($\mu_{pref}$) luminance contrast levels (AC), overall and within age groups, testing the hypothesis $H_0$: $\mu_{thr} = \mu_{pref}$ ($H_a$: $\mu_{thr} \neq \mu_{pref}$) [45].

|  | Mean difference (%) | T | df | Significance |
| --- | --- | --- | --- | --- |
| Overall | 47.6 | 77.8 | 2853 | <0.001 |
| Elderly | 60.9 | 62.4 | 1297 | <0.001 |
| Young | 36.6 | 55.9 | 1555 | <0.001 |

Reproduced with permission from *Displays*.

lower than the preferred (AC$_{pref}$) contrast level ($p < 0.001$). Performing the $t$-test again, between AC$_{thr}$ and AC$_{pref}$ but now within the separated age groups, the differences are still significant, $p < 0.001$ for both the young and the elderly group. But in case of elderly observers, the mean difference was greater than that for young observers (60.9 against 36.6).

The mean contrast values of the AC experiment are plotted in Figure 3.20, as a function of spatial frequency.

For the AC$_{thr}$ and AC$_{pref}$ data presented in Figure 3.20, separate analyses of variance (ANOVAs) were performed, within the subjects (for the spatial frequency and luminance level of the test images) and between the subjects (for the observer's age). In case of the threshold contrast (AC$_{thr}$), significant main effects were found for (1) age ($p < 0.001$), younger observers being more sensitive (as indicated by their lower threshold values) to luminance contrast; (2) luminance level ($p < 0.001$), observers being more sensitive in brighter conditions; and (3) spatial frequency ($p < 0.001$), observers being more sensitive to the mid spatial frequencies than for either low or high spatial frequencies.

All interactions were found to be significant: between luminance level and frequency ($p < 0.001$), age and luminance level ($p < 0.001$), age and frequency ($p < 0.001$), and age times spatial frequency times luminance level ($p < 0.001$). Observers were less sensitive to high and low spatial frequencies with the decrease of luminance level. Elderly observers became less sensitive with the reduction of luminance level. Elderly observers were also less sensitive to high spatial frequencies than young observers. They were less and less sensitive for not only the high but also the mid spatial frequencies as the luminance level decreased (see Figure 3.20).

For the preferred contrast condition (AC$_{pref}$), the main effects of age ($p < 0.001$) and spatial frequency ($p < 0.001$) were significant but there was no significant effect of luminance level ($p = 0.206$). Elderly observers had a preference of higher contrast, and generally, the preferred contrast of observers varied with spatial frequency. Similar to the threshold results, mid spatial frequencies needed less contrast to be preferred and high and low spatial frequencies requested higher preferred contrast values.

All interactions were found to be significant. There was a significant interaction between frequency and luminance level ($p < 0.001$), observers preferring higher contrast for mid and low spatial frequencies with the reduction of luminance level. There was a significant interaction between age and luminance level ($p < 0.001$), elderly observers preferring higher contrast levels with the decrease of the luminance level of the test images. There was also a significant interaction effect between age and spatial frequency ($p < 0.001$), elderly observers adjusting remarkably higher preferred contrast levels for the mid and low spatial frequencies.

The interaction effect among all three factors was significant ($p = 0.012$), elderly observers preferring very high values of contrast for the lower spatial frequencies, and, interestingly enough, young observers chose lower preferred contrasts as the luminance level of the test images dropped down to the darkest luminance level (see Figure 3.20).

**128** | *3 Ergonomic, Memory-Based, and Preference-Based Enhancement of Color Displays*

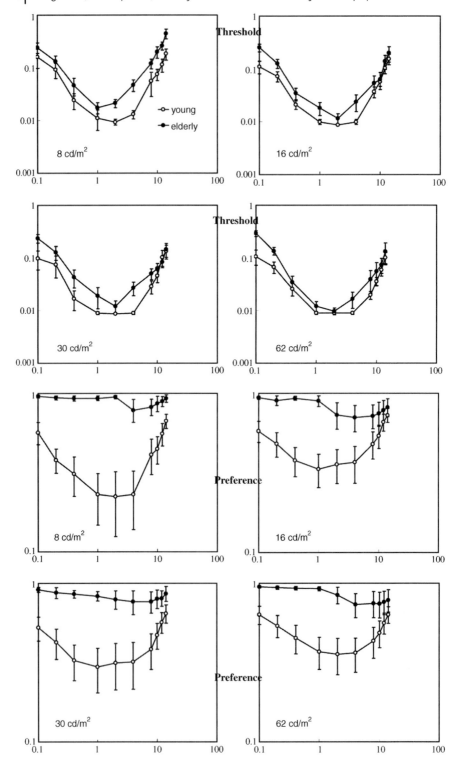

**Table 3.4** Paired-sample comparisons between the mean threshold ($\mu_{thr}$) and the mean preferred ($\mu_{pref}$) chromaticity contrast levels (CC), overall and within age groups, testing the hypothesis $H_0$: $\mu_{thr} = \mu_{pref}$ ($H_a$: $\mu_{thr} \neq \mu_{pref}$) [45].

|  | Mean difference (%) | T | df | Significance |
| --- | --- | --- | --- | --- |
| Overall | 41.9 | 122.2 | 2799 | <0.001 |
| Elderly | 39.8 | 100.3 | 1999 | <0.001 |
| Young | 47.1 | 73.8 | 799 | <0.001 |

Reproduced with permission from *Displays*.

In Table 3.4, paired-sample t-tests are listed to investigate the significance of the overall mean difference between the *threshold chromaticity* contrasts and the *preferred chromaticity* contrasts ($CC_{thr}$ and $CC_{pref}$) found by the observers. Similar to the achromatic results, threshold values and preference values differed significantly on average but the difference was greater for younger observers than for elderly observers (41.68 against 33.04) (compare with Table 3.3).

The mean threshold chromaticity contrast values are plotted in Figure 3.21, while the mean preferred chromaticity contrast values are plotted in Figure 3.22.

For the $CC_{thr}$ and $CC_{pref}$ data presented in Figures 3.21 and 3.22, ANOVAs were carried out within the subjects for spatial frequency and hue angle of the chromatic test images and between the subjects for the observer's age. In case of the *threshold chromaticity* contrast ($CC_{thr}$), significant main effects were found for the CIELAB hue angle of the chromatic test patterns ($p < 0.001$), observers being more sensitive for some hue angles than for other hue angles, and spatial frequency ($p = 0.016$), observers being more sensitive to mid spatial frequencies than for either low or high spatial frequencies.

Other effects including mixed effects were not found to be significant but it can be seen from the average results shown in Figures 3.21 that elderly observers tend to have lower chromaticity threshold sensitivities than young observers. For the preferred chromaticity contrast condition ($CC_{pref}$), the main effects of hue angle and spatial frequency ($p < 0.001$) were significant but there was no significant main effect of age ($p = 0.218$). Observers preferred less chromaticity contrast of the hues 0° and 45° (i.e., the red and the orange tones) and more for other chromaticities.

**Figure 3.20** Average achromatic contrast as a function of spatial frequency (log–log axes): threshold (first two rows) and preference (last two rows) values comparing the mean of young and elderly age groups under different luminance levels. Note that the scale of the preference diagrams in the lower two rows is different from the threshold diagrams in the upper two rows. Open circles show the results of the young observers and black dots represent the data of the elderly observers. The scale of the vertical axes is the log contrast of Equation 3.3. The value of 1 corresponds to 100% contrast. Error bars represent 95% confidence intervals of the mean values calculated from 5 observers × 5 repetitions of both age groups [45]. Reproduced with permission from *Displays*.

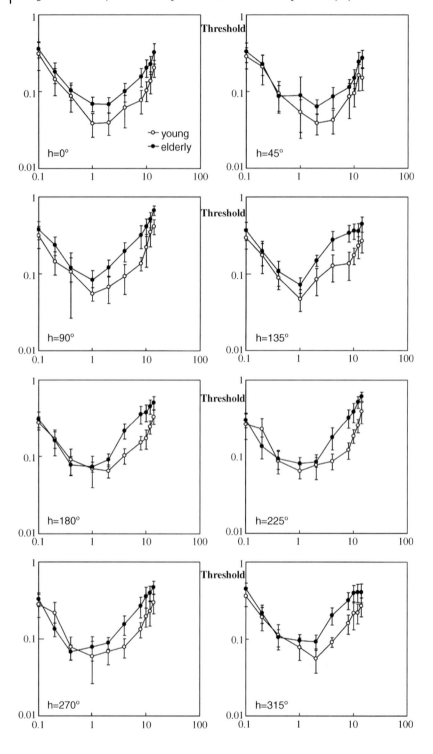

Similar to the chromatic threshold results, mid spatial frequencies needed less chromaticity contrast to be preferred and high and low spatial frequencies requested higher values. There was a significant interaction between hue angle and spatial frequency ($p < 0.001$) and there was a significant mixed effect of age times hue angle times spatial frequency ($p = 0.026$).

The observers' preferred chromaticity contrast was similar for each hue angle, for lower spatial frequencies. But from approximately 2 c/deg onward, different contrasts were preferred for different hues. Except for blue–green hues (180°, 225°, and 270°) where young and aged observers' preference coincided, elderly observers preferred greater contrast for lower spatial frequencies than young observers. The minima of the preferred chromaticity contrast curves of certain hue angles (0° and 45°) appeared at different spatial frequencies for young and elderly observers. This finding is corroborated by the interaction of all three factors (i.e., age, hue angle, and spatial frequency).

Summarizing the above findings, it can be stated that aging has an effect not only on luminance contrast *thresholds* or chromaticity contrast *thresholds* but also on the *preferred* luminance contrast and the *preferred* chromaticity contrast. Results indicate that, beyond elderly observers' expected luminance contrast sensitivity decline, the difference between the preferred contrast of the elderly and the preferred contrast of the young is even more significant than the threshold contrast difference.

The peak achromatic sensitivity (i.e., the inverse threshold) of the human visual system to mid spatial frequencies did not change but the sensitivity to high spatial frequencies dropped significantly with increasing age [53, 57], in accordance with the above results. This is actually a shifting of the peak sensitivity to lower frequencies during aging but the peak still remains in the domain of one to four cycles per degree of spatial frequencies.

*Preferred* luminance contrast curves follow a similar spatial frequency trend as the threshold curves; namely, for mid spatial frequencies, observers require less contrast to perceive it comfortably. The relatively large deviation values (see the error bars in Figure 3.20) are attributable to the more subjective nature of this task compared to the threshold task. As can be seen from Figure 3.20, the achromatic preferred contrast of elderly observers is significantly higher than that of young observers.

For lower and higher spatial frequencies (except for the mid-frequency region between 1 and 12 c/deg), elderly observers set almost the possible maximum for each luminance situation. Apart from the 30 cd/m$^2$ condition where the transition was smooth between the neighboring spatial frequencies, there were extraordinarily low

**Figure 3.21** Average chromaticity contrast as a function of spatial frequency (log–log axes). *Threshold values* comparing young and elderly observers with different hue angles of the test images (CIELAB *h* numbers are shown in degrees). Open circles show the mean values of the young observers, while black dots represent the mean of the elderly observers. The scale of the vertical axes is the log contrast of Equation 3.4, the value of 1 corresponding to 100% contrast. Error bars represent 95% confidence intervals of the mean values calculated from 5 observers × 5 repetitions for aged observers and 5 observers × 2 repetitions for young observers [45]. Reproduced with permission from *Displays*.

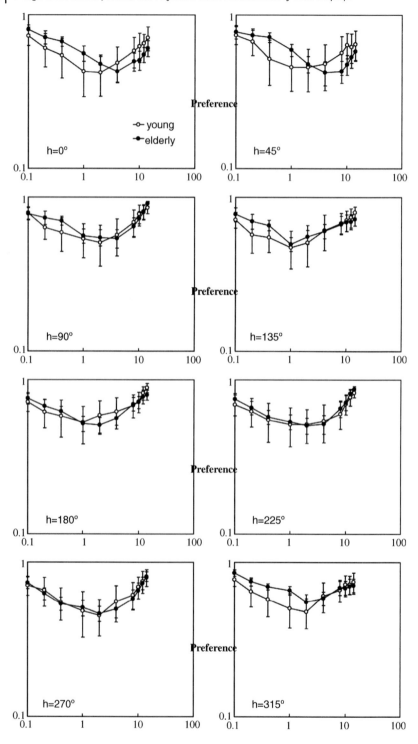

contrast preferences for mid spatial frequencies compared to low or high spatial frequencies. A possible answer to the increased difference between the two age groups for suprathreshold luminance contrasts is that they may depend on the optical properties of the opaque aged eye lens.

It was pointed out [62] that the reciprocal of human retinal contrast to any input image can be modeled by the sum of the reciprocal of the contrast value of the contributors in the system, that is, display contrast and eye contrast (see Equation 3.5).

$$(1/C_{\text{ret}}) = (1/C_{\text{disp}}) + (1/C_{\text{eye}}) \qquad (3.5)$$

The contrast contribution of the eye can be approximated by convolving the point spread function of the human eye (strongly dependent on age) with the image [63]. An aged (i.e., obscure) crystalline lens degrades contrasts considerably more than a young lens and this effect causes significantly less retinal contrast for aged persons than for young observers.

In the same study [60], luminance CMFs were concluded to have band-pass characteristics while chromatic CMFs exhibited low-pass style (at least in the 0.5–4 c/deg region they studied), though, for most of the chromatic cases, the resulting curves tended to exhibit a sensitivity loss at 0.5 c/deg compared to the 1 c/deg sensitivity result. The results (see also Section 4.4) suggest that the chromatic contrast sensitivity functions do have the general band-pass fashion for the whole spatial frequency domain analyzed, only the minima of the functions tend to shift toward lower spatial frequencies, as it was pointed out for chromaticity threshold compared to luminance threshold.

The small preference differences between the age groups for chromaticity contrast compared to achromatic contrast suggest that while – with increasing age – both chromatic and achromatic contrast sensitivities drop, suprathreshold perception stays more stable for chromaticity than for luminance. In Figure 3.23, the data of Figure 3.22 are replotted as a function of the hue angle of the test images.

In the top-left graph of Figure 3.23, the two curves representing young and elderly results overlap, which can be attributed to the relatively small spatial frequency (0.1 c/ deg). But, for higher spatial frequencies, preferred contrast levels diverge more with hue angle. The most outstanding change corresponds to the red and orange regions ($h = 0°$ and $h = 45°$). Here, both the young and the elderly observers set less chromaticity contrast than for the other hues, and, interestingly, these were the

**Figure 3.22** Average chromaticity contrast as a function of spatial frequency (log–log axes). *Preferred values* comparing young and elderly observers with different hue angles of the test images (CIELAB h numbers are shown in degrees). Open circles show the mean values of the young observers, while black dots represent the mean of the elderly observers. The scale of the vertical axes is the log contrast of Equation 3.4, the value of 1 corresponding to 100% contrast. Error bars represent 95% confidence intervals of the mean values calculated from 5 observers × 5 repetitions for aged observers and 5 observers × 2 repetitions for young observers [45]. Reproduced with permission from *Displays*.

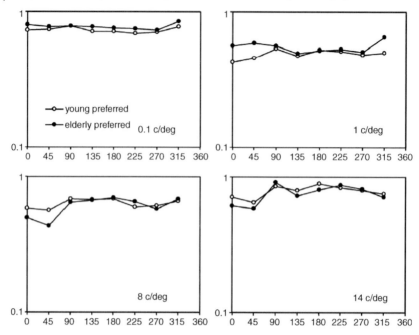

**Figure 3.23** Preferred chromaticity contrast of young (open circles) and elderly (black dots) observers replotted from Figure 3.22 as a function of CIELAB hue angle $h$ in degrees. The four plots represent the spatial frequencies of 0.1, 1, 8, and 14 c/deg [45]. Reproduced with permission from *Displays*.

only cases where elderly observers' preference was typically less than that of the young observers' for high frequencies.

The above finding is in accordance with the fact that the transmission of the crystalline lens of aged people is almost unchanged in the higher wavelength domain, that is, in the region of red and orange spectral colors. The 10–14 c/deg frequency corresponds to an approximately 2 pixel wide line on a display from a viewing distance of 60 cm. This implies that the use of chromaticity contrast should be encouraged to encode visual information in addition to luminance contrast – especially by using red or orange colors.

## 3.4
### Long-Term Memory Colors, Intercultural Differences, and Their Use to Evaluate and Improve Color Image Quality

In this section, long-term memory colors of a set of important familiar objects (often seen in the visual environment and stored in human memory) are quantified in color space. Their intercultural differences are also shown. An application of memory colors to increase the perceived color quality of still images and motion images is presented.

### 3.4.1
### Long-Term Memory Colors for Familiar Objects

Short-term color memory is often required to compare an original image with its reproduction both in the laboratory and in everyday life situations, for example, a person purchasing gloves to match a hat at home, an artist in his studio mixing a color on his palette, and a photographer looking at his photo in a viewing booth and then at the reproduction of his photo on a color monitor or a color inspector comparing a color sample with a color standard at another location. Short-term memory colors are often shifted toward long-term memory colors; that is, observers tend to substitute their actual short-term memory colors by long-term memory colors (so-called *prototypes*) of familiar objects instead of remembering the actually seen color [64–67].

The presence or absence of different *image cues* in the field of view (like the shape or texture of the familiar objects, for example, the texture of grass or the shape of a banana) influences the extent *long-term memory* colors are activated when remembering a short-term memory color. Image cues tend to *remind* the observer of previous experiences about the past color impressions of similar objects. The CIELAB $L^*$, $a^*$, and $b^*$ values of six long-term memory colors are summarized in Table 3.5.

Recently, long-term memory colors were quantified by using a three-phase psychophysical method [68]. The actual experiment [68] was carried out in the viewing situation of a self-luminous color CRT display, which is a very important application of memory colors (see Section 3.4.3). A simplified description of this experiment is given below. The full description can be found in Ref. [68].

The experiment took place in a dark room where the display was the only light source [68]. First, the nationality of the observers was Hungarian. To obtain a hint about the intercultural differences of long-term memory colors, the same experiment was repeated by Korean observers (see Section 3.4.2). There were 11 Hungarian observers of normal color vision altogether. All observers were young adults, college students familiar with computers and computer-controlled displays.

The experiment consisted of three phases. In the first phase, the *method of choice colors* was applied. In this method, observers had to select a long-term memory color

**Table 3.5** CIELAB $L^*$, $a^*$, and $b^*$ values of six important long-term memory colors under illuminant C [68].

| Memory color | $L^*$ | $a^*$ | $b^*$ | Reference |
|---|---|---|---|---|
| Caucasian skin | 79.5 | 16.1 | 10.4 | [69] |
| Blue sky | 54.0 | −17.0 | −28.1 | [69] |
| Green grass | 50.0 | −33.7 | 29.8 | [64] |
| Oriental skin | 63.9 | 14.0 | 16.1 | [70] |
| Deciduous foliage | 33.6 | −18.5 | 12.8 | [69] |
| Orange | 71.6 | 26.7 | 75.72 | [71] |

Reproduced with permission from *Color Research and Application*.

from 16 constant color stimuli (choice colors). Four lines of four constant color patches (4 × 4 = 16 choice color patches) were displayed on a middle gray background. There was a white adapting border at the screen boundary outside the middle gray background.

At the top of the middle gray background, a color name was displayed as a label. There was a checkbox under every color patch. Observers had to check only one checkbox associated with that color patch that was perceived to best correspond to the color name shown at the top. Following six memory color names were displayed: non-tanned skin, blue sky, green grass, deciduous foliage, banana, and orange.

The 16 choice colors were shown to the observer on one screen for each of the six color names. The observer indicated his/her choice for the color name shown. Thirty repetitions have been carried out with a random array of choice colors for each of the six memory colors. Therefore, the output of each observer was a set of 30 choices per memory color.

Choice colors were computed randomly. They were located inside spheres around the six color centers shown in Table 3.6 in CIELAB color space. The radii of these spheres are also shown in Table 3.6. Choice colors are depicted in Figure 3.24.

The first phase of the experiment was repeated with a set of new color centers estimated from the mean long-term memory colors resulting from the choice colors of Figure 3.24. The new tolerance values were estimated as twice the standard deviations of the mean long-term memory colors. The new color centers and tolerance values are listed in Table 3.7. Figure 3.25 shows the new choice colors and the new tolerance radii.

As can be seen from Figures 3.24 and 3.25, the tolerance radii became smaller in the repetition of the first phase of the method (i.e., the phase of choice colors). Therefore, observers could choose their long-term memory colors from a more limited set of choice colors in the repetition.

The *second phase* of the method of obtaining long-term memory colors was the method of *reproducing a color name*. In this method, there was a square-shaped color patch of changeable color in the middle of the screen within a dark gray frame of a thickness of 2 mm. The color of its uniform background was middle gray. There was

**Table 3.6** Color centers and tolerance values [68] used to calculate the random choice colors in the first phase of the long-term color memory experiment with Hungarian observers (see also Figure 3.24).

| Color name | $L^*$ | $a^*$ | $b^*$ | Tolerance radii |
| --- | --- | --- | --- | --- |
| Non-tanned skin | 79.5 | 16.1 | 10.4 | 10 |
| Blue sky | 54 | −17 | −28.1 | 15 |
| Green grass | 55 | −33.7 | 29.8 | 15 |
| Deciduous foliage | 36 | −18.5 | 12.8 | 20 |
| Banana | 70 | 0 | 65 | 30 |
| Orange | 71.6 | 26.7 | 69.72 | 15 |

Reproduced with permission from *Color Research and Application*.

## 3.4 Long-Term Memory Colors, Intercultural Differences, and Their Use to Evaluate

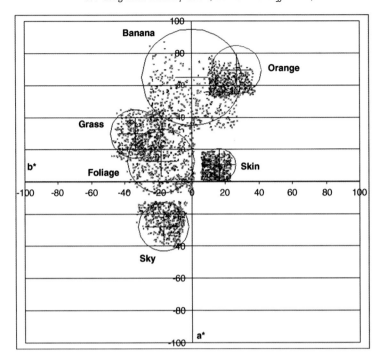

**Figure 3.24** CIELAB $a^*$ and $b^*$ values of the color centers of Table 3.6 (large plus signs) and actual random choice colors (small crosses) [68]. Large circles show the tolerance radii of Table 3.6. These choice colors were used in the first phase of the long-term color memory experiment with Hungarian observers. Reproduced with permission from *Color Research and Application*.

a white adapting border at the boundary of the screen, same as for the method of choice colors.

A color name was written by white characters above the changeable color patch. In the middle of the right half of the screen, there were three "sliders" to change hue,

**Table 3.7** New color centers and tolerance radii for the repetition of the method of choice colors (see text) [68].

| Color name | $L^*$ | $a^*$ | $b^*$ | Tolerance radii |
|---|---|---|---|---|
| Skin | 81.4 | 10.7 | 18.2 | 8 |
| Sky | 66.4 | −12.5 | −26.5 | 10 |
| Grass | 46.8 | −41.1 | 31.4 | 9 |
| Foliage | 44.8 | −40.1 | 29.2 | 11 |
| Banana | 81.5 | −10.8 | 64.7 | 11 |
| Orange | 68.0 | 26.1 | 62.7 | 8 |

Reproduced with permission from *Color Research and Application*.

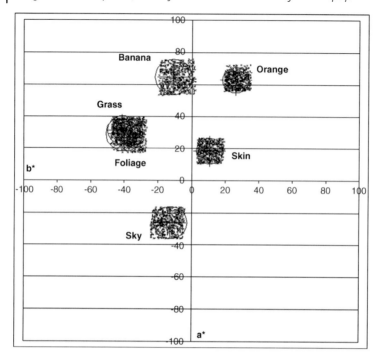

**Figure 3.25** CIELAB $a^*$ and $b^*$ values of the new color centers from Table 3.7 (large plus signs) and actual random choice colors (small crosses) in the repetition of the method of choice colors. Large circles show the tolerance radii from Table 3.7 [68]. Reproduced with permission from *Color Research and Application*.

saturation, and lightness in the display's native hue–saturation–value (HSV) color space. The HSV color space is very far from perceptually uniform but it was appropriate for just finding the long-term memory color.

Below the three sliders there was a "Ready" button. At the beginning, the color bar was dark gray with the first color name at the top. By the aid of the three sliders, the observer had to reproduce his/her memory color corresponding to a color name. Then the observer pushed the "Ready" button. This long-term memory color was stored. The output of each observer was a set of 10 colors per memory color, mixed for each of the six color names, the same names as used in the method of choice colors.

In the *third phase* of the experiment, observers carried out the method of *reproducing the most appropriate color* in a grayscale photo. The test image and the experimental procedure were similar to the method of "reproducing a color name" with the following differences: (1) the color patches of changeable color took place within different grayscale photorealistic images in such a part that could be described by a color name, for example, in the "grass" of a landscape image; and (2) color names were substituted by these grayscale images instead of the middle gray background.

By the aid of the three sliders, observers had to reproduce the "most suitable" colors corresponding to that part of the grayscale image where the changeable color patch was displayed. The output of each observer was a set of 10 colors per memory color, reproduced for each of the six grayscale photos containing non-tanned skin, blue sky, green grass, deciduous foliage, banana, and orange.

The mean long-term memory colors resulting from the three-phase psychophysical method [68] are described in the next section, in comparison with the Korean observer group.

### 3.4.2
### Intercultural Differences of Long-Term Memory Colors

Although serious efforts were taken to ensure the same experimental conditions for the Korean and Hungarian observers groups, minor differences could not be avoided (see Table 3.8).

As can be seen from Table 3.8, the Korean viewing conditions were very similar to the Hungarian viewing conditions. Experiments were carried out in separate laboratories, in Korea and in Hungary, respectively. The way of monitor characterization and the psychometric method were the same as in the Hungarian observations except that there was no repetition of the method of choice colors.

The Korean choice color centers and tolerance radii were the same as shown in Table 3.6 except that instead of Caucasian skin oriental skin was used (as specified in Table 3.5) with a tolerance radius of 10. In the Korean grayscale photo experiment, an image depicting an Oriental person was used instead of the Hungarian image depicting a Caucasian person. Overall mean long-term memory colors of all observers, all repetitions, and all three phases of the experiment are compared between Korean and Hungarian observers in Table 3.9 and depicted in Figure 3.26. A more detailed analysis can be found in Ref. [68].

The data of Table 3.9 and Figure 3.26 can be used for the dark/dim viewing condition and the 6500 K white point. Long-term memory colors corresponding to another viewing situation can be calculated by applying the CIECAM02 color appearance model (Section 2.1.9). Table 3.10 shows the significance of the differences between the Korean and the Hungarian long-term memory colors.

**Table 3.8** Comparison of the Korean and Hungarian long-term memory color experiments [68].

| Condition | Hungarian experiment | Korean experiment |
| --- | --- | --- |
| Viewing environment | Dark room | Dark room |
| Monitor white point | About 6500 K ($x=0.310$; $y=0.331$) | About 6500 K ($x=0.311$; $y=0.318$) |
| Peak white | 116 cd/m$^2$ | 117 cd/m$^2$ |
| Number of observers | 11 | 9 |
| Choice colors | 30 | 10 |
| Long-term memory color names | Non-tanned Caucasian skin, blue sky, green grass, deciduous foliage, banana orange | Oriental skin, blue sky, green grass, deciduous foliage, banana, orange |

140 | 3 Ergonomic, Memory-Based, and Preference-Based Enhancement of Color Displays

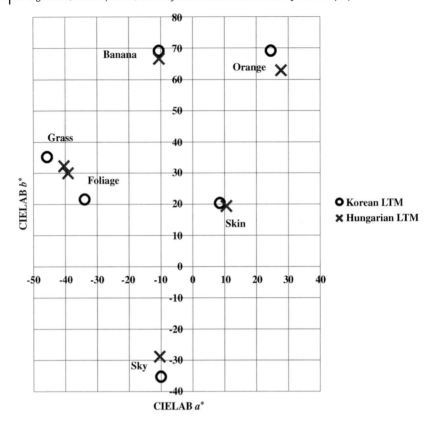

**Figure 3.26** Overall mean long-term memory colors (LT) of all observers in terms of mean CIELAB $a^*$ and $b^*$ values for all repetitions and all three experimental phases (see text). Comparison of Korean (K) and Hungarian (H) observers (see also Table 3.9) [68]. Reproduced with permission from *Color Research and Application*.

**Table 3.9** Overall mean long-term memory colors (LT) of all observers in terms of mean CIELAB values for all repetitions and all three experimental phases (see text).

| | LT | | | | | | | | | | | |
|---|---|---|---|---|---|---|---|---|---|---|---|---|
| | Skin | | Sky | | Grass | | Foliage | | Banana | | Orange | |
| | K | H | K | H | K | H | K | H | K | H | K | H |
| $L^*$ | 73.1 | 81.8 | 62.6 | 65.7 | 55.6 | 45.5 | 40.5 | 43.2 | 86.6 | 84.1 | 72.6 | 68.2 |
| $a^*$ | 8.6 | 10.5 | −10.0 | −10.4 | −45.7 | −40.4 | −33.9 | −39.1 | −10.5 | −10.6 | 24.5 | 27.7 |
| $b^*$ | 20.3 | 19.3 | −35.4 | −28.9 | 35.1 | 32.2 | 21.4 | 30.0 | 69.1 | 66.7 | 69.1 | 62.9 |

Comparison of Korean (K) and Hungarian (H) observers (see also Figure 3.26) [68].

## 3.4 Long-Term Memory Colors, Intercultural Differences, and Their Use to Evaluate

**Table 3.10** Significance of the differences between the long-term memory colors of Korean and Hungarian observers.

|  | \multicolumn{6}{c}{p} |  |  |  |  |
|---|---|---|---|---|---|---|
|  | Skin | Sky | Grass | Foliage | Banana | Orange |
| $L^*$ | **0.000** | 0.397 | **0.000** | 0.306 | **0.019** | **0.000** |
| $a^*$ | **0.037** | 0.835 | 0.060 | **0.038** | 0.715 | 0.150 |
| $b^*$ | 0.147 | 0.166 | 0.240 | **0.007** | 0.324 | **0.000** |

t-tests; bold numbers indicate significant differences at $p = 0.05$ [68]. Reproduced with permission from *Color Research and Application*.

As can be seen from Tables 3.9 and 3.10 as well as from Figure 3.26, Hungarian grass and foliage long-term memory colors were similar to each other but the Korean ones were different. Korean grass was lighter and much more saturated than Korean foliage. Korean green grass was lighter and more saturated than Hungarian green grass. Korean foliage was less saturated than Hungarian foliage.

Oriental skin and Caucasian skin had similar chroma. Oriental skin contained more yellow and Caucasian skin was found to be significantly lighter than Oriental skin. Korean sky had similar lightness as Hungarian sky but Korean sky was somewhat more saturated than Hungarian sky. Korean banana was somewhat lighter than Hungarian banana. Finally, Korean orange was lighter than Hungarian orange and Korean orange contained more yellow than Hungarian orange.

### 3.4.3
### Increasing Color Quality by Memory Colors

As mentioned above, mean long-term memory colors of familiar objects are also called *prototypical colors* in the literature [66, 67]. The perceived *naturalness* of a color image is a result of a comparison between the immediately perceived (so-called *original*) color O and the prototypical color. Prototypical colors have maximum naturalness [66, 67]. Going away from prototypical colors naturalness decreases [66, 67].

Observers tend to increase the perceived naturalness of O in their memory by *shifting* M toward the prototypical color [72]. An important finding is that observers tend to *shift* the original color toward the prototypical color and *not to replace it* by the prototypical color. If the original color O is located within a *tolerance volume* [66, 67] of perceived naturalness around the prototypical (or long-term memory) color in color space, then naturalness will become acceptable and the hue, chroma, and lightness shifts will not be significant.

An important application of long-term memory colors and their tolerance volumes is the enhancement of the perceived color quality of digital photorealistic images presented on a color display. Single color attributes such as hue can also be enhanced separately. For example, the hue angle values in a specific zone in the image (e.g., the "hue gamut" in a zone containing Caucasian skin) can be mapped into the hue tolerance interval (in the above sense) for Caucasian skin (which is equal to [58°; 80°] in terms of CIELAB $h$ [64]) in the following way.

In the first step, replace the *mode* of the above "hue gamut" by the middle point of the hue tolerance interval (69°), the maximum of the "hue gamut" by the upper limit of the hue tolerance interval (80°), and the minimum of the "hue gamut" by the lower limit of the constant hue interval (58°). In the second step, use linear estimation for the intermediate hue angle values of the "hue gamut." The chroma gamut and the lightness gamut can be mapped similarly.

This mapping procedure mimics the color shifts in human color memory toward long-term memory colors. Mapping can be improved by representing the color gamut of a specific zone of the image in a three-dimensional color space and then mapping it onto the tolerance volume.

Alternatively, the above procedure can also be used to evaluate the perceived naturalness of digital images presented on a self-luminous display. A naturalness index was recommended to predict the perceived naturalness of digital images [66, 67]. It can be calculated by means of a Gaussian function of the differences between average and prototypical *saturation* values.

An alternative way of calculation was also suggested [72]: divide the number of such pixels whose color is in a hue, chroma, or lightness *tolerance interval* by the total number of pixels of the digital image. The resulting *hue, chroma, or lightness naturalness indices* can be added to provide a *general naturalness index*. The difficulty with this way of calculation is that it is troublesome to decide to which type of image context a pixel belongs.

## 3.5
## Color Image Preference for White Point, Local Contrast, Global Contrast, Hue, and Chroma

This section describes an apparatus and a method to obtain a color image preference data set [73]. Color image transforms influencing color image preference are presented together with the results of color image preference related to the preferred white point, local contrast, global contrast, hue, and chroma. The important question of color image preference differences between young and elderly observers is also dealt with.

In modern society, elderly people become enthusiastic users of self-luminous displays optimized for the visual properties of young adults ignoring the specific attributes of the vision of the elderly though the existence of differences is evident despite long-term adaptation recompensating for some of the age-related changes in the human visual system.

Visual preference is a *high-level* psychological factor having a very important impact on the user acceptance of color displays. In this section, preference differences between young and elderly observers are discussed for the case of color photorealistic images presented on self-luminous displays, in terms of global and local contrast, white point, average chroma, and the effect of several standard image color transform algorithms.

It will be seen that there are significant color image preference differences between young and elderly observers. Some of these differences can be explained with neurophysiological changes while other changes may be attributed to cultural implications [73]. A novel image processing algorithm to get a preferred color image as a function of observer age is presented in Section 3.6.

Experiments on the color image *preference* differences between young and elderly observers are important because reconstructing *color appearance* to the elderly observers is *not* equivalent to design the elderly observer's preferred color image display since their color preference can be different from presenting perceptually correct color, that is, the appearance corresponding to the young adults'.

The focus of this section is concentrated on typical image enhancement methods expected to enhance images to be more preferred by young or by elderly observers. Classically, perceived color image quality was modeled with respect to a reference image, for example, by S-CIELAB, iCAM, and others (see also Section 4.4.3.2) [74–77]. These image quality metrics either are based on a *Minkowski metric* in image space using various decorrelating transforms or models of the human visual system or use a structural *similarity-based* approach that takes the direction of image distortion into account [78].

Other image quality modeling methods [79–81] consider the possibility of a no-reference image quality assignment via user preference utilizing the *cognitive representation* of the scene as a reference image. This section uses the latter concept with an image enhancing algorithm in mind. This image enhancing algorithm is enhancing the image without any reference image with respect to the age of the user. Several interesting questions are described elsewhere, for example, gender differences or cultural differences of image preference [80] or the spatial distribution of image features influencing preference [82].

### 3.5.1
**Apparatus and Method to Obtain a Color Image Preference Data Set**

An accurately characterized CRT display was used to display the color pictorial test images [73]. The mean characterization model prediction equaled $\Delta E^*_{ab} = 0.95$. For any pixel on the display in the *i*th row and the *j*th column, CIECAM02 $J$, $C$, and $h$ values were determined from the *rgb* values. These values are denoted by ($J_{ij}$, $C_{ij}$, $h_{ij}$) below. Eight elderly (average age: 69.1 years) and five young (average age: 25.8 years) color normal observers (males and females) took part in the experiments.

The method of image pair comparisons was used [83]. The resulting color image *preference scores* ($z$) were stored for further analysis. Comparisons were carried out by presenting image pairs selected from the set of original and transformed versions of the images [73]. Only one image appeared on the screen and the observer was able to switch between the two images of a pair by pressing a button. No time limit was established upon observation and the observer was allowed to switch between the two images unlimited times.

### 3.5.2
### Image Transforms of Color Image Preference

Several different types of color image transforms were tested for observer preference [73]. Four transforms were found to exhibit high relevance for preference. Only the latter four transforms are described here. The other transforms are reported elsewhere [73]. Image processing transforms were designed so that the amplitude or "strength" of their effect was characterized by a single scalar value as a parameter. The transformations modified the three perceptual correlates of CIECAM02 color space, $J$, $C$, and $h$. There was a further transform modifying the white point of the image.

Two transforms influenced lightness ($J$). The first transform was a local, convolution-based lightness contrast enhancement (denoted by LE). The second transform was a global lightness contrast enhancement (TC, standing for *tone curve*). A further transform influenced chroma. This was the so-called *hue-dependent chroma boost* (CH) enhancing the value of CIECAM02 $C$ uniformly in each relevant hue range (such as sky or skin tones, reds, or oranges) with different amounts of chroma in each hue range.

*Hue transforms* did *not* yield significant results. Finally, the fourth relevant transform (WP) changed the white point of the image. Correlations among these transforms were also tested to investigate the cross-effects. The importance of the *order* in which the transforms are applied to the test images was also pointed out [73].

The transforms TC and CH were designed to avoid out-of-gamut pixels by taking the gamut boundary into account, that is, the maximum $J$ and $C$ values at a given $h$ calculated from the colorimetric characterization model of the display. However, such a constraint was not feasible for the case of LE and WP; hence, a clipping at the gamut boundary was necessary. Since LE affected only lightness, this was accomplished simply by taking the largest displayable $J$ value for any given $C$ and $h$ values. For WP, $C$ values were limited in the same way.

In spite of the above gamut considerations, a small number of pixels still fell out of gamut, for example, due to rounding. Such pixels could be set to black or to their original value. Such a method, however, is not suitable for a consumer display. It was meant for the purpose of the experiment only.

The above four transforms and the corresponding color image preference results are described in Sections 3.5.3–3.5.6, respectively. The general finding was that elderly observers exhibited lower image preference scores than young observers. They were more indefinite and inconsequent in their preference judgments than young observers [73]. The mean preference score of the young observers was almost twice as much as that of the elderly (1.40 versus 0.79).

### 3.5.3
### Preferred White Point

With this transform (WP), the white point of an input image (assumed to captured and viewed under D65) was converted to a desired correlated color temperature (white tone, CCT) using the CIECAM02 color appearance model. The input

$$\begin{bmatrix} R \\ G \\ B \end{bmatrix} \xrightarrow{CRT \text{ model}} \begin{bmatrix} X \\ Y \\ Z \end{bmatrix} \xrightarrow[CIECAM\,02]{D\,65} \begin{bmatrix} J \\ C \\ h \end{bmatrix} \xrightarrow[(CIECAM\,02)^{-1}]{W} \begin{bmatrix} X \\ Y \\ Z \end{bmatrix} \xrightarrow{CRT \text{ model}} \begin{bmatrix} R \\ G \\ B \end{bmatrix}$$

**Figure 3.27** Workflow of the WP transform. The white point to be achieved by the transform is denoted by W [73]. Reproduced with permission from *Color Research and Application*.

parameter (so-called *external parameter*) was the white point to be achieved by the transform specified in Kelvin units. From this white point, the CIE *XYZ* tristimulus values of the transformed image were calculated. Various gamut compression methods were included to avoid those pixels falling off the gamut of the CRT display. The processing [73] is illustrated in Figure 3.27.

The white point of 12 images of different scenes (landscapes, artificial and natural surroundings, faces, and indoor photos) was transformed between approximately 3000 and 46 000 K (the minimal and maximal CCTs on the experimental display without serious image artifacts). These images were presented to the observers. The preference scores of the images were averaged for the observers. Thus, for each image, a mean observer preference curve was established for the different transformed versions of the image. A sample image is shown in Figure 3.28.

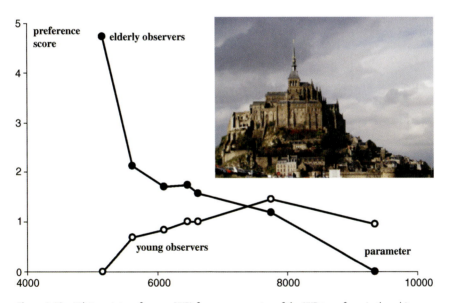

**Figure 3.28** White point preference (WP) for a sample test image (right) depicting the Mont Saint Michel Abbey in Bretagne, France after rainfall with cloudy sky. Black dots: elderly observers. Open circles: young observers. The parameter of the WP transform is the white point in Kelvin units [73]. Diagram reproduced with permission from *Color Research and Application*. The photo was taken by the authors.

As can be seen from Figure 3.28, elderly observers preferred the 5200 K version of this image while young observers preferred about 7800 K. The *optimum preference parameter* ($p_{opt}$) was defined in the following way. It was calculated from the preference function (i.e., from the preference score as a function of the transform parameter) by weighting a set of parameter values $\{p_i\}$ sampling the whole parameter range by the corresponding preference scores $\{z_i\}$ (see Equation 3.6). The value of $p_{opt}$ expresses *optimum preference* in the sense that the computing method of Equation 3.6 takes the entire preference function into account instead of simply taking the maximum of the preference function.

$$p_{opt} = \frac{\sum_i z_i p_i}{\sum_i z_i} \qquad (3.6)$$

To express optimum white point preference in the $u'$, $v'$ chromaticity diagram, now let $p_i$ (as a two-dimensional "vector" quantity) denote the $(u', v')$ values of the $i$th tested white point and let $z_i$ denote the corresponding preference score in Equation 3.6. The resulting optimum preference parameter ($p_{opt}$) is now a vector of the optimum $u'$ and $v'$ values computed by Equation 3.6. The latter result is shown in Figure 3.29 for the 12 test images.

As can be seen from Figure 3.29, preferred white point ranges between 5800 and 7000 K around a mean value of 6500 K. Image content influenced the preferred white point of the image significantly for both aged and young observers [73]. The lower

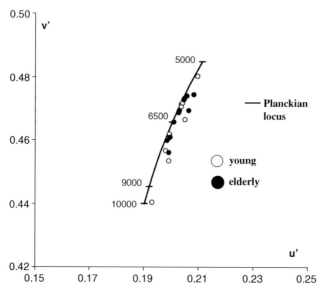

**Figure 3.29** Optimum preference parameter ($p_{opt}$) results of the white point preference (WP) experiment in the $u'$, $v'$ chromaticity diagram computed by Equation 3.6 (see text). Black dots: white point preference of elderly observers for each one of the 12 test images. Open circles: same result for young observers [73]. Reproduced with permission from *Color Research and Application*.

CCT preference of aged observers was confirmed. Elderly observers reported in several cases informally that they preferred warm tones to cold tones since this made the image more colorful to them. A reason may be that more light intensity is delivered at longer wavelengths than at shorter wavelengths due to the aging of the eye lens.

### 3.5.4
### Preferred Local Contrast

The achromatic local contrast enhancement (called LE) algorithm used the *Wallis filter* applied pixel-wise on the entire image. The following equation was used for the *ij*th pixel:

$$J'_{i,j} = (J_{i,j} - m_{i,j}) \cdot p + m_{i,j} \tag{3.7}$$

where

$$m_{i,j} = \frac{1}{(2 \cdot w + 1)^2} \cdot \sum_{k=i-w}^{i+w} \sum_{l=j-w}^{j+w} J_{k,l} \tag{3.8}$$

The new CIECAM02 ($J'_{ij}$, $C_{ij}$, $h_{ij}$) values of the *ij*th pixel were then transformed back to the display's native *rgb* values. Note that only the value of *J* was changed. The parameter *p* was the (positive) input parameter of the transform. The transform results in *blur* with $p < 1$ and local contrast *enhancement* with $p > 1$. With $p = 1$, the transform left the image intact. The parameter *w* denotes the radius of the convolution window in which the average was calculated. After an empirical analysis, this value was set to 19 pixels [73].

In this experiment, the same set of 12 test images was used as in Section 3.5.3. Color image transform parameter values (Equation 3.7) were restricted to a reasonable parameter range. For the LE transform, the resulting preference functions could be *averaged* among the test images (see Figure 3.30).

Figure 3.30 shows that young observers dislike any local lightness contrast enhancement above the very slight parameter value of 1.33. Note that parameter values under 1.0 in Equation 3.7 mean *blurring* of the original image. This result was independent of the image content. It can also be seen from Figure 3.30 that aged observers prefer a slight local contrast enhancement depending on the image content [73].

If the original image contains many details of high spatial frequencies, then aged observers tend to prefer higher local lightness contrast enhancement parameter values. Faces, skin, and noise seem to inhibit the preference of contrast enhancement. In typical skin images, neither age group preferred a harsh local contrast enhancement [73].

### 3.5.5
### Preferred Global Contrast

The global contrast enhancement algorithm (TC from the abbreviation of *tone curve*) modifies the CIECAM02 *J* value of all pixels in the image according to a sigmoid function $f_{TC}$, that is, $J'_{ij} = f_{TC}(J_{ij})$. This function is illustrated in Figure 3.31.

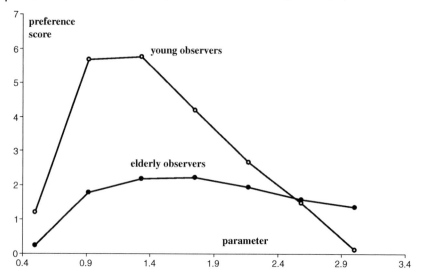

**Figure 3.30** Result of the local contrast enhancement (LE) color image preference experiment. The image transform parameter value of 1 corresponds to the original image. Black dots: elderly observers. Open circles: young observers [73]. Reproduced with permission from *Color Research and Application*.

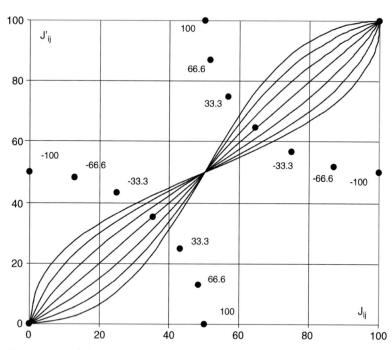

**Figure 3.31** Global contrast enhancement algorithm. Transfer function ($f_{TC}$) was of the transform TC. Numbers at the control points (black dots) refer to the parameter value $p$ of the transform [73]. Reproduced with permission from *Color Research and Application*.

## 3.5 Color Image Preference for White Point, Local Contrast, Global Contrast, Hue, and Chroma

The function $f_{TC}$ in Figure 3.31 is a continuous Bézier spline with four control points [73]. This function resulted in higher lightness for light colors and darker lightness for dark colors (compared to the mid-tones) enhancing the global contrast of the input image. The input parameter of the algorithm determined the position of the two changing control points labeled by the changing input parameter values in Figure 3.31 [73].

In this experiment, the same set of 12 test images was used as in Sections 3.5.3 and 3.5.4. Color image transform parameter values (Equation 3.7) covered the entire parameter range. For the TC transform, the resulting preference functions could be *averaged* among the test images (see Figure 3.32).

As can be seen from Figure 3.32, young observers preferred a slight positive global contrast enhancement parameter value while aged observers prefer a higher positive value. Since the TC transform's sigmoid distortion (see Figure 3.31) of the lightness level of the input image renders dark shades deeper and highlights lighter, the dynamic ranges of these regions *shrink*. This means that the image *loses* fine spatial details in its dark and light regions. The fact that young observers prefer no change or a slight enhancement indicates the importance of fine image details in every tone region for young observers [73].

Due to the aging of their visual system, elderly observers are more unconcerned with such fine image details. Therefore, they prefer more global contrast enhancement. In test images with dominating high spatial frequencies, young observers preferred the unchanged version in order not to lose fine image details but elderly observers preferred a moderate increase of global contrast [73].

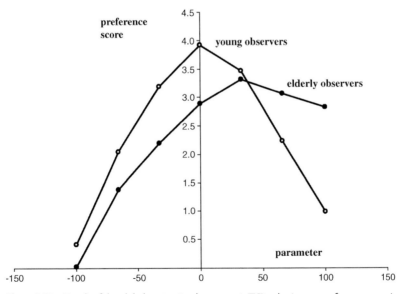

**Figure 3.32** Result of the global contrast enhancement (TC) color image preference experiment. The parameter value of 0 corresponds to the original image. Black dots: elderly observers. Open circles: young observers [73]. Reproduced with permission from *Color Research and Application*.

### 3.5.6
### Preferred Hue and Chroma

In the hue-dependent chroma boost transform (CH), the chroma values ($C_{ij}$) of a pixel were increased or decreased by $\Delta C^{\text{rel}}$ ($C_{ij}^{\text{rel}'} = C_{ij}^{\text{rel}} + \Delta C_{ij}^{\text{rel}}$). The latter value was computed by a Bézier spline function $f_{\Delta C}$: $\Delta C_{ij}^{\text{rel}} = f_{\Delta C} C_{ij}^{\text{rel}}$ if the hue of the pixel was in a given hue interval. The value of $f_{\Delta C}$ depended on the value of $C_{ij}$, relative to the maximum chroma ($C_{\max,j}$) of the display for $J_{ij}$ and $h_{ij}$.

The four control points ($P_0$, $P_1$, $P_2$, $P_3$) of the Bézier spline (see Figure 3.33) were constrained so that the slope of the curve in point $P_3$ was less than or equal to $-45°$ in order to have a monotonic transform. $P_0$ and $P_1$ were set to zero. The input parameter $p$ of the CH algorithm stood for the overall magnitude of $\Delta C$, that is, the height of control point $P_2$.

The idea of $f_{\Delta C}$ is that nearly achromatic colors were given their original chroma (i.e., no chroma change) and pixels with higher $C_{ij}$ values were boosted. But chroma boost was carried out carefully in order not to generate out-of-gamut colors, as mentioned earlier. With the CH transform, it was possible to apply different amounts

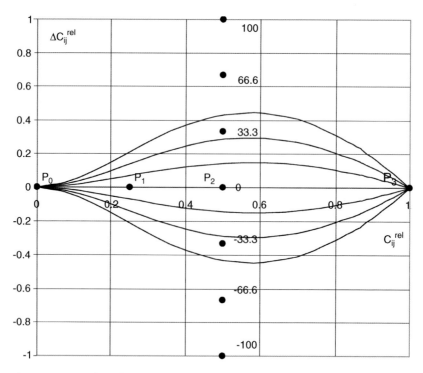

**Figure 3.33** Hue-dependent chroma boost transform (CH). $\Delta C_{ij}^{\text{rel}} = f_{\Delta C}(C_{ij}^{\text{rel}})$ function. The numbers at the control point $P_2$ refer to the parameter $p$ expressing the overall magnitude of $\Delta C$, that is, the height of control point $P_2$ [73]. Reproduced with permission from *Color Research and Application*.

of chroma enhancement in each segment of the hue circle, that is, in each hue interval [73].

The following hue intervals were introduced: red, orange, yellow, green, cyan, blue, and purple, as well as the hue intervals of some important long-term memory colors (Section 3.4), skin, sky, grass, and foliage. These hue intervals were defined by their CIECAM02 hue angle ranges quantified by using the values of focal colors (i.e., color stimuli that represent best examples for a given color category) [84] and long-term memory colors (Table 3.9, columns H). Only the following hue ranges turned out to be significant for color image preference: red, yellow, green (hence grass and foliage), and blue, as well as skin and sky [73].

In the hue-dependent chroma boost experiment, each one of the above six hue ranges was represented by seven test images containing many pixels inside the corresponding hue ranges. Figure 3.34 shows the color image preference results for the above six hue ranges.

As can be seen from Figure 3.34, the color image preference scores of young observers were again higher than those of elderly observers implying that young observers were more determined in their preference answers than elderly observers. The main tendency of both observer groups was to prefer a chroma up to a specific level of chroma. For high chroma levels, chroma preference begins to decrease [73, 85].

As can be seen from Figure 3.34, the course of the color image preference function depends on the *main* hue range of the object depicted in the image (red, yellow, green, blue, skin, or sky) and also on the observer's age. In case of green (including grass and foliage), sky, and skin colors, both age groups prefer a certain amount of chroma enhancement above which their color image preference begins to decrease or remains constant.

In the other three hue ranges (red, yellow, and blue), the preference function was monotonic, that is, observers preferred as high chroma values as possible inside the color gamut of the experimental display [73]. Comparing the images containing sky and other blue objects, the explanation may be that sky images let the observer recall the prototypical color of *blue sky* from long-term memory (Section 3.4.1) and this prototype has a specific chroma. Images with higher chroma are not preferred.

## 3.6
## Age-Dependent Method for Preference-Based Color Image Enhancement with Color Image Descriptors

In this section, a set of *image descriptors* (or simply *descriptors*) are defined [73] to enhance color image quality based on the age-dependent color image preference results of Section 3.5. The method of enhancement is described. This method yields preferred color images depending on the observer's age. This method is based on the image processing *transforms* themselves [73]. These transforms are expected to modify the image until it is preferred over the original by either elderly or young observers.

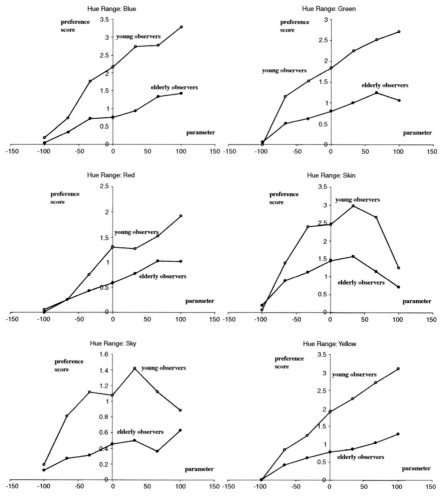

**Figure 3.34** Result of the hue-dependent chroma preference experiment. The parameter value of 0 corresponds to the original image. Black dots: elderly observers. Open circles: young observers [73]. Reproduced with permission from *Color Research and Application*.

In most cases, the amount of color image enhancement needed to achieve the preferred color image depends on the pictorial content and certain *measurable quantities* of the original (input) image itself. The equations presented in this section predict the preferred parameters of the image processing transform for any input image to get its preferred output version.

For the white point (WP) transform (see Section 3.5), the original image is transformed by the CIECAM02 color appearance model to the most preferred white point of the two age groups (see Figure 3.27). For the other three transforms (LE, TC, and CH), a more complex computational scheme shall be applied [73] where first a (scalar) *image descriptor* ($\zeta$) is defined for the input image. The value of $\zeta$ is changing

## 3.6 Age-Dependent Method for Preference-Based Color Image Enhancement

when the transform is applied to the input image. The change of the descriptor characterizes the effect of the image transform.

Second, the output value of the image descriptor ($\zeta^{out}$) depends on the value of the image transform parameter $p$. This dependence has to be parameterized by the input value of the descriptor $\zeta^{inp}$, as well. The mathematical relationship between the output values of the image descriptor $\zeta^{out}$ is modeled as a function of the parameter of the transform $p$ and the input value of the descriptor $\zeta^{inp}$. This is the so-called *descriptor input–output function* [73].

Third, from the preference functions of Section 3.5, the ideal value of the descriptor $\zeta^{opt}$ is calculated for each test image by using the weighting of Equation 3.6. Finally, the ideal value of the descriptor $\zeta^{opt}$ is approximated as a function of the input value of the descriptor. This is the so-called *ideal descriptor function*.

The above computational scheme can be used to achieve the output image as follows. A specific parameter value is calculated from the input value of the descriptor for the transform converting the value of the descriptor into the ideal value of the descriptor. For the global lightness transform (TC), the descriptor value $\zeta_{TC}$ was the weighted sum of the $J$-histogram $H(J)$ of the image. The weights were determined by a fourth-degree polynomial $J_p(J)$. This polynomial (see Figure 3.35) was positive at the extremities of the $J$ scale, that is, at the lower and upper fourth of the 0–10 scale. It was negative for mid-tones. So if the input image contains many pixels in the extremities then the value of the descriptor will be positive but if mid-tones dominate then it will be negative.

The computational formula of the image descriptor $\zeta_{TC}$ is given by Equation 3.9.

$$\zeta_{TC} = \sum_{0}^{J_{max}} \bar{H}(J) J_p(J) \qquad (3.9)$$

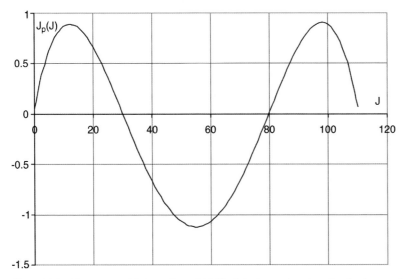

**Figure 3.35** Histogram weighting polynomial $J_p(J)$ of the image descriptor $\zeta_{TC}$ for the transform TC [73]. Reproduced with permission from *Color Research and Application*.

In Equation 3.9, the quantity $\bar{H}(J)$ is defined by Equation 3.10.

$$\bar{H}(J) = \frac{H(J)}{\sum_{0}^{J_{max}} H(J)} \tag{3.10}$$

The descriptor input–output function (predicting the value of the descriptor of the output image after transforming with parameter $p$) was approximated by the quadratic formula of Equation 3.11 [73].

$$\zeta_{TC}^{out} = \zeta_{TC}^{out}(p, \zeta_{TC}^{inp}) = a\left(\frac{p}{100}\right)^2 + b\left(\frac{p}{100}\right) + c$$
$$a = -0.2115\zeta_{TC}^{inp} - 0.0743 \tag{3.11}$$
$$b = 0.2618$$
$$c = \zeta_{TC}^{inp}$$

The ideal descriptor function is shown in Equation 3.12.

$$\zeta_{TC}^{opt}(\zeta_{TC}^{inp}) \cong \begin{cases} 0.9267\zeta_{TC}^{inp} + 0.044 & \text{for the aged} \\ 0.909\zeta_{TC}^{inp} + 0.0027 & \text{for the young} \end{cases} \tag{3.12}$$

For the LE algorithm, the descriptor value ($\zeta_{LE}$) is the spatial frequency dominating the image. This was calculated by weighting the spatial frequencies of the input image with the energy that the image contained at any given spatial frequency by using a two-dimensional fast Fourier transform (2D FFT) as defined by Equation 3.13. The term "energy" is used in the sense of the amount of signal energy present in the image in a given small frequency range [73].

$$f(r) = \oint_{f_x^2 + f_y^2 = r^2} |\Phi(f_x, f_y)| ds$$
$$\zeta_{LE} = \int_0^\infty rf(r) dr \tag{3.13}$$

The descriptor input–output function was approximated by the logarithmic formula of Equation 3.14.

$$\zeta_{LE}^{out}(p) \cong a \ln(p) + b, \qquad a = a(\zeta_{LE}^{inp}), \quad b = b(\zeta_{LE}^{inp}) \tag{3.14}$$

The quantities $a$ and $b$ in Equation 3.14 are defined by Equation 3.15.

$$a \cong -0.0621(\zeta_{LE}^{inp})^2 + 0.885\zeta_{LE}^{inp} - 1.9868$$
$$b \cong 0.9643\zeta_{LE}^{inp} + 0.197 \tag{3.15}$$

The ideal descriptor function was the linear function of Equation 3.16.

$$\zeta_{LE}^{opt}(\zeta_{LE}^{inp}) \cong \begin{cases} 1.0260\zeta_{LE}^{inp} + 0.2393 & \text{for the aged} \\ 0.9793\zeta_{LE}^{inp} + 0.4255 & \text{for the young} \end{cases} \tag{3.16}$$

For the chroma enhancement transform (CH), the descriptor ($\zeta_{CH}$) was the mean chroma in the given hue range of the image. The descriptor input–output function

**Table 3.11** Coefficients for calculating the ideal value of the image descriptor in the case of the CH transform [73].

| | Hue range (i) | | | | | |
|---|---|---|---|---|---|---|
| | Blue | Green | Red | Skin | Sky | Yellow |
| $a_i$ (young) | 1.05 | 0.90 | 1.18 | 0.68 | 0.83 | 0.97 |
| $\beta_i$ (young) | 1.73 | 5.68 | −6.87 | 6.81 | 6.47 | 3.90 |
| $a_i$ (aged) | 1.12 | 0.94 | 1.02 | 0.84 | 0.87 | 0.95 |
| $\beta_i$ (aged) | 0.62 | 4.51 | 3.36 | 4.00 | 3.87 | 4.04 |

Reproduced with permission from *Color Research and Application*.

was approximated by the linear formula of Equation 3.17.

$$\zeta_{CH,i} = a_i(p_i - 100) + b_i \tag{3.17}$$

The quantities $a_i$ and $b_i$ in Equation 3.17 are defined by Equation 3.18.

$$\begin{aligned} a_i &= 0.0011 \zeta_{CH,i}^F + 0.0415 \\ b_i &= 0.8751 \zeta_{CH,i}^{inp} - 3.1302 \end{aligned} \tag{3.18}$$

The ideal descriptor function is the linear function of Equation 3.19.

$$\zeta_{CH,i}^{opt}(\zeta_{CH,i}^{inp}) \cong \begin{cases} a_i^{aged} \zeta_{CH,i}^{inp} + \beta_i^{aged} & \text{for the aged} \\ a_i^{young} \zeta_{CH,i}^{inp} + a_i^{young} & \text{for the young} \end{cases} \tag{3.19}$$

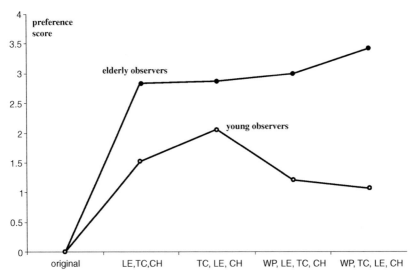

**Figure 3.36** Image preference results in the verification experiment. Black dots: elderly observers. Open circles: young observers [73]. Reproduced with permission from *Color Research and Application*.

The coefficients $\alpha_i$ and $\beta_i$ of Equation 3.19 depend on hue range and on the age of the observer (see Table 3.11).

Generally, the image enhancement algorithm calculates the parameter values for every transform depending on the setting of a two-stage *age dial* [73] (elderly or young). Then, the transforms are applied to the input image. It was tested in a verification experiment whether the enhanced image was more preferred by the observer than the input image [73]. The effect of the application order of the transforms (e.g., first WP, second LE, third TC, fourth CH, etc.) was also examined.

In the verification experiment, four test images, four aged and four young subjects, and four transform combinations were studied. Averaged image preference results (across observers and images) are shown in Figure 3.36.

Figure 3.36 shows that both age groups preferred the enhanced images over the original images. Aged observers preferred the transformations to be done in the following order: WP, TC, LE, CH. Young observers preferred TC, LE, CH without any white point transform.

## References

1 Oetjen, S. and Ziefle, M. (2009) A visual ergonomic evaluation of different screen types and screen technologies with respect to discrimination performance. *Appl. Ergon.*, **40**, 69–81.

2 ISO 9241-303:2008 (2008) *Ergonomics of Human–System Interaction. Part 303. Requirements for Electronic Visual Displays*, International Organization for Standardization.

3 Kraiss, K. and Moraal, J. (eds) (1976) *Introduction to Human Engineering*, Verlag TÜV Rheinland GmbH, Köln.

4 Shneiderman, B. (1998) *Designing the User Interface*, 3rd edn, Addison-Wesley.

5 Helander, M., Landauer, T., and Prabhu, P. (eds) (1997) *Handbook of Human–Computer Interaction*, 2nd edn, Elsevier.

6 Dix, A.J., Finlay, J.E., Abowd, G.D., and Beale, R. (1998) *Human–Computer Interaction*, 2nd edn, Prentice Hall Europe.

7 Nielsen, J. and Mack, R.L. (eds) (1994) *Usability Inspection Methods*, John Wiley & Sons, Inc.

8 Hick, W.E. (1952) On the rate of gain of information. *Q. J. Exp. Psychol.*, **4**, 11–26.

9 Hyman, R. (1953) Stimulus information as a determinant of reaction time. *J. Exp. Psychol.*, **45**, 188–196.

10 Fitts, P.M. (1954) The information capacity of the human motor system in controlling the amplitude of movement. *J. Exp. Psychol.*, **47**, 381–391.

11 Fitts, P.M. and Peterson, J.R. (1964) Information capacity of discrete motor responses. *J. Exp. Psychol.*, **67**, 103–113.

12 Rovamo, J. and Raninen, A. (1988) Critical flicker frequency as a function of stimulus area and luminance at various eccentricities in the human cone vision: a revision of Granit–Harper and Ferry–Potter laws. *Vis. Res.*, **28** (7), 785–790.

13 Kurtenbach, A., Mayser, H.M., Jägle, H., Fritsche, A., and Zrenner, E. (2006) Hyperoxia, hyperglycemia, and photoreceptor sensitivity in normal and diabetic subjects. *Vis. Neurosci.*, **23**, 651–661.

14 Bergquist, J. (2003) Visual ergonomics challenges in information-intensive mobile displays, Nokia Research Center. 10th International Display Workshops, December 3–5, Fukuoka, Japan.

15 Marumoto, T., Jonai, H., Villanueva, M.B.G., Sotoyama, M., and Saito, S. (2003)

A case report of ophthalmologic problems associated with the use of information technology among young students in Japan. Proceedings of the XVth Triennial Congress of the International Ergonomics Association, August 24–29, Ergonomics Society of Korea, Seoul, Korea.

16  Hasegawa, S., Omori, M., Watanabe, T., Matsunuma, Sh., and Miyao, M. (2009) Legible character size on mobile terminal screens: estimation using pinch-in/out on the iPod touch panel, in *Human Interface. Part II. HCII 2009 (Lecture Notes in Computer Science 5618)* (eds M.J. Smith and G. Salvendy), Springer, Berlin, pp. 395–402.

17  Lee, M.Y., Son, C.H., Kim, J.M., Lee, C.H., and Ha, Y.H. (2007) Illumination-level adaptive color reproduction method with lightness adaptation and flare compensation for mobile display. *J. Imaging Sci. Technol.*, **51** (1), 44–52.

18  Hunt, R.W.G. (2004) *The Reproduction of Color (Wiley-IS&T Series in Imaging Science and Technology)*, 6th edn, John Wiley & Sons, Ltd.

19  CIE 175:2006 (2006) *A Framework for the Measurement of Visual Appearance*, Commission Internationale de l'Éclairage.

20  CIE 195:2011 (2011) *Specification of Color Appearance for Reflective Media and Self-Luminous Display Comparisons*, Commission Internationale de l'Éclairage.

21  Roufs, J.A.J. and Boschman, M.C. (1997) Text quality metrics for visual display units. I. methodological aspects. *Displays*, **18**, 37–43.

22  Boschman, M.C. and Roufs, J.A.J. (1997) Text quality metrics for visual display units. II. An experimental survey. *Displays*, **18**, 45–64.

23  Jackson, R., MacDonald, L., and Freeman, K. (1994) *Computer Generated Color*, John Wiley & Sons, Inc., New York.

24  Bodrogi, P. (1999) On the use of the Ware and Cowan conversions factor formula in visual ergonomics, in *Proceedings of the CIE Symposium'99: 75 Years of CIE Photometry*, Akademiai Kiado, Budapest.

25  Carter, R.C. and Carter, E.C. (1988) Color coding for rapid location of small symbols. *Color Res. Appl.*, **13** (4), 226–234.

26  Travis, D. (1991) *Effective Color Displays*, Academic Press, New York.

27  Ware, C. and Cowan, W.B. (1983) Specification of heterochromatic brightness matches – a conversion factor for calculating luminances of stimuli which are equal in brightness. NRC Publ. No. 26055.

28  Bodrogi, P. (2003) Chromaticity contrast in visual search on the multi-color user interface. *Displays*, **24** (1), 39–48.

29  Knoblauch, K., Arditi, A., and Szlyk, J. (1991) Effect of chromatic and luminance contrast on reading. *J. Opt. Soc. Am. A*, **8** (2), 428–439.

30  Legge, G.E., Parish, D.H., Luebker, A., and Wurm, L.H. (1990) Psychophysics of reading. XI. Comparing color contrast and luminance contrast. *J. Opt. Soc. Am. A*, **7** (10), 2002–2010.

31  Spenkelink, G.P.J. and Besuijen, J. (1996) Chromaticity contrast, luminance contrast, and legibility of text. *J. SID*, **4** (3), 135–144.

32  Durrett, H.J. (ed.) (1987) *Color and the Computer*, Academic Press, New York.

33  Bauer, B. and McFadden, Sh. (1997) Linear separability and redundant color coding in visual search displays. *Displays*, **18**, 21–28.

34  Carter, R.C. (1982) Visual search with color. *J. Exp. Psychol. Hum. Percept. Perform.*, **8**, 21–28.

35  Carter, E.C. and Carter, R.C. (1981) Color and conspicuousness. *J. Opt. Soc. Am.*, **71**, 723–729.

36  Healey, C.G. (1999) Preattentive processing. Visualization using the low-level human visual system. SIGGRAPH'99, 26th International Conference on Computer Graphics and Interactive Techniques.

37  Monnier, P. and Nagy, A.L. (2001) Uncertainty, attentional capacity and chromatic mechanisms in visual search. *Vis. Res.*, **41** 313–328.

38  Nagy, A.L. (1999) Interactions between achromatic and chromatic mechanisms in visual search. *Vis. Res.*, **39**, 3253–3266.

39  Nagy, A.L. and Sanchez, R.R. (1990) Critical color differences determined with a visual search task. *J. Opt. Soc. Am. A*, **7**, 1209–1217.

40 Nagy, A.L. and Winterbottom, M. (2000) The achromatic mechanism and mechanisms tuned to chromaticity and luminance in visual search. *J. Opt. Soc. Am. A*, **17**, 369–379.

41 Bauer, B., Jolicoeur, P., and Cowan, W.B. (1996) Visual search for color targets that are or are not linearly separable from distractors. *Vis. Res.*, **36**, 1439–1465.

42 D'Zmura, M. (1991) Color in visual search. *Vis. Res.*, **31**, 951–966.

43 Carter, R.C. and Carter, E.C. (1982) High-contrast sets of colors. *Appl. Opt.*, **21**, 2936–2939.

44 Smallman, H.S. and Boynton, R.M. (1993) On the usefulness of basic color coding in an information display. *Displays*, **14** (3), 158–165.

45 Kutas, G., Kwak, Y., Bodrogi, P., Park, D.S., Lee, S.D., Choh, H.K., and Kim, C.Y. (2008) Luminance contrast and chromaticity contrast preference on the color display for young and elderly users. *Displays*, **29** (3), 297–307.

46 Sekuler, R. and Sekuler, A.B. (2000) Visual perception and cognition, in *Oxford Textbook of Geriatric Medicine*, 2nd edn (eds J.G. Evans, T.F. Williams, B.L. Beattie, J.P. Michel, and G.K. Wilcock), Oxford University Press, Oxford, pp. 874–880.

47 Derefeldt, G., Lennerstrand, G., and Lundh, B. (1979) Age variations in normal human contrast sensitivity. *Acta Ophthalmol. (Copenh.)*, **57** (4), 679–690.

48 Weale, R.A. (1988) Age and transmittance of the human crystalline lens. *J. Physiol.*, **395** (1), 577–587.

49 Shinomori, K., Scheferin, B.E., and Werner, J.S. (2001) Age-related changes in wavelength discrimination. *J. Opt. Soc. Am. A*, **18**, 310–318.

50 Sheferin, B.E. and Werner, J.S. (1990) Loci of spectral unique hues throughout the life span. *J. Opt. Soc. Am. A*, **7**, 305–311.

51 Owsley, C. and Sloane, M.E. (1990) Vision and aging, in *Handbook of Neuropsychology* (eds F. Boller and J. Grafman), Elsevier Science Publishers, Amsterdam, pp. 229–249.

52 Sloane, M.E., Owsley, C., and Jackson, C.A. (1988) Aging and luminance adaptation effect on spatial contrast sensitivity. *J. Opt. Soc. Am. A*, **5** (12), 2181–2190.

53 Higgins, K.E., Jaffe, M.J., Caruso, R.C., and de Monasterio, F.M. (1988) Spatial contrast sensitivity: effects of age, test–retest, and psycho-physical method. *J. Opt. Soc. Am. A*, **5** (12), 2173–2180.

54 Kelly, D.H. (1974) Spatio-temporal frequency characteristics of color-vision mechanisms. *J. Opt. Soc. Am.*, **64**, 983–990.

55 Kelly, D.H. (1983) Spatiotemporal variation of chromatic and achromatic contrast thresholds. *J. Opt. Soc. Am.*, **73**, 742–750.

56 Mullen, K.T. (1985) The contrast sensitivity of human color vision to red–green and blue–yellow chromatic gratings. *J. Physiol.*, **359**, 381–400.

57 Owsley, C., Sekuler, R., and Siemsen, D. (1983) Contrast sensitivity throughout adulthood. *Vis. Res.*, **23** (7), 689–699.

58 Peli, E. (1995) Suprathreshold contrast perception across differences in mean luminance: effects of stimulus size, dichoptic presentation, and length of adaptation. *J. Opt. Soc. Am. A*, **12**, 817–823.

59 Valberg, A. (2005) *Light Vision Color*, John Wiley & Sons, Ltd., Chichester, UK.

60 Delahunt, P.B., Hardy, J.L., Okijama, K., and Werner, J.S. (2005) Senescence of spatial chromatic contrast sensitivity. II. Matching under natural viewing conditions. *J. Opt. Soc. Am. A*, **22**, 60–67.

61 Hardy, J.L., Delahunt, P.B., Okijama, K., and Werner, J.S. (2005) Senescence of spatial chromatic contrast sensitivity. I. Detection under conditions controlled for optical factors. *J. Opt. Soc. Am. A*, **22**, 49–59.

62 de Wit, G.C. (2005) Contrast of displays on the retina. *J. SID*, **13** (2), 177–178.

63 CIE 135-1999 (1999) *CIE Collection 1999: Vision and Color, Physical Measurement of Light and Radiation. 135/1: Disability Glare*, Commission Internationale de l'Éclairage.

64 Bodrogi, P. and Tarczali, T. (2001) Color memory for various sky, skin, and plant colors: effect of the image context. *Color Res. Appl.*, **26** (4), 278–289.

65. Bodrogi, P. and Tarczali, T. (2002) Chapter 2: Investigation of color memory, in *Color Image Science: Exploiting Digital Media* (eds L.W. MacDonald and M.R. Luo), John Wiley & Sons, Ltd., Chichester, UK, pp. 23–47.
66. Yendrikhovskij, S.N., Blommaert, F.J.J., and de Ridder, H. (1999) Color reproduction and the naturalness constraint. *Color Res. Appl.*, **24**, 52–67.
67. Yendrikhovskij, S.N., Blommaert, F.J.J., and de Ridder, H. (1999) Representation of memory prototype for an object color. *Color Res. Appl.*, **24**, 393–410.
68. Tarczali, T., Park, D.S., Bodrogi, P., and Kim, C.Y. (2006) Long-term memory colors of Korean and Hungarian observers. *Color Res. Appl.*, **31** (3), 176–183.
69. Bartleson, C.J. (1960) Memory colors of familiar objects. *J. Opt. Soc. Am.*, **50**, 73–77.
70. Commission Internationale de l'Éclairage (CIE) (1996) Color Rendering. Specifying Color Rendering Properties of Light Sources. Report of CIE TC 1-33.
71. Siple, P. and Springer, R.M. (1983) Memory and preference for the colors of objects. *Percept. Psychophys.*, **34**, 363–370.
72. Bodrogi, P. (1998) Shifts of short-term color memory. Ph.D. thesis, University of Veszprém, Veszprém, Hungary.
73. Beke, L., Kutas, G., Kwak, Y., Sung, G.Y., Bodrogi, P., Park, D.S., Lee, S.D., Choh, H.K., and Kim, Ch.Y. (2008) Color preference of aged observers compared to young observers. *Color Res. Appl.*, **33** (5), 381–394.
74. Zhang, X. and Wandell, B.A. (1996) A spatial extension of CIELAB for digital color reproduction. SID'96 Digest, pp. 731–735.
75. Fairchild, M.D. and Johnson, G.M. (2002) Meet iCAM: a next-generation color appearance model. IS&T/SID 10th Color Imaging Conference, Scottsdale, pp. 33–38.
76. Sheikh, H.R. and Bovik, A.C. (2006) Image information and visual quality. *IEEE Trans. Image Process.*, **15** (2), 430–444.
77. Taylory, Ch.C., Pizloz, Z., Allebach, J.P., and Boumany, Ch.A. (1997) Image quality assessment with a Gabor pyramid model of the human visual system. Proceedings of the 1997 IS&T/SPIE International Symposium on Electronic Imaging Science and Technology, San Jose, CA, pp. 58–69.
78. Wang, Zh., Bovik, A.C., and Simoncelli, E.P. (2005) Structural approaches to image quality assessment, in *Handbook of Image and Video Processing* (ed. A. Bovik), Academic Press, New York, pp. 1–33.
79. Bringier, B., Richard, N., and Fernandez-Maloigne, C. (2006) Local contrast for no-reference color quality assessment. Proceedings, 75 Years of the CIE Standard Colorimetric Observer, Ottawa (ed. A. Carter).
80. Fernandez, S.R., Fairchild, M.D., and Braun, K. (2005) Analysis of observer and cultural variability while generating 'preferred' color reproductions of pictorial images. *J. Imaging Sci. Technol.*, **49**, 96–104.
81. Yoshida, A., Mantiuk, R., Myszkowski, K., and Seidel, H.P. (2006) Analysis of reproducing real-world appearance on displays of varying dynamic range. Proceedings of Eurographics, Vienna (eds E. Gröller and L. Szirmay-Kalos).
82. Babcock, J.S., Pelz, J.B., and Fairchild, M.D. (2003) Eye tracking observers during rank order, paired comparison, and graphical rating tasks. IS&T PICS Conference, Rochester, NY, pp. 10–15.
83. Guilford, J.P. (1954) *Psychometric Methods*, McGraw-Hill, New York.
84. Boynton, R.M. and Olson, C.X. (1987) Locating basic colors in the OSA space. *Color Res. Appl.*, **12** (2), 94–105.
85. Fedorovskaya, E.A., de Ridder, H., and Blommaert, F.J.J. (1998) Chroma variations and perceived quality of color images of natural scenes. *Color Res. Appl.*, **22** (2), 96–110.

# 4
# Color Management and Image Quality Improvement for Cinema Film and TV Production

In this chapter, the workflow in cinema film and TV production is described with all system components. Also, requirements for a high-end color management are pointed out. Workflow components such as the film or digital cinema camera, the film scanner, and the laser film recorder are dealt with. The application of color management for digital camera and for postproduction and related problems such as color gamut mapping and the nonadditivity of the monitor or the digital cinema projector are analyzed.

Before the digital film content is distributed to cinemas worldwide, it has to be compressed and watermarked in order to protect the copyright of the filmmakers. Several image compression and watermarking methods based on human image perception have been developed and evaluated. These methods are also the subject of this chapter – together with the application of the spatiotemporal characteristics of the human visual system to optimize the optical performance of imaging optics for digital cinema cameras. Finally, the color rendering potential of the light sources used for illuminants in TV and cinema production and the emotional aspects of cinema film will be described.

## 4.1
### Workflow in Cinema Film and TV Production Today – Components and Systems

### 4.1.1
### Workflow

Today, there are various possibilities for the production of digital cinema and TV content [1] (see Figure 4.1).

In the conventional form, the object scene is shot with an analog film camera and the negative film material is developed chemically. After the development of the negative film, there are two methods for the subsequent image processing:

1) The negative film (master negative film) is copied on the intermediate (so-called intermed) positive (IP) film with high-quality optics and three color channels in

---

*Illumination, Color and Imaging: Evaluation and Optimization of Visual Displays*, First Edition.
Peter Bodrogi and Tran Quoc Khanh.
© 2012 Wiley-VCH Verlag GmbH & Co. KGaA. Published 2012 by Wiley-VCH Verlag GmbH & Co. KGaA.

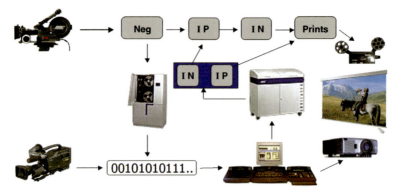

**Figure 4.1** Production chain in the modern TV and cinema industry [1].

the copying apparatus. The color density of the IP film is influenced by the variation of the light intensity in each color channel. This light intensity adjustment is the most important task of an analog film laboratory. From the main IP film, different IN (intermed negative) films are copied for the film content distribution to different countries. In each country, the IN film is copied to many positive films (print film) for the cinemas of the particular country.

2) Alternatively, the master negative film can be scanned either in a data scanner for high-quality cinema applications in 14 bits per channel resolution or in a Telecine scanner with 10 or 8 bits per channel resolution for TV applications. The density values of the negative film are converted into the digital signals of the CCD or CMOS sensors of the scanner that are made available for further image correction in postproduction on a digital platform.

Recently, the original object scene can also be shot digitally with a digital cinema camera in 12 bits per channel for cinema films and 8 or 10 bits per channel for HDTV applications. The digital images can also be corrected in the postproduction. There exist three technical systems with three very different color generation processes:

1) The color process of the film camera with the spectral sensitivities of the negative film layers, the film scanner, the light sources illuminating the negative film (xenon lamp or LED radiation), and the spectral sensitivity of the scanner sensor.
2) The color process of the digital camera with the spectral sensitivities of the sensor pixels and the color algorithms processing the color pixel signals.
3) The color process that generates color images on the monitor or with the digital projector. For high-end cinematographic applications, only the conventional CRT monitor and DLP projection technology are suitable for the presentation of high-quality images.

The signal transfer behavior of the different technical systems is not identical. In the best case, the digital camera, the digital projector, and the monitor have a linear behavior between the input and output signals. But the behavior of the negative film is highly nonlinear.

After the correction of the digital images, the final film content can be projected in a public digital cinema theater with a digital projector. But this case does not represent a complete digital penetration today because several cities do not have a digital cinema due to the high price of the digital projector and the limited distribution possibilities of digital film content such as film transfer with data recorder, glass fiber, or satellite signal transfer. Therefore, the widespread appearance of digital cinemas can be expected only in the future while there are very different estimations about the year of complete digital penetration. Today (2012), a popular solution often remains to be the recording of the digital images back on the intermed negative film (IN film) that is then chemically developed and copied onto the IP or print film for the analog cinema rooms worldwide.

By analyzing the modern production chain in the cinema and TV industry with the above hybrid structure, one can conclude that many different technical systems are used with different ways of color signal generation and conversion. Hence, TV and cinema technology is confronted with different device color spaces and, in turn, the principle of color transparency becomes an important criterion. The latter criterion can be formulated in the following way: In every stage of the color image processing workflow, the image has to be processed so as to yield the same color appearance of the colored objects in the real original scene during the filmmaking process as either the analog film or the digital cinema camera. Thus, the reference of the color reproduction is the color appearance of the real object scene. In order to control the color reproduction in every step of the workflow, a color image processing process (hardware and software) is necessary in order to transform the color signals between the individual technical systems.

Accordingly, an optimal color image processing process has to contain the following features:

1) Implementation of a device-independent color system (color space) predicting perceived color attributes (brightness, hue, and saturation; see Section 1.1.3) under the concrete viewing condition. The specification of the viewing condition includes the spectral power distribution (or chromaticity coordinates) of the illuminant, the adaptation luminance (or brightness) of the scene, the luminance (or brightness) distribution of the colored objects in the scene, and the viewing angle from the position of the observer. Remind that, according to Chapter 1, human color perception is the product of the activity of the photoreceptors on the retina, the adaptation state, the signal transfer and compression on the way from the receptors to the brain, and color image processing at the later stages of the workflow in the visual brain.

2) A transformation algorithm converting the system-dependent features (device color attributes) in a system-independent color space. Hereby, the speed of image processing plays an important role because TV and cinema contents are motion pictures with a minimum frame rate of 24 frames per second (f/s) and a minimum bit depth of 8 bits per color channel at a high spatial resolution, for example, HDTV resolution of $1920 \times 1080$ pixels per image.

Due to the very different color reproduction methods, it is obvious that the color monitor image and the image after recording on the IN negative film and projecting onto the cinema screen shall have a different color appearance. The real color scene is also different from the monitor image. It is therefore necessary to transform the system-dependent color attributes into a system-independent color space. In 1993, the International Color Consortium (ICC) was established with the membership of the most important companies involved in color reproduction technology. The aim of the ICC is to develop and maintain an open and manufacturer-independent standard for the exchange of color data that must also be operation system independent.

### 4.1.2
### Structure of Color Management in Today's Cinema and TV Technology

The ICC describes the system-dependent color space by a profile. This means that every system has its own profile (so-called ICC profile) that has to be checked and renewed from time to time because the system may be changing. There are principally three types of ICC profile [2–4] (see Figure 4.2):

1) **The input profile (digital camera or scanner)**: the input signals are the color values of the input image and the output signals are the system signals from the sensor electronics.
2) **The display profile**: certain input signals (analog or digital) are used for the reproduction of a digital image with well-defined color coordinates on the monitor or with the digital projector.
3) **The output profile**: depending on the input digital signal, a well-defined color coordinate is generated on the output medium, for example, on the negative film.

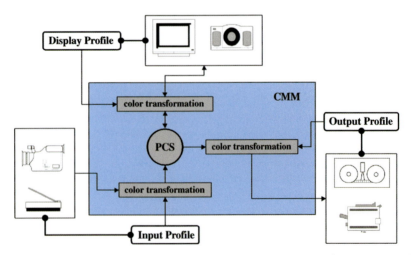

**Figure 4.2** Architecture of color management according to the principles of the ICC. PCS: profile connecting space; CMM: color management module [1].

As mentioned above, the profile remains constant only in that case if the system described by the profile does not change. This aspect requires a system calibration on a regular basis or a control feedback mechanism in order to keep the system constant, for example, against changes of the scanner lamp. Environmental operating conditions can also influence the behavior of the system and therefore its profile. There are some important guidelines for the cinema industry to keep the working conditions constant:

- All monitors for the feature film correction have to be calibrated colorimetrically every week.
- The colorimeter and the luminance meters have to be calibrated yearly.
- The chemical processes of the film laboratory have to be checked two to three times within 24 h.
- The reference cinema room for the final control of film image quality shall be measured monthly in order to take the degradation of the projection lamps and the yellowness of the cinema screen into account.
- The film scanner and the laser film recorder have to be set up in a temperature-stabilized room with a fluctuation of maximum 0.5 °C.

Numerical computations based on ICC profiles are generally not very difficult. A set of predefined color samples are measured colorimetrically or spectroradiometrically and the numerical correlates of their color attributes are then compared with the system output data. In practical color management, profiling software can be used in order to calculate the system-dependent profile. In the generated profiles, the information for the transformation is included either as a $3 \times 3$ matrix or as a 3D look-up table (3D-LUT).

The profile connection space (PCS) represents a system-independent color space, either as the CIELAB coordinates or as the CIE $XYZ$ tristimulus values. With these values, the system-dependent color data can be transformed from system to system. For the calculation and transformation, a computational module is needed, the so-called color management module (CMM) (see Figure 4.2). There are different CMM software packages on the market that generally yield different results. The best CMM software is the one delivering the color data corresponding to the visual perception of color image experts.

### 4.1.3
**Color Management Solutions**

The color space transformation converts the data of a system-dependent color space into a system-independent color space. The ICC defined the color spaces CIE $XYZ$ and CIELAB as system-independent color spaces. In practice, CIELAB is mostly preferred due to its correspondence to the perceived attributes of human color vision (unlike $XYZ$) (see Section 1.1.3). Also, the issues of color gamut mapping can be understood more intuitively in terms of the CIELAB color space. Generally, there are three possibilities for the color transformations [2]:

1) The color transformation is carried out with a 3 × 3 matrix. It can only be used if the dimensions of the source and target color spaces are the same and if the systems are linear. The 3 × 3 matrix has been used for many years in order to transform the TV camera signal into the monitor data while the attainable quality of color rendering quality is moderate.
2) Transformations with a matrix with nonlinear parameters. Every color value in the target color space is a polynomial function of the RGB color values of the source space in the following form:

$$R_{target} = aR + bRG + cRB + dGB + eR^2 + f \tag{4.1}$$

In Equation 4.1, $R_{target}$ represents the R (red) color value in the target space. Similar relationships hold for the G and B values. The procedure of Equation 4.1 is applied in high-end professional digital cameras for still and motion picture imaging. Equation 4.1 takes the nonlinearity and nonadditivity aspects of the camera sensors and the monitors into account. For the digital sensor, a further effect is also accounted for by Equation 4.1, that is, the internal transport of a part of the signal of a particular sensor pixel to the neighboring pixels.

3) The transformation with the 3D-LUT is time-consuming. But the 3D-LUT can be used for all possible color spaces. It is also usable for nonlinear systems. A LUT encodes the procedure of transforming the color values of the source space into the color values of the target space. For an accurate transformation, the LUT has to contain numerous grid values and requires a higher amount of dataflow. Hence, a real-time computation requires an enormous computational effort. For the color transformation from the film color space to the monitor color space or from the monitor color space to the film projector space in the path of the film laser recorder, a 3D-LUT with, for example, 33 × 33 × 33 grid values can be used. Color values between the grid values can be calculated by a linear interpolation. This grid size is feasible in the postproduction within one color correction session in real time provided that a fast hardware is used. For a high-end digital motion picture camera with 10 bits per color channel resolution and with a frame rate of 25 frames per second, a 3D-LUT with a grid size of 17 × 17 × 17 is the best compromise between processing time and color rendering quality.

## 4.2
### Components of the Cinema Production Chain

### 4.2.1
#### Camera Technology in Overview

Camera technology started at the beginning of the twentieth century and had a very dynamic and intensive development in the past 15 years. The key features influencing the image quality captured by cameras are described below.

The first key feature is the optical system imaging the object plane onto the negative film plane. The most important requirements for the optics are the highest

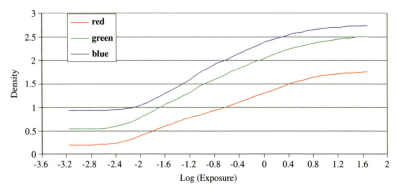

**Figure 4.3** Density of the film as a function of the logarithm of exposure.

possible modulation transfer function (MTF) and the lowest optical aberration such as astigmatism, distortion, or chromatic aberration. In order to have the best image quality for every distance to the object plane, many high-end optical systems of fixed focal length have been developed and manufactured.

The second issue concerns the contrast transfer functions of the film material. Negative films have three layers for the red, green, and blue color channels and the contrast transfer functions of the layers have to be optimized. In addition, the spectral sensitivity functions of the three layers should optimize the color fidelity of the colored objects in the captured scene.

Third, the exposure sensitivity of the film materials is also important. The dynamic range of the film (i.e., the ratio of the maximum luminance of the scene highlights and the minimum luminance of the darkest objects rendered by the same film) lies between 600 : 1 and 1000 : 1 for the most professional HDTV cameras for TV films. With the motion picture film material of current film industry, a dynamic range of more than 15 000 : 1 can be achieved. Figure 4.3 shows the density of the film as a function of the logarithm of exposure ranging between −2.6 and 1.6 for the case of a typical film material.

Characteristics such as the ones depicted in Figure 4.3 shall be the aim for the next generation of digital motion picture cameras in the near future. A high dynamic range should mean in the language of filmmakers that both the highlights and the dark parts of the scene including dark objects can be imaged on the film and on the cinema canvas with all the fine details and delicate color shadings (see Figure 4.4). This feature is very important to enhance perceived image quality.

In Figure 4.4, a church interior is illustrated as an example of a scene with a very high dynamic range. The windows on the left and right sides have a high luminance but due to the high dynamic range of the cinema camera, the modulation and structure of these windows can be clearly seen in the rendered image. At the same time, the dark and colored objects of the altar maintain all perceivable color and spatial details.

In Figure 4.5, the contrast transfer functions of a negative film material (type 200T with the sensitivity of 200 ASA for tungsten light exposure) and a typical high-end

**Figure 4.4** Example of a church interior for a scene with a very high dynamic range.

film camera optics system are illustrated. At a spatial frequency of more than 80 cycles/mm, the contrast transfer value is still remarkable with more than 40% for the blue and green film layers. Because the film width is about 24.89 mm, which is projected on the canvas in the cinema room, one should have 24.89 mm × 80 cycles/mm = 1991 cycles per image or 3982 pixels per image width. The best film is therefore able to present the image resolution of about 4000 pixels (so-called 4k resolution). This resolution is much better than the resolution of the recently often mentioned high-definition television (HDTV) with the resolution of 1920 × 1080 or 1080 × 720 pixels.

In Figure 4.6, the spectral sensitivity of a new negative film (type 500D, sensitivity of 500 ASA for daylight exposure) and that of another (conventional) negative film (type 250D) are illustrated.

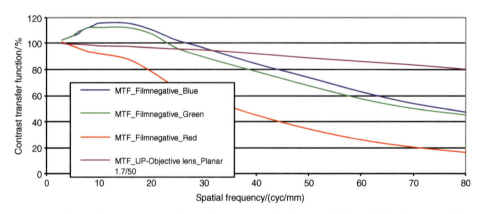

**Figure 4.5** Contrast transfer functions of a negative film material (type 200T with the sensitivity of 200 ASA for tungsten light exposure) and a typical high-end film camera optics system.

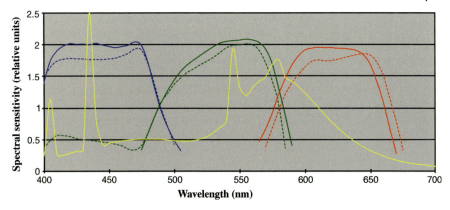

**Figure 4.6** Spectral sensitivity of a new negative film (type 500D, continuous line) in comparison to a conventional negative film (type 250D, dotted line) and a fluorescent lamp spectrum (yellow line).

In Figure 4.6, the curves for the green and red layers of the new film type are shifted closer together in comparison to the conventional film. The crossing point of the two sensitivity curves is at 580 nm so that the color fidelity for the skin tone can be improved by using the new film. Skin tones are very important colors for cinematography because they represent possibly the most important long-term memory colors (see Section 3.4.1). They have a typical spectral reflectance function with a remarkable edge at 580 nm independent of whether they are Asian or European skin tones. This can be seen from the two skin tones in Figure 6.13.

The spectral reflectance functions of the skin of several persons from different countries were measured and this is depicted in Figure 4.7, where the edge at 580 nm is well visible for all persons.

Since 2000, film industry has tried to develop a generation of digital cameras for cinematographic applications. In the research phase, many HDTV cameras designed

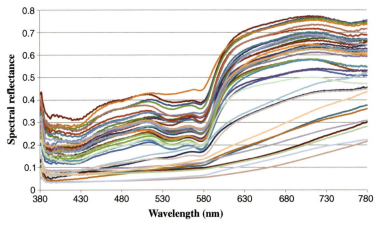

**Figure 4.7** Spectral reflectance of different (international) skin tones measured at the Technische Universität Darmstadt (Germany).

for TV application have been tested whether they can be applied for cinema purposes. Specifically, cinema digital cameras have to fulfill the following requirements:

1) Possible use of the optics of the current analog film cameras with their already optimized optical quality.
2) The sensor should have the same format as the negative film area (about 24.89 mm × 18.76 mm) in order to maintain the depth focus of the current film. The sensor has to be positioned at the same position as the film.
3) A high dynamic range of more than 15 000 : 1 should be maintained but at least 4000 : 1.
4) Color rendering capacity has to be better than the one of the negative film. It is possible to achieve this goal if the spectral sensitivity of the sensor pixels is optimized.
5) The frame rate of the analog film camera is up to 150 f/s so that a high-quality digital cinema camera should be optimized and designed for at least 96 f/s. This high frame rate is necessary for shooting the fast motion of the objects such as motion of the vehicles or the sport events such as football or baseball.

After the first test study of HDTV cameras worldwide, it was pointed out that the cameras examined did not fulfill the above-described cinematographic requirements (1–5). Therefore, a new generation of cinema cameras had to be developed. From the point of view of image quality, the following optimizations have been carried out:

1) Sensor technology was changed to CMOS (complementary metal-oxide semiconductor) technology from CCD sensor technology because CMOS technology allowed a fast readout of the signal from the pixels after the exposure in order to have the frame rate of up to 96 f/s. In addition, CMOS sensors avoid some image artifacts such as blooming or smear taking place at the overexposure of the sensor pixels. The sensor area is 24 mm × 18 mm; hence, it covers the same size as the conventional negative film. Because the optics of the film camera has to be maintained, only one sensor has to be used for all three color channels (RGB). Therefore, the pixels for the red, green, and blue information are located periodically on the same sensor.
2) The spectral sensitivity of the whole optical chain has become the product of the spectral transmittance of the optics, the filters eliminating the infrared and ultraviolet radiation, the low-pass filter avoiding the alias structure of the sensor, the spectral transmittance of the color lacquer for the RGB pixels, and, finally, the spectral sensitivity of the monochrome CMOS sensor that is generally a semiconductor silicon sensor.

Figure 4.8 shows the spectral sensitivity curves of the whole chain of a newly developed digital cinema camera for the three color channels (RGB).

As can be seen from Figure 4.8, the spectral sensitivity of the green channel is similar to $V(\lambda)$, the human luminous efficiency function for photopic vision with a peak wavelength of 550 nm. The blue channel has a wide spectral bandwidth with more sensitivity. This is very important for special blue effects in TV and cinema production. The curves of the red and green channels highly overlap in the

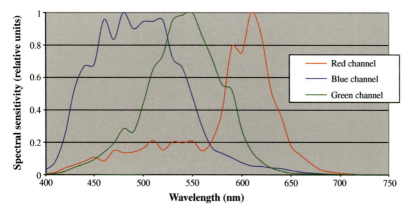

**Figure 4.8** Spectral sensitivity of the whole optical chain of the digital cinema camera.

wavelength range around 580 nm and therefore allow a very good skin tone reproduction. The spectral sensitivity of the red channel is high between 560 and 650 nm in order to guarantee a good color rendering of the important red objects such as red textiles, red food, and red flowers. For the wavelengths above 690 nm, the spectral sensitivity of the color channels has to be low because all artificial and natural objects have a high reflectance between 690 nm and the infrared range and this signal would falsify the color rendering of the camera. Figure 4.9 shows a test image for the camera containing skin tone, colored test charts, and a spatial pattern to test the spatial resolution of the camera.

**Figure 4.9** Test image for the camera containing skin tone, colored test charts, and a spatial pattern to test the spatial resolution of the camera.

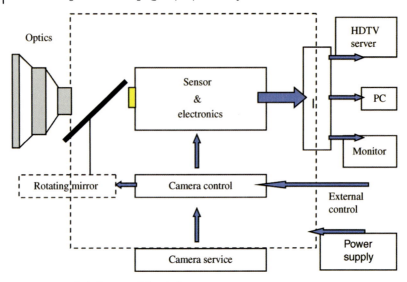

**Figure 4.10** Block diagram of the whole camera system.

Figure 4.10 shows the block diagram of the whole camera system. As can be seen from Figure 4.10, the optics of the camera images the object plane onto the sensor plane in the time slot in which the rotating mirror lets the light beam pass. In the other time slot, the mirror blocks the light beam and reflects it toward the incoming light into an optical viewfinder so that the cinematographer can observe the object plane during the film shooting sequence. The photons are absorbed in the sensor pixel generating a photocurrent that is converted into a photovoltage by the transistors inside each pixel. The voltage is digitalized by an A/D converter at a color resolution of 12 bits per color channel.

The signal of the digital pixels can be processed in two ways depending on the two different application fields:

1) **The live mode**: This mode is applied if the image has to be used for a HDTV application, for example, a live HDTV sports event or a HDTV feature film with a low-cost budget. For this mode, the image is downsampled to the HDTV resolution, that is, $1920 \times 1080$ pixels per image and 8 bits per color channel. Image processing steps such as black value correction, linearizing the signal, dead pixel correction and interpolation, and color management (see below) are performed in real time at frame rates of 24, 25, or 30 f/s and with a moderately demanding data processing algorithm.
2) **The raw mode**: Every cinematographer would like to create a high-quality cinema film with excellently rendered brilliant colors. To achieve this, the work time to be invested is not the key factor. Therefore, unprocessed raw digital signals can be read out in the highest possible bit depth (mostly 12 bits per channel) and transferred to a postproduction company. During postproduction, time-consuming high-end image processing algorithms are executed in high bit depth and high spatial image

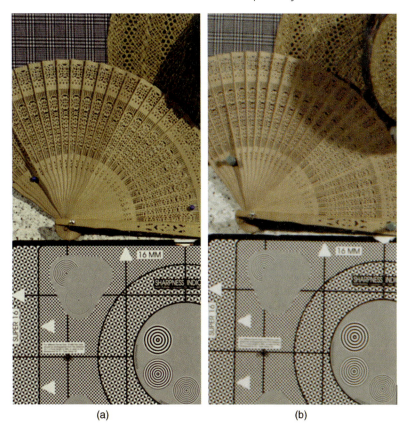

**Figure 4.11** Comparison of the spatial resolution of a digital cinema camera (a) and a HDTV camera (b).

resolution. Special color experts (so-called colorists) have the possibility to correct and/or process every tiny image sequence according to the goal of the film producer.

In Figure 4.11, the spatial resolution of a real digital cinema camera (a) is compared with the resolution of a high-end digital camera for HDTV applications (b).

As can be seen from Figure 4.11, the spatial resolution of the digital cinema camera is higher than the resolution of the high-end digital camera for HDTV applications. Observe in Figure 4.11 that the digital cinema camera can resolve even delicate spatial details.

Independent of the application (live mode or raw mode), the aim of color processing is to adjust the color appearance of the colored objects on the monitor to the color appearance of the same real objects at the location of making the film. This is illustrated in Figure 4.12.

As can be seen from Figure 4.12, to obtain the same color appearance, the color transformation unit coverts the $RGB$ values of the camera ($R_K$, $G_K$, $B_K$) to the RGB values of the monitor ($R_M$, $G_M$, $B_M$).

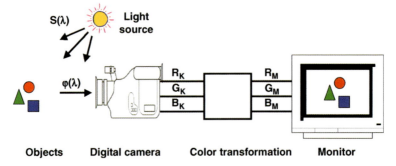

**Figure 4.12** Color processing: adjusting the color appearance of the colored objects on the monitor to the color appearance of the same real objects at the location of making the film.

## 4.2.2
### Postproduction Systems

For cinema film productions of highest image quality, analog film cameras have to be used due to their high dynamic range and high (4k) resolution. After shooting the scene, the negative film can be chemically developed in the film laboratory and scanned in order to have the digital images in high quality. These images can be then corrected and processed in the digital platform.

The film scanner for cinema motion picture films consists of an LED illumination unit with three LED systems for the blue, green, and red channels (see Figure 4.13).

As can be seen from Figure 4.13, the blue, green, and red LED radiations are projected through the first entrance opening into the inside of an integrating sphere. The diffuse radiation is used for the uniform illumination of the film transported along the second opening of the sphere. In each scan, the film is illuminated in transmission mode with blue, green, or red radiation and imaged by an optimized

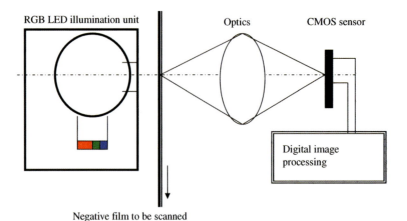

**Figure 4.13** Block scheme of a cinema film scanner.

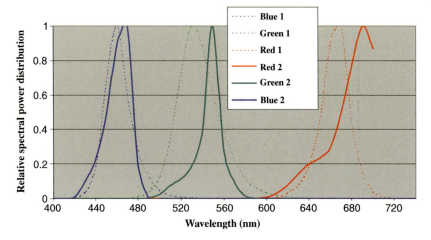

**Figure 4.14** Relative spectral power distribution of the RGB LED illumination unit of the film scanner for cinema motion picture films. R, G, and B LED radiations have to be optimized (see the Red 2, Green 2, and Blue 2 curves compared to Red 1, Green 1, and Blue 1) by the cautious selection of the LEDs and the band-pass and cutoff filters so that the resulting spectral power distributions do not overlap as this could cause a desaturation of the resulting RGB images.

optical system onto the active surface of an area semiconductor sensor of CMOS technology.

This monochrome sensor has the typical spectral sensitivity of silicon materials. The resolution of this sensor is $3112 \times 2048$ pixels. This resolution can be expanded by the microscanning algorithm up to $6048 \times 4000$ pixels per image. The blue, green, and red images of the negative film are then processed in order to correct the black value (i.e., the dark current), to interpolate death pixels, to linearize the signals, and to carry out ICC color management. The images are then stored and transferred to postproduction for further creative color corrections and image improvements, as mentioned above.

Figure 4.14 shows the relative spectral power distribution of the RGB LED illumination unit of the film scanner for cinema motion picture films.

Note that the R, G, and B LED radiations have to be optimized (see the Red 2, Green 2, and Blue 2 curves (optimized) in Figure 4.14 compared to Red 1, Green 1, and Blue 1 (not optimized)). Optimization can be carried out by the careful selection of the LEDs and the band-pass and cutoff filters so that the resulting spectral power distributions do not overlap as this could desaturate the resulting RGB images.

Finally, the improved and colorimetrically corrected digital images have to be prepared for the worldwide distribution of film content. The first possibility is to project these images with a high-end digital cinema projector. Today (2012), this option may only be possible in some reference cinema rooms worldwide (as mentioned earlier, there are different opinions about the extent and speed of digital penetration). It may be still not popular for mass public film presentation due to the high prices of the digital projectors and the limited possibilities to distribute digital

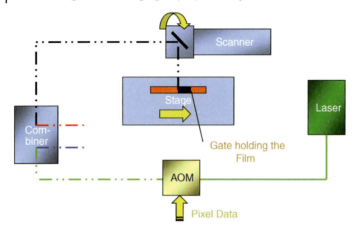

**Figure 4.15** Block scheme of the optical part of the film recorder for the green channel.

content. Today, a still more widespread possibility is to use conventional analog film projectors in the cinema room. Analog projectors are stable on a long-term basis and not expensive. This option requires a conversion of the final digital images back to the film medium to be performed by a laser film recorder.

The resolution of digital cinema images is $3112 \times 2048$ pixels or $1920 \times 1080$ pixels for HDTV cinema applications. Each pixel has a color resolution of 14 bits per color channel for high-end cinema purposes or 10 bits per color channel for HDTV. The laser recorder contains three solid-state laser units of high long-term stability (red: 650 nm; green: 532 nm; blue: 450 nm). The advantage of lasers is that they emit monochromatic radiation with a high beaming quality – imaging onto an AOM (acousto-optical modulator) unit for each color channel (see Figure 4.15).

The AOM unit receives the RGB image data and modulates the constant incoming laser radiation so that the output laser radiation has a degree of modulation ranging between 0% (for a dark pixel) and 100% (for the maximum *rgb* pixel value). Therefore, the modulated laser radiation carries the image information. The laser beam is deflected and imaged by the high-end optics. Finally, it is focused on the film at the position corresponding to the original pixel position of the digital image (see Figure 4.16).

The modulated intensity of the laser causes – after the exposure – a varying optical density in the corresponding film layer, red, green, or blue. The film is developed in the film laboratory after the recording process and can be copied to the IP film or print film; thus, film projection in the cinema room becomes possible. In Figure 4.17, the conversion function of the transformation from digital image values to film density is shown.

4.2.3
### CIELAB and CIEDE 2000 Color Difference Formulas Under the Viewing Conditions of TV and Cinema Production

The film production chain contains the following components: film acquisition with digital or analog film camera, film postproduction in which the images are corrected

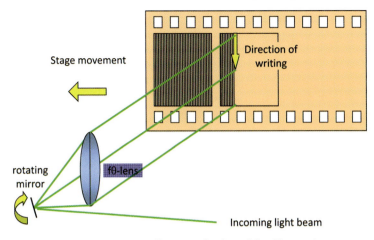

**Figure 4.16** Writing process on film material with modulated laser radiation.

and compared with the color objects of the real scenes, and the projection of the corrected film in which the color images on the cinema room canvas have to be compared to the images on the monitor used in the earlier postproduction. In each phase of this comparison, colorists compare the color images of different digital

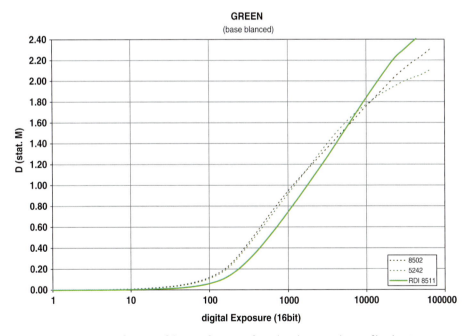

**Figure 4.17** Conversion function of the transformation from digital image values to film density. 8502: a commercial intermed film material; 5242: another commercial film material; RDI 8511: a new commercial film material developed especially for the laser technology.

imaging systems and media and evaluate the results of ICC color management visually, based on their empirical color vision experience and know-how [5–10].

For image engineers and system developers, the difference of the color images in different technical media can be expressed and analyzed by different color difference formulas based on different color spaces such as CIELUV, CIELAB, or CIECAM02 (see Sections 1.1.3 and 1.1.4). In the process of application of ICC color management and the empirical selection of the most appropriate color difference formula according to visual experience, some important questions have to be answered. These questions are not only of interest for cinema or TV industry but also essential for printing, lacquer, and textile industry and recently also for LED industry to be able to bin the LEDs into different categories after the LED packaging process. These questions can be formulated as follows:

1) Which color space and which color difference formulas can be used to predict the color difference perception of the human visual system?
2) How can we build a categorical color difference scale (e.g., with the categories "just noticeable", "very good", "good", "tolerable", and "bad") by the aid of the various color difference formulas?
3) Is there a difference in color difference perception between normal subjects who use the color products as normal consumers and trained persons and experts who evaluate color product quality everyday as a part of their job, for example, the color quality of an image sequence, a printed picture, a new fashion collection, or a color paint on the wall of a new building?

This section describes a series of visual experiments under the typical viewing conditions of TV and cinema production to answer at least a part of the above questions. In this context, the CIELAB and CIEDE2000 color difference formulas were the subject of the study. For further details on these formulas, see Section 1.1.4 and Refs. [5, 7, 8].

#### 4.2.3.1 Procedure of the Visual Experiment

The experimental setup was built in an optical laboratory and had the design of a typical monitor-based color correcting office in cinema postproduction (see Figure 4.18).

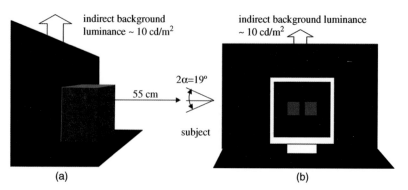

**Figure 4.18** Experimental setup of the color difference experiment. (a) Global view of the setup. (b) A test image presented on the CRT monitor.

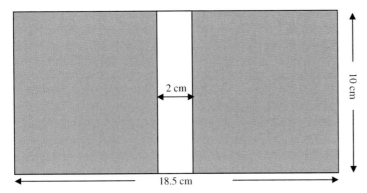

**Figure 4.19** Size and arrangement of the two color patches in the color difference experiment.

As can be seen from Figure 4.18, subjects sat at a distance of 55 cm to the CRT monitor used in the experiment. Subjects had to evaluate their perceived color difference between the test color sample on the right and the reference color sample on the left side of the monitor. The two homogeneous color patches were centered on the screen of the monitor with a distance of 2 cm between them. Only still images were displayed. The surround of the monitor had an average luminance of 10 cd/m². The viewing angle ($2\alpha$) of the two color patches was about 19° in the horizontal direction. Note that, for professional cinema film color correction, only CRT monitors are used because commercial LCD monitors are not suitable for this high-end professional application due to their insufficient image quality. Figure 4.19 shows the size and arrangement of the two color patches.

The luminance of the white point of the monitor was 96 cd/m² and that for the black point was about 0.4 cd/m². These values are typical for the viewing conditions of most postproduction companies. Colorimetric calibration of the monitor to the white point of illuminant D65 was performed by using commercial software. The monitor was switched on 1 h before the experiment. Test color patches were generated digitally in Adobe Photoshop®. In order to limit the number of test color samples, only eight color centers were considered: the $L^*$, $a^*$, and $b^*$ values of peak red (i.e., with 100% intensity of the red color channel), peak green, peak blue (100%), and cyan, magenta, and yellow as well as a skin tone and a gray tone were selected as color centers. Color differences had to be evaluated around these eight color centers in color space. The $L^*$, $a^*$, and $b^*$ values of cyan, magenta, yellow, and skin tone were derived from the data of the corresponding colors of the Macbeth ColorChecker® chart.

The $L^*$, $a^*$, and $b^*$ values of the above eight color centers were varied around each color center in order to obtain a set of test colors around each color center (see Table 4.1).

To obtain the gray test color samples, the values of $a^*$ and $b^*$ were varied around the gray color center by using every possible pair combinations of 1, 0, and $-1$ as of $\Delta a^*$ and $\Delta b^*$, for example, $\Delta a^* = 1$ and $\Delta b^* = -1$, $\Delta a^* = 0$ and $\Delta b^* = -1$, and so on. The $L^*$, $a^*$, and $b^*$ values of the color centers and the test colors were imported into

**Table 4.1** Selection of the test color samples for the peak blue, peak red, peak green, skin, cyan, magenta, and yellow color centers (CC) in the color difference experiment (except gray).

| CC | −5 | −4 | −3 | −2 | −1 | CC | A | 1 | 2 | 3 | 4 | 5 |
|---|---|---|---|---|---|---|---|---|---|---|---|---|
| Blue | Void | Void | Void | 28 | 29 | 30 | $L^*$ | 31 | 32 | 34 | 38 | Void |
|  | 43 | 52 | 60 | 64 | 66 | 68 | $a^*$ | 70 | 71 | 72 | Void | Void |
|  | Void | −128 | −120 | −116 | −114 | −112 | $b^*$ | −111 | −110 | −109 | −108 | Void |
| Red | Void | Void | 51 | 52 | 53 | 54 | $L^*$ | 55 | 56 | 57 | 58 | Void |
|  | Void | Void | 73 | 77 | 79 | 81 | $a^*$ | 83 | 85 | 89 | 97 | 106 |
|  | 45 | 54 | 62 | 66 | 68 | 70 | $b^*$ | 72 | 74 | 78 | Void | Void |
| Green | Void | 84 | 85 | 86 | 87 | 88 | $L^*$ | 89 | 90 | 92 | 96 | 100 |
|  | Void | Void | Void | −81 | −80 | −79 | $a^*$ | −78 | −77 | −75 | −71 | Void |
|  | Void | Void | 73 | 77 | 79 | 81 | $b^*$ | 83 | 85 | 89 | 97 | 106 |
| Skin | Void | Void | Void | 64 | 65 | 66 | $L^*$ | 67 | 68 | Void | Void | Void |
|  | Void | Void | 23 | 24 | 25 | 26 | $a^*$ | 27 | Void | Void | Void | Void |
|  | Void | Void | Void | 20 | 21 | 22 | $b^*$ | 23 | 24 | 25 | Void | Void |
| Cyan | Void | 70 | 71 | 72 | 73 | 74 | $L^*$ | 75 | 76 | 77 | Void | Void |
|  | −40 | −39 | −38 | −37 | −36 | −35 | $a^*$ | −34 | −33 | −32 | −31 | −30 |
|  | Void | Void | −2 | −1 | 0 | 1 | $b^*$ | 2 | 3 | Void | Void | Void |
| Magenta | Void | Void | Void | 51 | 52 | 53 | $L^*$ | 54 | 55 | Void | Void | Void |
|  | 63 | 64 | 66 | 65 | 66 | 67 | $a^*$ | 68 | 69 | 70 | 71 | 75 |
|  | Void | Void | −18 | −17 | −16 | −15 | $b^*$ | −14 | −13 | −12 | −11 | Void |
| Yellow | Void | Void | 83 | 84 | 85 | 86 | $L^*$ | 87 | 88 | Void | Void | Void |
|  | Void | −4 | −3 | −2 | −1 | 0 | $a^*$ | 1 | 2 | 3 | 4 | Void |
|  | 77 | 81 | 82 | 83 | 84 | 85 | $b^*$ | 86 | 87 | 88 | 89 | Void |

First row: numbering of the test colors. Column CC: color centers. Column A: attribute being varied ($L^*$, $a^*$, or $b^*$). Other columns (the background indicated the test color, for example, peak red on red background, etc.): test colors, either only $L^*$ or only $a^*$ or only $b^*$ was varied compared to the color center while the other two coordinates were the same as in the central column. Void: not included in the experiment.

Photoshop® to convert them into *rgb* values. Conversion errors from CIELAB to RGB due to colorimetric characterization inaccuracies were irrelevant because all color samples were measured spectroradiometrically *in situ* on the monitor for the prediction of color differences. The spectroradiometer had a holographic grating allowing for the measurement in the visible wavelength range between 380 and 730 nm. The step width was 3.5 nm and the bandwidth was 10 nm. From the measured spectral power distributions, the XYZ values of the two color samples (color center and test color) were computed and then the color difference between them.

After the selection of a panel of observers with normal color vision, two groups were defined among them, an expert group (5 women and 7 men) and a nonexpert group (5 women and 10 men). The expert group included the colorists and image engineers dealing with color correction and color quality in the departments of postproduction and image processing. The nonexpert group included camera engineers, electronic engineers, and technicians who did not have a close professional contact to the color images though motivated to carry out this experiment due to their professional orientation. Subjects were between 20 and 60 years of age (mean age: 36 years).

For each test color–color center combination, observers indicated their answer on a sheet of paper. Observers were asked whether they perceived a color difference. They could answer with "Yes", "No", or "No decision" (if they were not sure). The experiment took about 2 h and observers were allowed to have a break of 5–6 min. At the beginning, subjects adapted for 10 min to the viewing conditions in the test room while they were taught about the aim and procedure of the experiment. The experiment started with the gray samples and then the blue, green, red, cyan, magenta, yellow, and skin colors followed.

#### 4.2.3.2 Experimental Results

Color difference values were calculated with both formulas (CIELAB and CIEDE2000) for every test color–color center combination, for the variation of the $L^*$, $a^*$, and $b^*$ values separately. Percentages of the subjects who gave "Yes", "No", and "No decision" answers can be seen in Figures 4.20 (as a function of CIELAB color difference) and 4.21 (as a function of CIEDE2000 color difference) for the blue color center, as an example.

In the experiment, a test sequence was started with a test sample of high color difference that was reduced stepwise to a minimum color difference and then the color difference was increased again until the highest color difference (see the columns −5 to −1 and then the columns 1–5 of Table 4.1). Accordingly, in Figures 4.20 and 4.21, the light blue columns are higher for higher color differences as more subjects responded with "Yes, I can see a color difference."

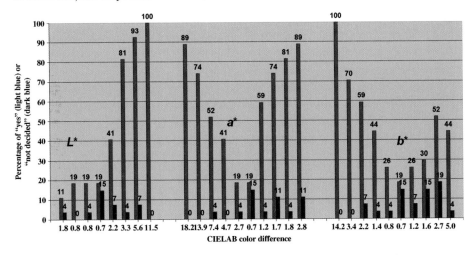

**Figure 4.20** Abscissa: CIELAB color difference values between the blue test color samples and the blue color center according to Table 4.1 by changing the $L^*$, $a^*$, and $b^*$ values of the test color sample, respectively. Ordinate: percentage of the subjects who answered "Yes, I can see a color difference" (light blue columns) and percentage of the subjects who were not sure or could not decide about whether they perceived a color difference (dark blue columns).

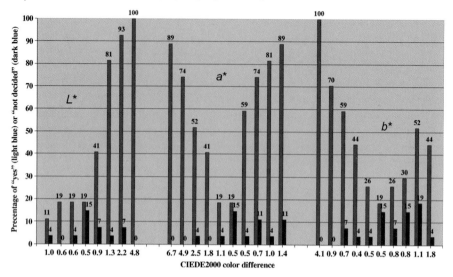

**Figure 4.21** Same as Figure 4.20 but for the CIEDE2000 color difference values.

From Figures 4.20 and 4.21, the computed color difference corresponding to the 50% threshold was estimated. This threshold value was defined by the criterion of 50% of the subjects perceiving a color difference. These threshold values are listed in Table 4.2.

As can be seen from Table 4.2, for red, green, and blue, CIELAB color differences are very different from CIEDE2000 color differences. CIEDE2000 color differences correlate better with perceived color differences than CIELAB color differences. It seems to be more or less possible to assign the CIEDE2000 color difference value of 1.0 to the visibility threshold except for the blue and red color centers (see the bold values in Table 4.2). Generally, the CIELAB color difference values of the primary colors (red, green, and blue) scatter more than the values of the other (mixed) colors; see, for example, the green threshold value of about 3.5 but a similar tendency can be observed for red and blue.

It is interesting that subjects were highly sensitive to the changes of the skin tone. It was easy to perceive a color difference between two skin tones and it was very difficult to design skin tone color samples on the monitor that were not perceived as different. A further observation was that the magnitude of threshold color difference depended on whether the color difference was increased or decreased during a given test sequence. Color difference perception was apparently affected by the color difference observed before. This means that if a well-visible color difference had been seen first, then a smaller color difference was no more visible at the next observation. If, however, a very small but perceptible color difference was presented first, then the next higher color difference was visible. This is a possible learning effect in color difference perception that can be improved by long time training.

This visual color difference experiment showed that the CIELAB color difference formula applied to predict the criterion of just noticeable color differences does not

**Table 4.2** Values of 50% visual threshold color differences in terms of the CIELAB and CIEDE2000 color difference formulas for the test color–color center pairs of Table 4.1 by decreasing (D: −5 to −1 in the first row of Table 4.1) and increasing (I: 1 to 5 in the first row of Table 4.1) the color difference along the $L^*$, $a^*$, and $b^*$ axes.

| Color center | Variation | CIELAB | | | CIEDE2000 | | |
|---|---|---|---|---|---|---|---|
| | | $L^*$ | $a^*$ | $b^*$ | $L^*$ | $a^*$ | $b^*$ |
| Blue | D | Void | >5.0 | 1.8–2.0 | Void | 2.5 | 0.7 |
| | I | 2.6–2.8 | 1.2 | 2.0 | 1.1 | 0.5 | 1.1 |
| Green | D | >3.4 | 3.0–3.2 | 3.6 | >1.1 | 1.1 | 1.0 |
| | I | 3.5 | Void | 3.0 | 0.8 | Void | Void |
| Red | D | 3.0 | 3.0 | 1.4 | 2.1 | 1.1 | 1.2 |
| | I | 1.5 | 1.5 | 1.2 | 0.6 | 1.1 | 1.2 |
| Cyan | D | 2.0 | 2.0 | 2.0 | 1.5 | 1.3 | 1.2 |
| | I | 2.5 | 1.8 | 2.0 | 1.6 | 0.6 | 1.3 |
| Magenta | D | 1.0 | 1.0 | 2.0 | 0.5 | 0.3 | 0.5 |
| | I | 1.0 | 1.3 | 1.2 | 0.9 | 0.3 | 0.5 |
| Yellow | D | 2.0 | 1.7 | 1.7 | 1.0 | 0.7 | 0.6 |
| | I | 1.2 | 1.7 | 1.5 | 0.7 | 0.7 | 0.8 |
| Skin tone | D | 1.5 | 1.5 | 1.0 | 0.9 | 0.9 | 0.7 |
| | I | 1.0 | 1.5 | 1.0 | 0.7 | 0.8 | 0.9 |
| Gray | D | 0.9 | 1.0 | 1.0 | 0.9 | 0.9 | 1.0 |
| | I | 0.5 | 0.7 | 0.9 | 0.8 | 1.5 | 1.0 |

deliver constant and reliable color difference values as it does not correspond to the perception of just noticeable color differences. The CIEDE2000 color difference formula, however, turned out to be more stable across the different color attributes and color centers. There was a tendency of yielding the ideal constant value of 1.0 for just noticeable color difference. Let us mention that, in the current ICC color management system, CIELAB color space is used. It would be advisable to apply a more uniform color space such as CIECAM02-UCS (see also Section 6.2) instead of CIELAB in the future.

For the color correction of new high-end cinema films in highly qualified postproduction companies, the criterion "just noticeable" is preferred. Instead of this criterion, the criterion "just tolerable" allowing higher perceived color differences can also be used later in the film distribution chain worldwide on the way to the public cinema rooms.

A further aim of the experiment was to point out whether the group of color experts can perceive color differences better than the group of nonexperts. The percentages of "Yes, I can see a color difference" answers of experts and nonexperts are compared in Figure 4.22 for the skin color center in a similar way as in Figures 4.20 and 4.21.

As can be seen from Figure 4.22, there is only a negligible difference between the expert and the nonexpert groups of subjects. Possibly, experts do not have a superior color difference perception ability compared to nonexperts. Nevertheless, it is

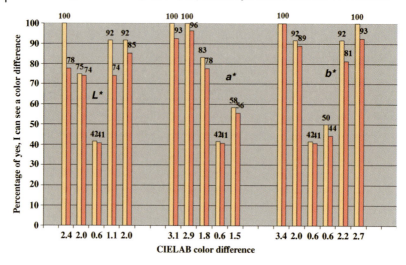

**Figure 4.22** Comparison of the answers of color experts with the answers of all observers in the color difference perception experiment. Percentages of "Yes, I can see a color difference" answers of experts (orange columns) and all observers (red columns). The way of data presentation is similar to Figures 4.20 and 4.21.

probable that they can interpret and explain their perceived color differences better than nonexperts.

### 4.2.4
### Applications of the CIECAM02 Color Appearance Model in the Digital Image Processing System for Motion Picture Films

In TV and cinema film production, several different technical systems with different color spaces and different viewing conditions are applied. In the image generation phase, the digital motion picture camera and the film camera have completely different spectral sensitivities due to the difference between the digital sensor and the analog film layers. At the locations of film production, the viewing condition is subject to changes from outdoor illumination conditions (with illuminance levels of up to 100 000 lx) to indoor conditions in dim or dark rooms (with very low illuminance levels of about 3–10 lx).

Luminaires used at the location vary between daylight simulating lamps and tungsten lamps with correlated color temperatures ranging between 2500 and 6500 K. Raw film pictures are transported to postproduction companies and corrected/enhanced by viewing them with a digital projector of a maximum luminance of 48 cd/m² or on so-called reference monitor, mostly a CRT display with a maximum white point luminance of 80–100 cd/m², a viewing angle of 40°, and a white point correlated color temperature of 6500 K. The final cinema film is then evaluated in a so-called reference cinema room with a screen luminance of maximum 55 cd/m² (on average about 10 cd/m²).

According to the above-mentioned changes of native device color space and viewing conditions, the color image appearance of the scenes of the film also changes (see Section 2.1.8) – especially the visual attributes of brightness, hue, saturation, and perceived dynamic luminance range. Color appearance changes can be diminished by the use of an appropriate color management system. Such systems have been tested and proven in the printing and display industries and, since 2002, also in the motion picture industry. The current ICC color management system has many advantages but also some essential deficiencies as enumerated below:

1) The profile connecting color space is CIELAB color that does not take viewing conditions such as surround luminance level into account when trying to predict the color appearance of the colored objects.
2) CIELAB's chromatic adaptation formula is not usable because the reference white of chromatic adaptation is the $XYZ$ value of the white point – instead of the retinal cone receptor signals (LMS) adapted to the white point.
3) CIELAB turned out to be perceptually inaccurate, especially in the blue hue range.

Therefore, professional cinematography has tried for many years to apply a better color system for digital color value conversion taking the different viewing conditions into account. One possibility is the CIECAM02 color appearance model (see also Sections 1.1.3 and 2.1.9) that was tested in a visual experiment to be described in this section. Note that the CIECAM02 color appearance model contains several advantages compared to former color appearance models and includes several further improvements. It implements a visually relevant chromatic adaptation formula and calculates color attributes according to the viewing conditions (also depending on background and surround luminance level). The most important improvement in CIECAM02 is the consideration of the influence of the background and the surround (see also Section 2.1.8) on the color appearance of the color stimulus (see Figure 4.23).

Inverting the calculation of the numerical correlates of color attributes (e.g., lightness, hue, and chroma) by the aid of the CIECAM02 model gives the users a chance to determine the $XYZ$ values of a certain object so that the color appearance of this object perceived by the human visual system is the same under different illuminants and viewing conditions. The following example taken from the printing industry may clarify this important application:

- **Situation 1**: Color perception of a digital picture on a self-luminous monitor in a well-illuminated office workplace. The monitor is characterized colorimetrically with a white point of 6500 K (D65).
- **Situation 2**: The same digital picture is printed by a color-managed professional printer and viewed in a dim living room illuminated by a tungsten halogen lamp with a color temperature of 3200 K.

The question is what relative $XYZ$ values the printed picture should contain at each image point so that the color appearance of this picture in the living room is the same as on the monitor in the office. The calculation scheme is illustrated in Figure 4.24.

In Figure 4.24, the tristimulus values $X_1$, $Y_1$, and $Z_1$ represent – at every pixel – the color stimulus of an object displayed on the monitor. These $XYZ$ values can be

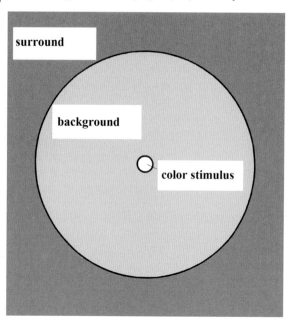

**Figure 4.23** Field of view in the CIECAM02 color appearance model: color stimulus or color element considered (about 2°), background (a 10° field immediately surrounding the color element), and surround (outside the background) (see also Section 2.1.8).

measured *in situ* or calculated by using the colorimetric characterization model of the monitor (see Chapter 2). The white point of the monitor has the tristimulus values $X_{W1}$, $Y_{W1}$, $Z_{W1}$. By applying a matrix transformation in the CIECAM02 model, the tristimulus values $X_1$, $Y_1$, and $Z_1$ are converted to retinal cone signals, $L_1$, $M_1$, and $S_1$. From latter data, the workflow of the CIECAM02 model calculates the numerical values of the perceptual correlates of the color attributes (see Section 2.1.8) by considering chromatic adaptation and signal compression in the human visual system.

Finally, a color space is also defined with the three orthogonal axes $J$ (lightness), $a_c$ (red–green axis), and $b_c$ (yellow–blue axis), similar to CIELAB $L^*$, $a^*$, and $b^*$ color space. These numerical correlates of the above-mentioned color attributes describe the color appearance of the colored object displayed on the monitor. These color attributes have to be kept constant for the case of the printed picture viewed in the darker room illuminated by the tungsten halogen lamp. Therefore, the numerical correlates of these color attributes constitute the input data for the CIECAM02 backward model (see Figure 4.24). The new cone signals ($L_2$, $M_2$, and $S_2$) can be computed by substituting the new viewing parameters characterizing the dim living room and the new white point, that is, the light source chromaticity of tungsten halogen. From the new cone signal values, the new tristimulus values ($X_2$, $Y_2$, and $Z_2$, the so-called corresponding color) can be calculated for the printed picture.

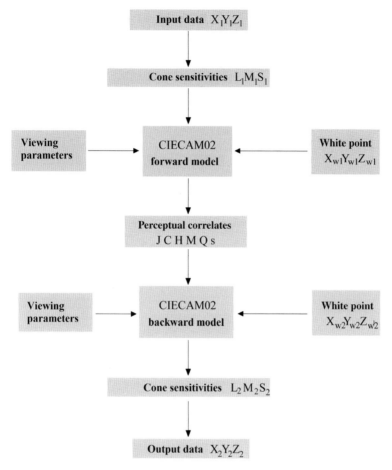

**Figure 4.24** Application of the CIECAM02 color appearance model to the calculation of XYZ values corresponding to different viewing conditions. Index 1: first viewing condition; index 2: second viewing condition.

As mentioned above, a visual experiment was performed to test the performance of CIECAM02 and to improve image color appearance in the different stages of the cinema production chain. The stimulus of the experiment is illustrated in Figure 4.25.

As can be seen from Figure 4.25a, the tabletop arrangement contained the Macbeth ColorChecker® chart, a set of colored textiles and printed products such as colored paper and books, and some other objects such as candles and an artificial flower. This tabletop arrangement was illuminated by tungsten halogen lamps with a correlated color temperature of 3200 K homogeneously. The tabletop was imaged by a 6 megapixel motion picture camera to display the image on the monitor (see Figure 4.25b).

These digital picture data were converted into HDTV format with 1920 × 1080 pixels per image. Each image pixel contained a set of *rgb* values that was in turn converted into *XYZ* values to establish the input data for the CIECAM02 model. The

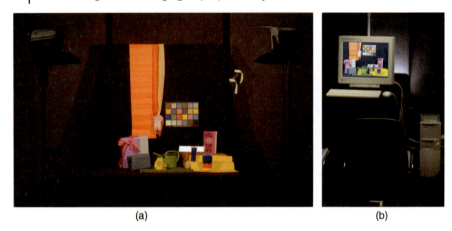

**Figure 4.25** Arrangement of the visual experiment to test the performance of CIECAM02. (a) A tabletop scene with real colored objects. (b) Monitor image of the tabletop.

CIECAM02 viewing condition parameters (see Section 2.1.9) of the tabletop arrangement and the monitor are compared as follows:

Tabletop side (Figure 4.25a):

- Relative background luminance: $Y_b = 22$.
- White point luminance: $L_W = 1150\,\text{cd}/\text{m}^2$.
- Surround: average; parameters: $c = 0.69$; $F = 1.0$; $N_c = 1.0$.
- Adapting field luminance: $L_A = 230\,\text{cd}/\text{m}^2$.
- White point of the light source: $X_{W1} = 105.1$; $Y_{W1} = 100.0$; $Z_{W1} = 46.6$.

Monitor side (Figure 4.25b):

- Relative background luminance: $Y_b = 20$.
- Luminance of the monitor's peak white: $L_W = 80\,\text{cd}/\text{m}^2$.
- Surround: dark; parameters: $c = 0.525$; $F = 0.8$; $N_c = 0.8$.
- Adapting field luminance: $L_A = 16\,\text{cd}/\text{m}^2$.
- White point of the monitor (calibrated to D65 and profiled): $X_{W2} = 95.3$; $Y_{W2} = 100.0$; $Z_{W2} = 109.2$.

For the purpose of the visual experiment, three versions of the monitor image were computed. Two images were calculated by the aid of the CIECAM02 model (dark: as in the real viewing condition of the monitor side; dim: the viewing condition parameters were changed to dim) and a further monitor image was computed with the original ICC color management algorithm used in the digital motion picture camera used to capture the photo of the tabletop.

Fourteen subjects carried out the experiment. All of them were experts of color image analysis like image engineers or cinema film colorists. The task of the subject was to compare the subjective color image appearance of the original tabletop scene with each one of the three image versions on the monitor. The method of memory matching was applied in which subjects could adapt to the illuminant completely. The reason is that the color temperature difference between the real tabletop scene with the tungsten halogen light ($T_c = 3200$ K) and the monitor ($T_c = 6500$ K) was too high for the purpose of a side-by-side comparison. Chromatic adaptation mechanisms of the human visual system cannot bridge such a large (but practically relevant) white point chromaticity difference in a side-by-side test.

First, the tabletop was observed for 5 min. Subjects were instructed to view and evaluate each colored object separately and also the total scene. Then, subjects looked at the monitor displaying a full-screen white patch without any image structure for 2 min. After that, subjects could see a version of the tabletop image on the monitor and compare it with the color appearance of the real tabletop recalled from their memory. Subjects filled a questionnaire with two different scales.

On the first scale, subjects judged the similarity of color appearance for every object individually and indicated their average impression on the scale. They were taught not to consider the entire tabletop and also not the relationship among the different colors. On the second scale, subjects judged the total color image appearance of the entire tabletop by paying attention to the luminance distribution and color appearance of all colored objects at the same time. Figure 4.26 shows the two scales of the questionnaire.

The above observation procedure was repeated for each one of the three versions of the monitor images (CIECAM02-dim, ICC, and CIECAM02-dark, see above). At the beginning of the comparison with each monitor image, subjects observed the real tabletop scene once again to enhance the effect of recalling from memory. Figure 4.27 illustrates the mean result of the second scale, that is, the overall similarity judgment of color appearance between the entire tabletop and each one of the three image versions.

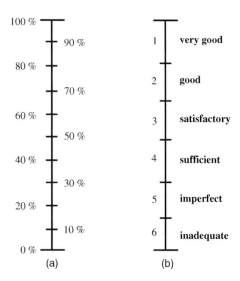

**Figure 4.26** Questionnaire used to compare the color image appearance of the tabletop with the monitor (see Figure 4.25). (a) First scale to evaluate the color appearance of every individual object (average impression). (b) Second scale to judge the total color image appearance of the entire tabletop by paying attention to the luminance distribution and color appearance of all colored objects.

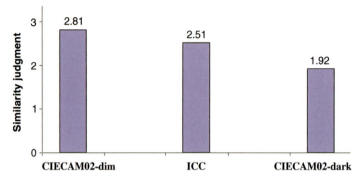

**Figure 4.27** Mean result of the second scale of Figure 4.26: overall similarity judgment of color appearance between the entire tabletop and each one of the three image versions.

As can be seen from Figure 4.27, the CIECAM02-dark version of the monitor image obtained the best similarity judgment and the CIECAM02-dim version had the worse evaluation although both images were processed by CIECAM02. This implies that CIECAM02 is only useful if its viewing condition parameters are chosen correctly (compare with Section 2.1.9). A similar finding can be seen from Figure 4.28 depicting the mean result of the first scale, that is, the similarity of color appearance of the individual objects.

As can be seen from Figure 4.28, the CIECAM02-dark version of the monitor image had the best judgment again.

The above experiment and also other experiments of the TV and cinema film industry have shown that the CIECAM02 color appearance model should be an improvement in certain circumstances. But the model is not perfect – it should not be the ultimate solution to predict image color appearance (see the hint to the image color appearance framework in Section 2.1.8). It is also necessary to test whether an extension of CIECAM02 would be usable in the mesopic (twilight) range of vision.

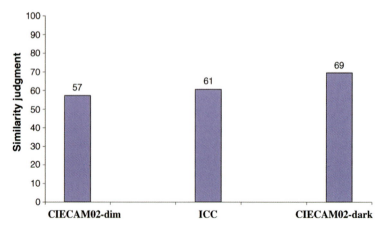

**Figure 4.28** Mean result of the first scale of Figure 4.26: similarity of color appearance between the tabletop and each one of the three image versions for every individual object.

Also, a new algorithm has to be developed in order to choose the viewing condition parameters correctly for real viewing situations. These aspects can be interesting subjects of research in the future.

## 4.3
## Color Gamut Differences

When color information is transferred from one color space to another, the problem of color gamut differences arises because the color gamut of the image source device and the gamut of the target device are not identical. In a color gamut study [3, 4], a cinema motion picture digital camera and a HDTV CRT monitor were characterized colorimetrically. The result showed that the gamut of the camera completely enclosed the gamut of the monitor independent of lightness and the hue angle (see Figure 4.29).

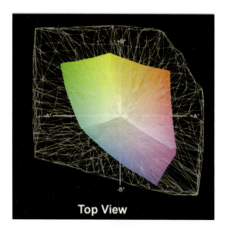
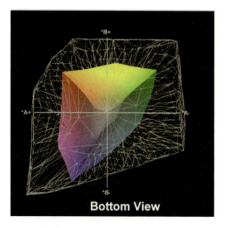
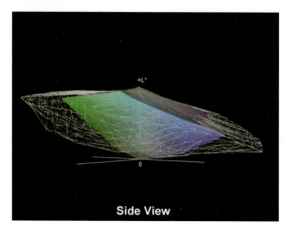

**Figure 4.29** Color gamut of a cinema motion picture digital camera (outer white mesh) and a HDTV CRT monitor (inner color solid). Top, bottom, and side views in CIELAB color space.

As can be seen from Figure 4.29, certain colored objects taken by the digital camera can occur between the outer white mesh representing the color gamut of the camera and the color solid of the monitor inside this mesh. Latter colored objects are so-called out-of-gamut colors and cannot be displayed on the CRT monitor.

The problem becomes more difficult if the colors have to be transformed from the monitor to the film or vice versa. Specifically, in the currently widely used hybrid production chain, the monitor is the medium of image color correction in postproduction while the film is the medium to shoot the real scenes with an analog film camera or to project the complete story in the cinema. Figure 4.30 illustrates this situation, that is, the gamut of the xenon lamp film projector together with the gamut of the CRT monitor.

As can be seen from Figure 4.30, the overlap of the film projector gamut and the CRT gamut is only partial and there exist out-of-gamut colors for both devices. A further color gamut issue is related to the use of digital cinema as a platform of film presentation on monitors in the postproduction for image correction and the use of digital projectors for cinema presentations. Namely, in the digital platform, the color gamut of the monitor and the color gamut of the digital projector are not identical (see Figure 4.31).

As can be seen from Figure 4.31, the overlapping regions of the two three-dimensional color solids depend on lightness and hue angle. The above findings imply that, beside the improvement of the accuracy of color transformations, the

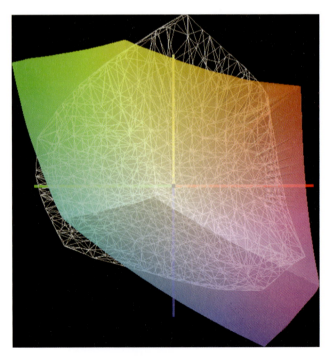

**Figure 4.30** Color gamut of a film projector (white mesh) and a CRT monitor (color solid) in CIELAB color space.

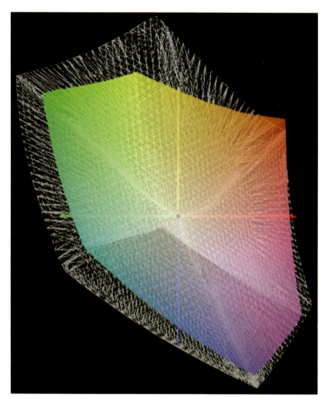

**Figure 4.31** Color gamut of a DLP projector (white mesh) and a CRT monitor (color solid) in CIELAB color space.

control of out-of-gamut colors is a further very important task of color management. In a recent study [3], two solutions were found to handle out-of-gamut colors in the cinema industry:

1) **"Hard clip" method**: The gamut of the medium having a greater color solid is hard clipped to the boundary surface of the smaller color solid. The advantage of this method is the high color fidelity of all colors within the smaller color body. The disadvantage is the loss of color details in the volume of out-of-gamut colors.
2) **"Soft clip" method**: The colors of the greater color solid are compressed to the gamut of the smaller color volume in order to transform the out-of-gamut colors without loss or with a small loss into the smaller (target) color space. The disadvantage is the loss of color fidelity for saturated colors (i.e., a loss of saturation). In the case of the color transformation from the monitor to film projection, colorists can choose different strengths for the soft clipping algorithm according to the film content (see also Section 4.6).

In practice, prior information (i.e., already in the greater color solid) of those colors that will undoubtedly become out-of-gamut in the smaller color solid makes the work of the colorist easier. For example, to record the large-gamut digital monitor image to the film of smaller gamut, the color management software of the recorder can be programmed to display a so-called out-of-gamut warning or out-of-gamut indicator image (see the example of Figure 4.32).

As can be seen from Figure 4.32 (middle image), the colors within the film gamut are masked by gray tones and just the out-of-gamut colors are colored in the out-of-gamut indicator image. Colorists working in postproduction have to process an original HDTV image and record it on the film of smaller color gamut, for example, in the orange–red hue angle region (see Figure 4.30). The missing orange–red region can also be seen in the middle (out-of-gamut indicator) image of Figure 4.32. The

**Figure 4.32** Method of out-of-gamut indicator image. Top: original HDTV image; middle: out-of-gamut indicator image; bottom: corrected (i.e., color compressed) image.

colorist scrutinizes the set of out-of-gamut colors and compresses them corresponding to film content and the imagination of the cinematographer.

## 4.4
## Exploiting the Spatial–Temporal Characteristics of Color Vision for Digital TV, Cinema, and Camera Development

Daily life objects and images projected in a cinema or displayed on a TV screen not only contain color and brightness information but also include spatial and temporal structures. There are both fine and rough spatial structures that can be very difficult or easy for the observers to perceive. Parallel to this aspect, in the temporal domain, there are processes and events containing both fast and slow motion. It is therefore necessary to study the spatial and temporal detection capacity of the human visual system including the stages of information processing in the eye and the visual brain. Results of this research help optoelectronic engineers develop high-quality products for printing industry, digital photography, and TV and cinema imaging.

### 4.4.1
### Spatial and Temporal Characteristics in TV and Cinema Production

Generally, until recently, visual acuity has been regarded as the most important factor of spatial vision used to assess the capacity of the human eye to resolve two objects in space or to identify the fine spatial structure of very small objects. In optical physiology, visual acuity can be tested with test patterns of high luminance contrast and varying size. The smallest test pattern size that the observer can just recognize yields a measure of visual acuity of a particular subject. Such test patterns are the so-called Snellen letters or the Landolt rings of different sizes.

In the past 20–30 years, optical physiology and lighting science have also dealt with the concept of contrast detection and analyzed the spectral contrast sensitivity of the human eye. The motivation for this research has been the knowledge that the spatial structure processing unit of the human visual system consists of many independent parallel channels. Each channel is sensitive to a test pattern of a certain size. The detection process is therefore a spatial integration process at which a certain threshold value of contrast and object size should be necessary.

For the measurement of contrast sensitivity, sine test patterns with a well-defined spatial frequency range and with variable contrasts are used (see Figure 3.19). The contrast of a sine pattern can be defined by Equation 4.2.

$$C = (L_1 - L_2)/(L_1 + L_2) \qquad (4.2)$$

$L_1$ and $L_2$ are the maximum and minimum luminance of the test pattern at the tested spatial frequency. If the experimenter reduces the contrast of the sine test pattern at a given frequency in defined steps until the observer can just perceive the contrast or the minimal modulation of luminance, then the threshold contrast value ($C_{th}$) is obtained at the limit of perception. The inverse value of threshold contrast is the

**Figure 4.33** Internal factors affecting human contrast sensitivity [11]. At the final stage, integration is performed over the whole test sample area of the size of $X_T$, $Y_T$ and over the total integration time $T$.

so-called contrast sensitivity (CS) (see Equation 4.3).

$$CS = 1/C_{th} \tag{4.3}$$

If the above test is repeated with a wide range of spatial frequencies, then a contrast sensitivity function can be calculated. The contrast sensitivity function tells how much contrast sensitivity (or threshold contrast) is there at a given spatial frequency. Contrast sensitivity is influenced by the number of sine periods in the pattern, test sample size, background luminance, and the orientation of the pattern, for example, vertical, horizontal, or diagonal. These factors are so-called external factors. In addition, contrast sensitivity also depends on a number of so-called internal factors [11] affected by the anatomy and condition of the human visual system, both the eye optics and the signal processing in the visual brain. Internal factors are illustrated in Figure 4.33.

As can be seen from Figure 4.33, there is a certain amount of noise (so-called external noise) at the object currently being observed, for example, the noise of a TV motion picture or the grain noise of a cinema film on the canvas. The object is imaged onto the human retina by the eye lens that is not an optimal optical imaging system and causes some optical aberration. Therefore, the optical modulation transfer function of the eye is deteriorated. In addition, the photoreceptor distribution on the retinal surface builds a discrete mosaic so that the spatial sampling period is not fine enough in most cases.

The radiant flux transported onto the retina exhibits a certain time-dependent fluctuation: the number of photons per time unit is not constant over time so that a so-called photon noise occurs. After the absorption of the photons in the receptors, photochemical signals are transferred over the ganglion cells to the central processing regions of the visual brain where a so-called lateral inhibition takes place. Within the process of signal transfer, a statistical fluctuation of the signal is registered that causes an internal noise. At the final stage, integration is performed over the whole test sample area and over a finite integration time (see Figure 4.33).

If the visual signal arrives at the central processing stages of the brain and the absolute signal strength is about three times the internal noise (i.e., the signal-to-noise ratio is equal to 3), then a contrast threshold value can be perceived and registered, or, in other words, the contrast detection process becomes possible. The contrast sensitivity function depends on two important factors:

1) **The eye lens optics**: this is not perfect and causes certain optical aberrations such as astigmatism and spherical aberration so that the modulation transfer function

## 4.4 Exploiting the Spatial–Temporal Characteristics of Color Vision for Digital TV

(see also Section 5.3.2.2) is strongly dependent on pupil diameter. A greater pupil diameter increases spherical aberration and therefore reduces the values of MTF. According to Ref. [12], the MTF of the human eye optics can be calculated by means of Equations 4.4 and 4.5.

$$\text{MTF}_{\text{opt}}(u) = \exp(-2\pi^2 \delta^2 u^2) \tag{4.4}$$

$$\delta = [(0.5 \text{ arcmin})^2 + (0.08 \text{ arcmin} \cdot d)^2]^{0.5} \tag{4.5}$$

In Equations 4.4 and 4.5, the symbol $u$ represents the spatial frequency in *line pair per degree* (Lp/deg) units or in other words cycles per degree (cpd) units (see Section 1.1.2), while the symbol $d$ stands for the pupil diameter of the human eye in mm units.

2) **Lateral inhibition**: the mechanism of lateral inhibition can be explained by the existence of antagonistic receptive fields. As already learned in Section 1.1.2, there are receptive fields on the retina, with a center that can be excited by light and with a periphery that can be inhibited by light (so-called on-center receptive fields) and the complementary receptive fields (so-called off-center receptive fields). A point spread profile on the human retina has the shape of a band-pass filter with smaller MTF values at low spatial frequencies [13]. The reduction of MTF at low spatial frequencies can be formulated [12] by Equation 4.6.

$$\text{MTF}_{\text{lat}}(u) = [1 - \exp(-(u/7)^2)]^{0.5} \tag{4.6}$$

Figure 4.34 illustrates the MTF curve for lateral inhibition and contrast sensitivity for different pupil diameters.

As can be seen from Figure 4.34, the MTF values of lateral inhibition increase until the spatial frequencies around 16 Lp/deg. Also, the MTF function of the whole human visual system including lens optics and lateral inhibition is limited at the side of higher spatial frequencies by the optics of the eye lens. Therefore, the resolution of the human eye is defined by the human eye lens. The crossing point of the MTF of

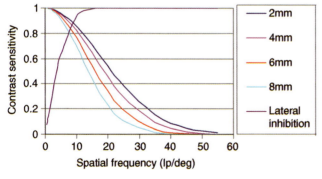

**Figure 4.34** MTF curve for lateral inhibition and contrast sensitivity for different pupil diameters. Lp/deg: line pairs per degree (the same as the cpd unit).

lateral inhibition and of the eye lens should be the position of the maximum of the MTF for the whole human visual system.

From all components that contribute to the contrast sensitivity of the whole human visual system, a physiological model of the MTF function can be established [12] with retinal illuminance as a parameter. This is shown in Figure 1.7 for three retinal illuminance levels, $E_1 = 2200$ Td (Troland) representing a well-illuminated working office, $E_2 = 600$ Td for the viewing situation of color image correction with TV monitors in the control departments of modern TV stations, and $E_3 = 314$ Td for a typical visual environment of cinema film presentation. These considerations are valid for the contrast sensitivity of the luminance channel only, that is, for the resolution of achromatic structure. If a red–green or a blue–yellow sine pattern is used instead of the grayscale pattern, then the contrast sensitivity of the chromatic (spectrally opponent) channels can be determined [11]. This is depicted in Figure 1.8 together with the contrast sensitivity of the luminance channel.

Psychophysical studies on the temporal behavior of the human optical signal processing system were performed by De Lange [14] and Kelly [15–17]. These studies focused on the temporal modulation of the luminance channel. In the last time, some physiological measurements were directly carried out at the stage of ganglion cells by exploring neural signals as a response to temporal chromatic modulations of the visual stimulus [18, 19]. From the physiological measurements at the ganglion cells and the psychophysical contrast sensitivity results, some relations can be established about the temporal filtering behavior of the different brain processing stages regarding both chromatic and achromatic modulations.

If the temporal frequency of periodical sine-shape modulated radiation is increased slowly in small steps, observers perceive the stimulus (i.e., the incoming radiation) in three different phases from a rough flicker form over the fine modulation flicker and finally a temporally constant perception. As introduced in Section 1.1.2, the frequency at which observers have a constant perception is called critical flicker frequency (CFF) or flicker fusion frequency (FFF). The following parameters influence the value of CFF:

1) **Luminance**: a higher luminance of the object causes a higher CFF (see Section 3.1).
2) **Size**: for small test field sizes, CFF increases with test field size.
3) **Position in the viewing field**: the maximal flicker fusion frequency is 50 Hz for a foveal object position and relatively high object luminances. At an extrafoveal object position (e.g., 30° nasal, 50° temporal), the value of CFF increases up to 70 Hz. This is the reason why in the SMPTE standard [20], the maximal luminance in the four corners of the screen in cinema rooms is limited to 80–90% of the luminance in the screen center – in order to avoid flicker perception for extrafoveal observations of motion pictures.

Figure 1.9 illustrates the temporal contrast sensitivity functions of the achromatic (luminance) and chromatic channels. According to these findings, some special reference cinema rooms apply a frame rate of 48 frames per second in order to minimize the flicker perception related to the film projection system while in most commercial cinemas worldwide, film projection frame rate is limited to 24 frames per second.

## 4.4.2
### Optimization of the Resolution of Digital Motion Picture Cameras

The optimization of a digital motion picture camera includes the optimization of the imaging system containing the optics (lens systems, filters) and the digital semiconductor sensors, the improvement of the electronics (bit depth, signal-to-noise ratio), and the formats to store and transfer picture data in digital form. At the final stage, TV or cinema film products (motion pictures) can be displayed on a TV monitor, on a home cinema screen, or in public cinema rooms. For professional premium applications, motion images are presented with the best digital projectors in digital formats without data compression and with a bit depth of 12 bit per color channel so that the image quality is mainly dependent on the quality of the digital motion picture cameras.

For home cinema and TV applications, the data format should have the bit depth of 8 bit with a high image compression factor. At the end of the chain of image generation, correction, formatting, transfer, and display, observers can perceive and evaluate image quality making use of their visual, optical, and psychophysical capabilities. To consider this, this section deals with the optimization of the resolution of imaging camera systems and investigates the effect of human contrast sensitivity and spatial resolution ability of the observers in a typical cinema room.

The SMPTE standard [20] defines for the illumination of cinema screens a maximal luminance of 55 cd/m² without the film material in the film projector. Because an average motion picture should have light and dark scenes, the average luminance of the whole film can be estimated to have about 30 cd/m² on which the cinema visitor has to adapt during the film presentation. In optical physiology, there is a well-known formula predicting pupil diameter as a function of adaptation luminance ($L_a$) (see Equation 4.7).

$$D = 5 - 3\tanh(0.4 \log L_a) \tag{4.7}$$

According to Equation 4.7, the pupil diameter of the observer is 3.4 mm at 30 cd/m² resulting in a circular pupil area of 9.1 mm². For this pupil area and the adaptation luminance of 30 cd/m², the retinal illuminance is equal to 274 Td. Therefore, a contrast sensitivity function can be calculated (see Section 4.4.1). This function is depicted in Figure 4.35.

As can be seen from Figure 4.35, the spatial frequency of maximum contrast sensitivity is equal to about 3 Lp/deg. The highest perceptible spatial frequency is about 32 Lp/deg while the spatial frequency for the contrast sensitivity of 0.5 (50%) is about 10 Lp/deg.

For further calculations and optimizations, one can assume that the observers take place at three different distances from the cinema screen: at a distance equal to the screen height and at the distance of three and five times the screen height. The

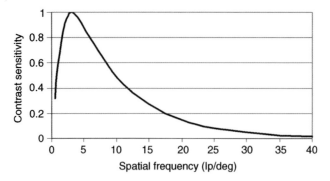

**Figure 4.35** Contrast sensitivity function for the retinal illuminance of 274 Td representing the average observer in a typical cinema.

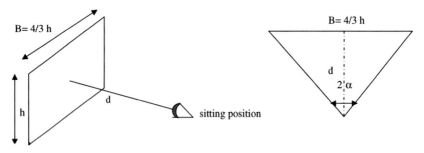

**Figure 4.36** Scheme of sitting positions in a cinema room.

cinema screen (canvas) has a format ratio of 4 (width) to 3 (height). The sitting position of the observer is shown in Figure 4.36.

With the above three sitting positions in mind, the corresponding three viewing angles ($2\alpha$) can be computed together with the number of periods (line pairs) within the whole viewing angle for the maximal, minimal, and 50% contrast sensitivity. The result of this computation can be seen in Table 4.3.

**Table 4.3** Viewing angle of the cinema screen (canvas) in three typical sitting positions in front of the canvas in the cinema (see Figure 4.36).

| SP | VA | NP for max. C | NP for 50% C | NP for min. C |
| --- | --- | --- | --- | --- |
| $d=h$ | 74.86 | 223.8 | 748.6 | 2395 |
| $d=3h$ | 27.8 | 83.4 | 278 | 890 |
| $d=5h$ | 16.87 | 50.6 | 168.7 | 540 |

The number of periods (line pairs, Lp) within this whole viewing angle for maximal, minimal, and 50% contrast sensitivity is shown in the last three columns. SP: sitting position; VA: viewing angle ($2\alpha$ in degrees); NP: number of periods (line pairs, Lp) within $2\alpha$; C: contrast.

**Table 4.4** Number of periods (line pairs) per millimeter at the film plane for maximal, minimal, and 50% contrast sensitivity (see also the legend of Table 4.3).

| SP | VA | NP for max. C | NP for 50% C | NP for min. C |
| --- | --- | --- | --- | --- |
| $d = h$  | 74.86 | 9.3  | 31.2 | 99.8 |
| $d = 3h$ | 27.8  | 3.47 | 11.6 | 37.0 |
| $d = 5h$ | 16.87 | 2.1  | 7.0  | 22.5 |

In this table, NP means the number of periods (line pairs, Lp) per millimeter.

The cinema screen is fully illuminated by an optically optimized projection lens of a film projector. The positive film of a width of 24 mm is inserted into the projector for motion picture projection. Because the number of periods (i.e., the number of line pairs) on the cinema canvas (i.e., at the image plane) must equal the number of periods of the film positive at the object plane, one can calculate the number of periods per millimeter at the film plane, for maximal, minimal, and 50% contrast sensitivity, by dividing the data of Table 4.3 by 24 mm. This calculation is shown in Table 4.4.

Table 4.4 implies some important requirements for the optical engineer working in film industry:

1) It is very important to optimize the imaging optics to the maximum of the modulation transfer function between 2 and 10 Lp/mm because human contrast sensitivity is very high in this frequency range.
2) Because the whole spatial frequency range of the contrast sensitivity of the human visual system is important, it is necessary to optimize the modulation transfer function of the imaging optics (for film camera, digital cameras, cinema film scanners, and recorders) until the spatial frequency of 31 Lp/mm.
3) In a typical case – in a good cinema room – observers can resolve spatial details up to 37 Lp/mm. But in a reference cinema room used to control the quality of a new film to be brought to the market, the engineers and technicians inspecting the film move typically very close to the cinema canvas in order to recognize delicate spatial details. Therefore, for reference films and premium applications, spatial frequencies up to 80 Lp/mm are interesting for the optical engineers to optimize their lens systems.

The above knowledge is not new – it was found in numerous experiments in the past in the worldwide optical industry. The above three requirements summarize this knowledge for a successful development of the optical imaging system. From these requirements and the experience of the optical laboratories of film industry that a loss of MTF of only 1% is already visible at the spatial frequencies between 5 and 10 Lp/mm, the minimal MTF values of the imaging lens can be computed. This is shown in Table 4.5.

Figure 4.37 shows the MTF function of a theoretical resolution limited lens for the wavelength of 550 nm and an aperture of $k = 2.3$ in comparison with a real camera

Table 4.5 Minimal MTF values of the imaging lens of the cinema camera for the blue, red, and green channels and for different spatial frequencies.

| Spatial frequency (Lp/mm) | Blue channel (450 nm) | Red channel (650 nm) | Green channel (550 nm) |
| --- | --- | --- | --- |
| 10 | 0.88 | 0.9 | 0.9 |
| 20 | 0.85 | 0.85 | 0.85 |
| 40 | 0.5 | 0.60 | 0.60 |

lens of a typical motion picture film camera. It can be noticed that the MTF values of these two lenses are practically identical in the spatial frequency range of up to 32 Lp/mm and the difference is very small up to 60 Lp/mm.

It is also known that aliasing artifacts (see Section 2.1.7) limit the resolution of digital cameras because the pixels are arranged periodically on the sensor sampling the original scene when taking the image. Regarding this sampling issue, the developers of digital motion picture cameras have to take the following aspects into account:

1) Excellent imaging lenses have a relative high MTF value at the spatial frequencies greater than 60–120 Lp/mm. Therefore, the potential of an aliasing effect is high from the optical point of view.
2) At a small sensor pixel (on the order of 4–6 μm), the sampling rate is high and this causes a very high MTF function. But the effective sensor area absorbing radiation

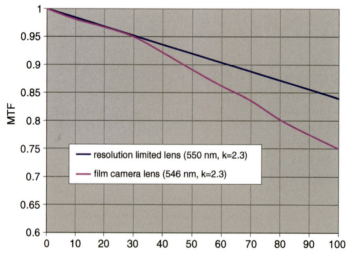

Figure 4.37 MTF functions of a theoretical resolution limited lens for the wavelength of 550 nm and an aperture of $k=2.3$ in comparison with a real film camera lens.

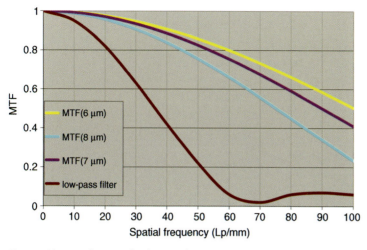

**Figure 4.38** MTF functions for three pixel sizes, 6, 7, and 8 μm, and the resulting low-pass filter.

is low and the resulting signal-to-noise ratio can result in a poor and not very brilliant image quality.

3) By the aid of a conservative design of the low-pass filter, one can keep the aliasing artifacts small. But in this case, MTF is relatively low in the range of up to 40 Lp/mm; hence, the image is not very sharp.

In Figure 4.38, MTF curves are shown for three pixel sizes, 6, 7, and 8 μm. These sizes are feasible with today's CMOS technology using the 0.35 or the 0.5 μm wafer process. In Figure 4.38, the resulting low-pass filter can also be seen.

Discussing Figure 4.38 further, the pixel size of 8 μm can be chosen as a good compromise between resolution and sharpness on one side and signal-to-noise ratio and brilliance of the image on the other side. For the pixel size of 8 μm, the Nyquist frequency at which the alias pattern can be observed for the first time is equal to 62.5 Lp/mm. The MTF of the low-pass filter (see Figure 4.38) allows for a relatively high MTF value of 0.4 (40%) at the spatial frequency of 40 Lp/mm. From 60 up to 100 and 120 Lp/mm, the MTF values of the low-pass filter are very low in order to reduce any higher order alias structure.

Besides the modulation transfer function as a measure of contrast transfer capacity, an excellent imaging optics should cause only a minimal lateral chromatic aberration. Chromatic aberrations are usually perceived when the observer approaches the screen being now able to scrutinize the edge of an object (e.g., a black letter or the edge of the roof of a house imaged on the screen) and now an unexpected colored contour is seen. Lateral chromatic aberration is acceptable for professional cinematographers if a visual acuity of or better than 1 arcmin, that is, 0.00029 rad, can be achieved. This means a lateral color aberration of a width of 11 μm observed from a distance of 1 m. Figure 4.39 illustrates the distribution of

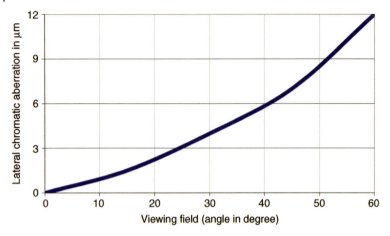

**Figure 4.39** Distribution of the lateral color aberration of a high-end motion picture camera optics over the viewing field.

lateral color aberration of a high-end motion picture camera optics over the viewing field.

If the imaging lens system, the digital sensor pixels, and the low-pass filtering of the system are optimized, then it is possible to develop high-quality digital motion picture cameras for cinema film production with MTF values better than those of other high-quality HDTV cameras, as illustrated in Figure 4.40.

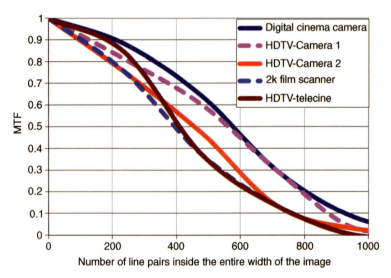

**Figure 4.40** MTF curve of a high-quality digital motion picture camera for cinema film production in comparison with high-quality HDTV cameras and a film scanner.

## 4.4.3
### Perceptual and Image Quality Aspects of Compressed Motion Pictures

#### 4.4.3.1  Necessity of Motion Picture Compression
In today's postproduction, TV and cinema industry receives digital motion pictures either from a highly qualitative HDTV camera or from a digital film scanner. For a modern professional HDTV camera, the dataflow at the 12 bit quantization level for each one of the RGB color channels and at a frame rate of 24 frames per second should have a bandwidth of 1.8 Gbit/s. At a frame rate of 60 frames per second, the bandwidth should be approximately 4.5 Gbit/s.

For the modern digital intermediate (DI) process, cinema engineers and technicians have developed a high-quality film scanner of 4k or 6k film resolution, that is, $4096 \times 3112$ or $6048 \times 4096$ pixels per film frame and with 14-bit digitalization per color channel. Therefore, the production of an average cinema film of 100 min should have a data capacity of 9.7 terabytes. Although data storage technology exhibited excellent progress in the last time, this dataflow is too high so that TV and cinema industry must reduce the amount of data for worldwide film distribution. To this end, several motion picture compression methods and infrastructure (both hardware and software) were and still have to be developed and validated.

Especially for data acquisition (e.g., the filmmaking process at the premises) and for highly qualitative postproduction (e.g., color and contrast correction, cutting process, or animation), dataflow reduction is not of highest priority. In these applications, the image quality of the digital master film is very important for further high-end distribution to premium cinemas and also for image storage in digital archives without loss of any visual information.

In the past 10 years, several motion image compression methods were developed and validated. Numerous tests in the past showed that the wavelet-based method according to MJPEG2000 is the most suitable tool – in comparison to the other compression methods – due to its especially efficient compression factors, scalability, loss-free data compression, and high image quality up to a high compression degree.

This section describes the results of an experiment on the MJPEG2000 method. The method itself, however, is described elsewhere [21]. In this section, tools of objective and subjective evaluation of digital images, selection of test sequences, and test arrangements are described. Subsequently, the result of the experiment is presented. The aim of this experiment was to point out the tolerance limits and threshold values of *image difference* perception and the resulting compression factor needed to conserve the very good image quality of the film to be able to distribute it worldwide.

#### 4.4.3.2  Methods of Image Quality Evaluation
Because the individual estimation of different subjects about image quality is very different, reliable criteria and methods have to be worked out and proven in order to rank and evaluate the performance of the different image compression tools. There exist objective and subjective methods for this purpose.

In the subjective method, visual tests are carried out with a number of observers to estimate the image quality of certain well-defined motion picture sequences in a viewing condition typical for TV and cinema rooms. The collection of subjective data is time-consuming and needs a careful logistic preparation and data collection. But subjective tests with scientific methods have the advantage that they can be planned flexibly according to the intended applications. Also, the collected data set is relevant and usable for real purposes because the complexity of the whole context of image quality and a variety of technical and scientific questions can be answered.

Objective methods are mostly not able to regard the complexity of visual perceptions but they can answer certain questions related to technical aspects. They can be done independent of human observers and they are more flexible concerning time and location. The evaluation and estimation of image quality and the loss of certain image properties in an image processing system can be carried out on the basis of some well-studied mathematical, physical, or psychophysical models.

There are two mostly used objective methods, the MSE (mean-square error) method and the PSNR (peak signal-to-noise ratio) method. The MSE method measures the difference between an original image $x(m, n)$ and the image $y(m, n)$ to be tested. This method is often applied due to the simplicity of its calculation (see Equation 4.8).

$$\text{MSE} = \frac{1}{MN} \sum_{m=0}^{M-1} \sum_{n=0}^{N-1} [y(m, n) - x(m, n)]^2 \tag{4.8}$$

In Equation 4.8, $y(m, n)$ is the image to be tested having the size (width × height) of $(M \times N)$ and $x(m, n)$ is the original image, also with the size of $(M \times N)$. The PSNR method is derived from the MSE method (see Equation 4.9).

$$\text{PSNR} = 20 \log_{10} \left( \frac{b}{\sqrt{\text{MSE}}} \right) \tag{4.9}$$

In Equation 4.9, the symbol $b$ represents the maximum possible pixel code value of the image, for example, 255 or 1023. PSNR data are given in dB units and usually range between 20 and 50 dB. Absolute PSNR values are not important in most cases because different images of different subjective image quality can have equal PSNR values. But a comparison of subjectively measured and calculated PSNR values under the same test condition is relevant – it should yield information about image quality. In image compression technology, the PSNR method is often applied. The Moving Pictures Experts Group (MPEG) uses an empirical value of 0.5 dB for PSNR as a measure of just noticeable difference between two images. Many compression methods optimize their output data with the aim of the highest possible PSNR values.

Concerning the subjective methods of testing image quality, in the double-stimulus continuous quality-scale (DSCQS) method [21], test subjects observe the original image and the quality-reduced image (test image) pairwise. To do the evaluation, subjects have to mark their perceived image quality on a linear interval scale. The image quality difference between the original image and the test image can be calculated.

In the DSIS (double-stimulus impairment scale) method, subjects evaluate the test image sequence in comparison to the original images. The original (error-free) sequence is shown as a first sequence and in the next step the compressed image sequence is presented. The aim of the DSIS method is not the evaluation of total image quality but rather the identification and estimation of an image error or image impairment caused by image compression in the test image.

Generally, test procedures have four phases. In the training phase, participants get an instruction in oral and written forms. It is explained what they should concentrate on and how image quality is to be evaluated. In the demonstration phase, a set of images is shown and evaluated by the participants informed about some special critical locations in those images. Results of this phase are not taken into account. In the pseudo test phase, the entire test procedure is launched but the images shown are not the same images that are to be tested in the regular phase. In the fourth (regular) phase, the test is carried out with the intended test image sequence. For the evaluation of image quality, every participant fills a questionnaire with a scale containing the judgment of image quality.

The SDS (simultaneous double-stimulus) method evaluates image quality similar to the DSIS method. The difference is that in the SDS method the test image sequence and the original image sequence are presented simultaneously; hence, the subjective evaluation becomes more critical. This method is often used by professional image engineers in premium TV and cinema film production companies. In the questionnaire, participants answer to the question of how much difference there is between the two presented image sequences.

### 4.4.3.3 The Image Quality Experiment

The experiment took place in the research laboratory of a cinematographic company. The test arrangement was built according to the recommendation of ITU BT.500 and BT.710. A high-end monitor was applied (often used as a reference monitor in cinema industry) with $1920 \times 1080$ pixels resolution at 50 Hz. The maximal luminance of the monitor was $70 \, cd/m^2$. A tubular fluorescent lamp with diffuse light intensity, a correlated color temperature of 6500 K, and a high color rendering index was used as illumination unit behind the monitor. From the observer's position, the background luminance was $7.0 \, cd/m^2$ (10% of the maximal luminance of the monitor).

The viewing field subtended 60° (vertical) and 90° (horizontal). The distance between the observer and the monitor was 54 cm (twice the image height of the monitor). In every test cycle, one, two, or three subjects could carry out the evaluation simultaneously. A disk recorder was used as playing equipment. Data sequences were stored in RGB format with 8 bits per color channel and with an image resolution of $1920 \times 1080$ pixels. For the subjective tests, a play list was preprogrammed with which the image sequences could be displayed in a certain time order with well-defined time durations so that the test conditions could be exactly reproduced.

The selection of the test sequences for the evaluation of the compression methods is very important for the relevance of the subjective and objective tests. In order to guarantee the comparability of the different tests, the implementation of the internationally agreed standard test sequences is useful. The Video Quality Experts

**Figure 4.41** HDTV test scene "tabletop".

Group (VQEG) defined some test sequences for similar compression evaluations. Because no test sequences were available in HDTV resolution at the time of this experiment, the experimenters generated a set of digital image sequences with a digital cinema camera and scanned some analog film materials.

There were two digital sequences made with the digital cinema cameras. The sequence "tabletop" (see Figure 4.41) shows a table with a black cloth on the table and as a background in the viewing field. There are several different colored objects containing different levels of contrast on the table. The corresponding motion picture sequence contains a slow camera motion horizontally from the left to the right side of the table.

In the test sequence "roof", many objects can be seen in fine spatial resolution including the hair and skin structures of the persons in the images, the leaf scene, the edge structure of the roofs of the houses, and the face expressions and mimic of the persons (see Figure 4.42).

Figures 4.43–4.45 show the test sequences captured by the analog film camera with a 35 mm cinema film material in 4k resolution (i.e., $4096 \times 3112$ pixels per film image) and scanned with 14 bits per color channel. By using cubic pixel interpolation and conversion, a new digital data film was generated in HDTV resolution ($1920 \times 1080$ pixels). In these scanned test sequences, the characteristic grain structure (of different spatial frequencies) of the film images can be clearly seen.

In the test sequence "sea" (Figure 4.43), subjects saw the entire panorama of a sea in a hill region where a slow vertical camera motion was carried out. During the camera motion, the total image content was moving. The fine structure of the water waves on the sea surface, the cloud structure on the sky, and the fine contrast transition from the sky side to the dark side of the hill constitute the content of this sequence.

**Figure 4.42**  HDTV test scene "roof".

In the sequence "church", the roof and the white walls contain many fine details in different spatial frequencies and the edges of the roof form the transition between the darker roof and the much lighter and more homogeneous blue sky (see Figure 4.44).

In the sequence "faces", close-up images of persons were shown. The skin tones, the fine details of their hair and their faces, and the textiles are the reasons for the selection of this sequence.

The first question of the experiment concerned the efficiency of wavelet-based compression methods. Several comparisons were performed between the different

**Figure 4.43**  Test scene "sea".

**Figure 4.44** Test scene "church".

compression methods about compression efficiency. The evaluation criterion was the peak signal-to-noise ratio. The reason for the choice of this criterion is its simple methodology and the relatively small time and organization effort.

For the analysis of the compression efficiency of the wavelet-based compression methods, some images of different nature were generated in different compression degrees (or compression factors) with the MJPEG2000 method. The following test images were used: the still picture of the motion picture sequence "tabletop", the SFR (sequential frequency response) test chart [22], the picture Lenna (the

**Figure 4.45** Test scene "faces".

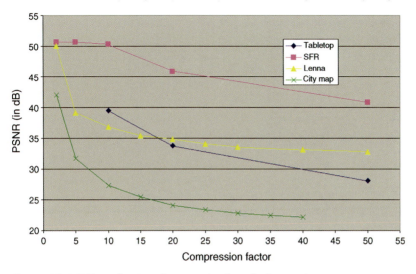

**Figure 4.46** PSNR as a function of compression factor for four test images.

well-known portrait of a woman), and a city map image. The comparison of the PSNR values of these still pictures confirmed the dependence of compression efficiency on image content and image size (see Figure 4.46).

As can be seen from Figure 4.46, the images such as "tabletop" or "Lenna" containing a variety of structures and color information but also the "SFR" image achieve a higher PSNR value for equal compression factors than the image "city map" containing only a limited number of spatial frequencies. Generally, one can say that natural images can be compressed more efficiently than artificial images because, in a natural image, many objects and structures can be found with a high number of different spatial frequencies.

The next question of the experiment was to determine how much compression is allowed by the observers for an image sequence compressed with MJPEG2000 to be judged so that the image quality is acceptable or tolerable. In this context, acceptability and tolerability mean that, although critical observers recognize image compression in the test image sequence, they do accept the quality of the image or do not consider the image depreciated.

To answer this question, the DSIS test method (see above) was applied. Altogether, 34 subjects were asked to evaluate three different sequences containing images of seven different compression factors together with the original images of the three motion pictures sequences "church", "roof", and "faces" (see Figures 4.42–4.45). These sequences were preprogrammed in a certain time order so that the test condition was the same for all subjects. Figure 4.47 shows the result of image quality evaluation, that is, the percentage of observers who answered acceptable as a function of compression factor.

As can be seen from Figure 4.47, at the compression factor of 16 : 1, more than 60% of all subjects evaluated image quality as "acceptable". The evaluation of the judgments of the observers showed that the perceived difference between the original

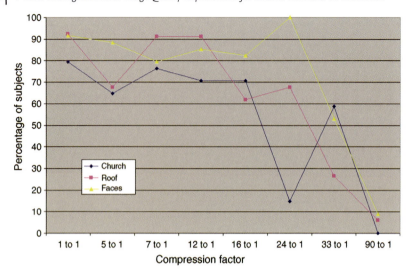

**Figure 4.47** Result of image quality evaluation: percentage of observers who answered "acceptable", as a function of compression factor.

and the test images increased with higher compression factors independent of image content. Note that, with the wavelet-based compression method of MJPEG2000, the perceptible image difference between the original and the compressed images is really small for compression factors up to 16:1.

Because the image quality of a cinema film is very important for the success of a film, the observers deciding whether the image quality is acceptable before going to the broader audience are not the final customers, that is, not the visitors of the cinema. The decision makers are the film producers, the cinematographers, and the image engineers who have a lot of experience in image evaluation. Therefore, in the next phase of the experiment, the SDS method was applied. As mentioned above, in this method, observers have the possibility to evaluate the simultaneous presentation of the reference and test images.

In order to utilize the full HDTV resolution, one half of every image consisted of the reference image and the other half of the test image with the same content. The monitor image was split vertically and the two halves were shown side by side (see Figure 4.48).

The image database for this phase of the study contained the sequences "church" and "roof" (see Figures 4.42 and 4.44) compressed with the MJPEG2000 method. A panel of 17 color image experts was invited to take part in the test. They were aware of the purpose of the experiment in detail and accurately instructed about the aim and time schedule of the test. The test sequences started with a higher compression factor at which the loss of image quality was visible and ended with the best image quality level. Subsequently, the test procedure was changed so that the sequences could be repeated or displayed at adjustable playing speeds; hence, the identification of possible image differences between the original and compressed images became

**Figure 4.48** Example of an image used in the SDS method in the so-called butterfly mode; that is, one half of the image shows the original sequence while the other sequence to be tested is viewed as a vertically mirrored version of the image.

easier for the subjects. The result of this phase of the experiment can be seen in Figure 4.49.

As can be seen from Figure 4.49, for the sequence "church" only 47% of all subjects recognized the difference at the compression factor of 7:1. The same percentage (47%) occurred for the case of the sequence "roof" at the compression factor of 12:1.

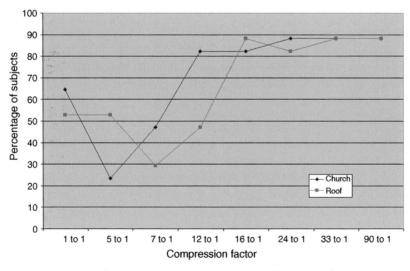

**Figure 4.49** Result of image quality evaluation: percentage of observers who answered "acceptable", as a function of compression factor.

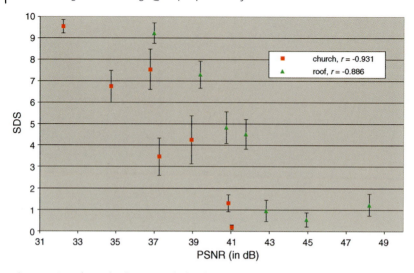

**Figure 4.50** Relationship between calculated PSNR values and subjective SDS data.

Conversely, one can say that, in this special test procedure – in which subjects could evaluate the images in direct comparison and with considerable time and effort – about 50% of the experts could not see the difference between the test sequence and the original sequence at a compression factor of 7:1.

In Figure 4.50, the relationship between the calculated PSNR values (objective method) and the subjective SDS data is illustrated.

In Figure 4.50, higher PSNR values and lower SDS data indicate better image quality. Therefore, the correlation coefficients ($r$) are negative. The correlation coefficient for the sequence "church" is $-0.931$ and for the sequence "roof" is $-0.886$. At a PSNR value of 43 dB, the image quality was evaluated to be very good with an SDS value of 1.0 or 1.5.

In summary, the results of the experiment described in this section suggest that, for image color correction and for digital mastering, the compression without loss has to be used with MJPEG2000. For the distribution to premium cinema events, a compression factor of 7:1 or 12:1 can be chosen, and for normal high-quality distribution, a compression factor of about 16:1 is acceptable.

### 4.4.4
### Perception-Oriented Development of Watermarking Algorithms for the Protection of Digital Motion Picture Films

#### 4.4.4.1 Motivation and Aims of Watermarking Development

Cinematography has used digital technology for many years. The traditional analog cinema data transfer chain has been replaced or expanded by a number of new digital systems including cinema film generation, postproduction, and distribution. The digital process has several advantages such as lower processing time, more possibilities for film content improvement, and a consistent and reproducible creation of image quality, for example, image contrast improvement and color conversion.

In the digital medium, the production and distribution of illegal copies of the new films is much easier than in the analog one because the digital data form allows a perfect film copy by the aid of today's digital technology at a very low technical effort. American film industry estimated the sale loss arising from the violation of film rights and illegal reproduction of video film carriers to be approximately 3 billion U.S. dollars per year. Most of the losses by illegal reproduction of film copies usually take place in the first weeks after publication. In this short time interval, only a few persons may have access right to the complete high-quality film material. Therefore, the aim of worldwide TV and cinema film industry is to develop a technology or a tool for a firm protection and reliable source identification in order to avoid production of illegal copies from the beginning of film postproduction onward.

In the process of guaranteeing digital media copyrights, the terminology "watermarking" has played an important role in the last time. A watermark is an apparently random image signal added to the regular image as additional information. It is not easy to remove this information because it is not part of the structure of the data format; for example, it is not added as an alphanumeric tag at the end of the image file. It is rather an integrated part of image information. In the past, several different watermarking systems have been used in order to transport this additional information. But not all watermarking systems are consistent and reliable. Sometimes, watermarking information is easy to find and simple to remove.

Most watermarking algorithms are only suitable for watermarking images of lower resolution, such as those from the Internet, standard TV, or DVD resolution. These algorithms add to the image only a pseudorandom noise pattern. They are not able to be used for video sequences in real HDTV or higher resolution such as the 4k resolution of $4096 \times 3112$ pixels per image. Because the random noise pattern mentioned above is independent of the individual image content, this pattern is perceived as an erroneous structure.

In Figure 4.51, schematic representations of the MTFs of four different video resolutions are illustrated, from 4k resolution over 2k, DVD down to SVCD resolution.

As can be seen from Figure 4.51, the transferable frequency range is rapidly reduced in case of the DVD and SVCD resolutions. In many of the current watermarking systems, the algorithm utilizes the whole available frequency range to implement the watermark signal. It is quite plausible to understand that the main part of this information will be lost if the original watermarked image has 4k or HDTV resolution and the illegal copy of the film is published in SVCD resolution. If the resolution is downsampled, then the watermark is automatically destroyed without using any additional tool to remove it. Therefore, one aim of the development of new reliable watermarking methods is to maintain the watermark after the downsampling process from high cinema film resolution to DVD or standard TV resolution.

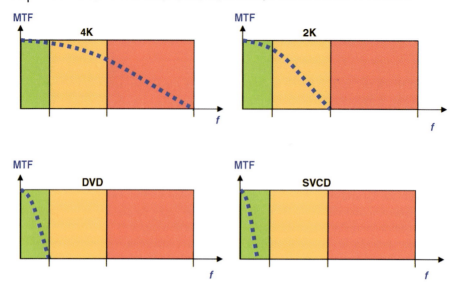

**Figure 4.51** Schematic representations of the MTFs of four different video resolutions: 4k and 2k resolutions, DVD, and SVCD. $f$: spatial frequency.

#### 4.4.4.2 Requirements for Watermarking Technology

The following requirements were formulated to develop new watermarking technologies:

- **Constant image quality**: The amount of dataflow, image quality, and robust watermarking are contradicting features of the system. The criterion of constant image quality is that the image change due to the implementation of the watermarks shall not be visible.
- **Robust watermarking**: the watermark as an image object must be robust against all image changing operations, for example, the addition of noise, downscaling, or image distortions. This requirement makes sure that the original watermark can be used for prosecution in case of copyright violation.
- **Moderate dataflow quantity**: if more data have to be transported, then more changes have to be done in the images. These changes tend to reduce image quality and data security. Therefore, the transferred dataflow should be as small as possible.

During the development of the new watermarking algorithm, several concepts were worked out to fulfill the above requirements. For constant image quality, the watermarking process can be done with the blue channel of the image because the changes and deformations in the blue channel are not easily visible compared to the red or green channel. The implementation of the watermark in the blue channel (at least up to a certain strength) does not reduce perceived image quality.

But this is not the only reason for using the blue channel as a watermarking platform [23]. The low contrast sensitivity of the human vision system in the blue

**Figure 4.52** A picture with a dynamic water mark (a) and the watermark as a structure (b). Observe that the watermark is added as a mask along the edges of picture (a) (hence it is image dependent or "dynamic", for example, the hair structure). Picture (b) is the difference picture between the watermarked image and the original image.

channel (see the blue curve in Figure 1.8) is utilized for a further improvement. This knowledge is essential for the development of modern image compression and watermarking algorithms. In order to avoid the loss of the watermark after downsampling the image from the original high resolution, the watermark is already designed in the blue channel of low resolution. By integrating this low-resolution watermark, the quality of the image in the blue channel is reduced but this reduction cannot be perceived by the human eye. In addition, the new method helps design a dynamic watermark that can be changed depending on image content – it is not only a static mask (see Figure 4.52).

As can be seen from Figure 4.52a, the watermark is added as a *mask* along the edges of the picture and hence it is image dependent or "dynamic"; observe, for example, the hair structure in Figure 4.52. For the requirement of robust marking, the so-called "not blind watermarking algorithm" is useful. This means that when an illegally attacking person searches for a watermark, then an original image without watermark would be needed to accomplish this search. Because the illegally attacking person does not have this original image, it cannot be checked whether the watermark removal was successful and this property improves the robustness of the watermarking method.

#### 4.4.4.3 Experiment to Test Watermark Implementations

The implementation of a watermark always results in an image error or image change. This error can be visible or disturbing depending on the type and intensity of the error. In order to evaluate and compare the different watermarking methods, objective and well-recommended criteria have to be used. In addition to the visual evaluations, robustness must also be taken into account in the assessment. For the visual and numerical evaluation of watermarking systems, an experiment was carried out by using the same test sequences as in Section 4.4.3.3 (used for compression judgment): "church", "faces", "sea", "roof", and "tabletop" in HDTV resolution (see Figures 4.41–4.45).

For the evaluation of image quality and the comparison of the different algorithms, the same objective quantities were used as described in Section 4.4.3.2, that is, PSNR and MSE. Note that a higher PSNR value means less change in the image. In addition to the PSNR values, the so-called Sarnoff JND (just noticeable difference in the context of *image differences*) [24, 25] metric for color video values was also taken into consideration. The value of JND is computed from an image model. It is intended to predict the image difference perception of the human visual system during the observation of similar images. In the above sense, the following JND scale was introduced:

- **JND 1**: The differences between the two images to be compared are not or scarcely visible for normal or trained subjects. JND 1 means a 75% probability for the perception of an image difference. If a pair of images has a JND value less than 1, then one can assume that the image difference is not visible, also not for the circumstance that the exact location and shape of the image impairment is known.
- **JND 3**: The differences are visible and perceivable for a detailed and careful observation. If two images have a JND value less than 3, then they are not clearly different. The difference is perceivable if the observer knows the location of the image change.
- **JND 5**: The differences between the two images are clearly visible. In other words, it is easy for the observer to see the difference between the original image and the test image.

In order to design a watermarking method resistant against different attacks, it is necessary to check – during the phase of algorithm development – whether the algorithm is able to generate such watermarks that are able to maintain their identity after applying different image operations. The concrete steps can be described as follows. With every watermarking method, an image or a sequence of images should be set up with a watermark or with various watermarks.

A number of operations should then be applied on these image sequences such as deformation, distortion, downscaling, or the implementation of various noise structures. For each algorithm under test, a number of images will be obtained and one should search for the watermarks in these images. The result is an overview of watermark recognition safety that is dependent on the algorithm and on the performed operations.

Recognition safety does not have a mathematical unit – it is a quantity introduced in order to answer the question of how reliably a watermark can be identified. There is a threshold value that has to be exceeded in order to state the condition of recognition security. The value of 1 means that the implemented watermark and the extracted watermark are equal.

Concerning the subjective tests, the evaluation criteria PSNR, JND, and robustness are suitable for the direct comparison of the different watermarking algorithms. To answer the question of how image impairment is perceived and judged visually, it is necessary and important to perform visual tests. These visual tests were carried out according to the ITU recommendation [26]. The same test conditions were applied as in Section 4.4.3.3 concerning the monitor, background luminance, and observation distance.

Because the changes among the images watermarked by the method under test are very small, the SDS test method was applied (see Section 4.4.3.2). In order to use full HDTV resolution, each test image obtained one of its two halves from the reference image and the other half from the test image with the same content. The monitor image was split vertically and the halves of the reference and test images were shown side by side (see Figure 4.48).

Observers did not know on which side of the monitor the original image was placed. The test sequences were the sequence "church", "sea", and "roof" (see Figures 4.42–4.44). For each sequence, six combinations of the test sequence and original sequence were presented. The first sequence was the sequence in which the original images were displayed on both sides of the monitor and this combination was repeated later in the test without informing the observers. Test subjects evaluated the perceived image difference by putting a cross on a continuous interval scale (i.e., a line on the questionnaire) between the two extremes "equal" (left, corresponding to the value of 0) and "different" (right, corresponding to the value of 10).

In order to provide an anchor stimulus for the subjects, the image combination "original–original" was first demonstrated to correspond to the "equal" reference point at the left-hand side of the scale. For the reference point "different" (on the right-hand side of the scale), the combination of the original image and the image watermarked with a modest watermarking method (different from the method under test) providing obvious image differences was shown. In the next step of the training, before the test, the subject was asked to describe the perceived difference orally. All 10 subjects were image experts in cinema postproduction. The experimenter explained in oral and written forms the content of the experiment and the types of possible image differences.

For the watermarking method under test, two different algorithms were developed, watermarking in the luminance channel and watermarking in the blue channel (of low spatial resolution). In the watermarking algorithm with the luminance channel, the following steps were performed (see Figure 4.53):

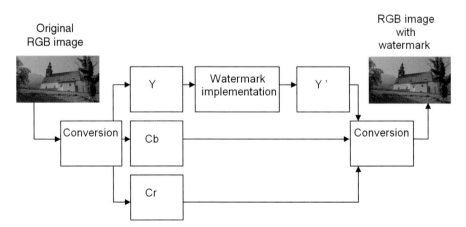

**Figure 4.53** Block scheme of the watermarking algorithm in the luminance channel.

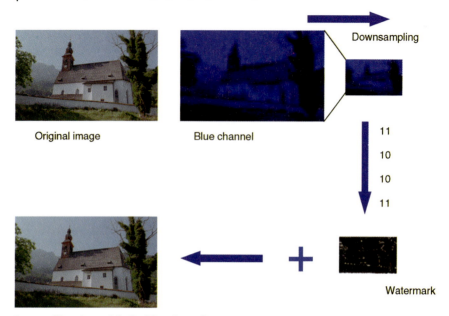

Figure 4.54 Block scheme of the watermarking algorithm in the blue channel.

1) Data transformation from RGB color space to YCbCr color space.
2) Watermark implementation into the luminance channel.
3) Data conversion back from YCbCr color space into RGB color space.

In the watermarking algorithm with the blue channel, the following steps were performed (see Figure 4.54):

1) Extraction of the blue channel from the RGB color image.
2) Cubic interpolation of the blue image down to a lower resolution.
3) Implementation of a watermark in this new blue image.
4) Superposition of the blue image of lower resolution to the original image.

The above-mentioned five test sequences comprised 650 motion images altogether. The borderlines between these test sequences are marked with red color in Figure 4.55 illustrating the calculated PSNR values as a function of video frame number.

As can be seen from Figure 4.55, the two algorithms under test (luminance and blue channels) deliver PSNR values 15–20 dB higher than the PSNR values of the other (modest) method. The differences in the results between the luminance and the blue channel algorithm can be explained not only by the image processing differences applied in the two different channels. Another reason of this difference is the lower spatial resolution of the blue channel image. In the first 100 images of the sequence "church", the PSNR values of the luminance channel algorithm were better than the PSNR values of the blue channel algorithm because, in this sequence,

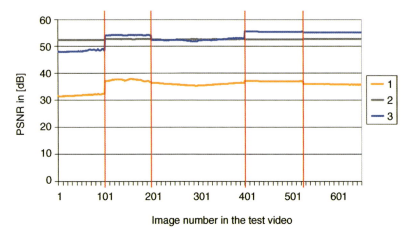

**Figure 4.55** Result of PSNR calculation in the test of three watermarking algorithms: (1) a modest method; (2) the algorithm under test in the luminance channel; (3) the algorithm under test in the blue channel. Red lines: borderlines between two consecutive test sequences.

the edge structures had a large high spatial frequency content changed intensively in the blue channel of lower spatial resolution. In the other sequences containing lower spatial frequencies, however, the PSNR values of the blue channel method were higher.

Figure 4.56 shows the result of computational image evaluation according to the JND metric [24]. As mentioned above, if a JND value is lower than 3, then no difference was noticed between the test image and the original image.

As can be seen from Figure 4.56, over all test sequences, the JND results with the blue channel method turned out to be the best because spatial image changes in the blue channel are less noticeable corresponding to human perception. The JND values

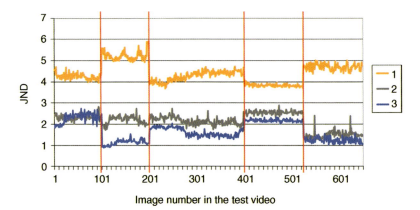

**Figure 4.56** Result of JND [24] calculation in the test of three watermarking algorithms: (1) a modest method; (2) the algorithm under test in the luminance channel; (3) the algorithm under test in the blue channel. Red lines: borderlines between two consecutive test sequences.

of both the luminance algorithm and the blue channel algorithm are lower than 3. In the test sequence "faces" (see Figure 4.45, from image no. 101 to image no. 201 in Figure 4.56), the JND values of the other (modest) method are higher than 5 because the pictures of this sequence do not have much spatial structure in the domain of higher spatial frequencies and most pictures of this sequence contained blurred scenes. The other (modest) method, however, utilized the whole range of spatial frequencies integrating fine details within the watermarks in these blurred scenes so that they were obviously noticed and perceived as image structure differences.

Concerning the result of the subjective test of perceived image differences on the scale between the two anchor points ("equal" on the left and "different" on the right), the other (modest) method was clearly recognized as "different" (see Figure 4.57).

As can be seen from Figure 4.57, the results of the two methods under test, that is, the luminance and blue channel algorithms, were judged in most cases with values being lower than 1 and gave the information on no or only little difference between the test image and the original image. Subject no. 5 gave the judgment of 1.5 for the monitor image containing the original image on both sides (see the yellow column of subject no. 5 in Figure 4.57). This finding also implies that the results with the two methods under test were below the threshold of image difference perception. The two watermarking methods summarized in Figures 4.53–4.54 indicate that watermarking is possible for high-resolution images, too. The method in the blue channel (Figure 4.54) improves robustness by keeping visual image quality unchanged.

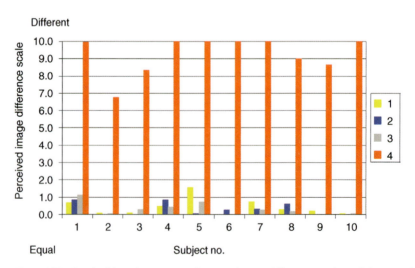

**Figure 4.57** Result of the subjective test of perceived image differences on the scale between the two anchor points "equal" (0) and "different" (10). Abscissa: subject number. (1) Original image on both sides; (2) blue channel algorithm; (3) luminance channel algorithm; (4) other (modest) method.

## 4.5
## Optimum Spectral Power Distributions for Cinematographic Light Sources and Their Color Rendering Properties

During the production of TV and cinema films, different light sources can be used depending on the locations and story of the film. In indoor productions in studios and halls, tungsten lamps and light sources with correlated color temperatures between 2700 and 3200 K are usually applied. In outdoor film production, light sources with correlated color temperatures in the range of 5400 and 5900 K can be used as additional light sources to natural skylight and sunlight.

In both types of location, two types of illumination are used, diffuse illumination and spotlight. Diffuse illumination with light sources on the order between 50 and 500 W (floodlight) should create a shadow-free illumination on the objects, for example, on the face of the artists. Spotlight using optics (reflectors, Fresnel lenses) concentrates light intensity in a certain direction. In the latter case, light sources on the order between 250 W and about 18 kW are used.

In most cases, the image quality of TV and cinema productions should be high to be able to communicate the story of the film correctly and have a chance on the TV and cinema market. A TV film or a cinema film expresses the ideas of filmmakers in the form of motion pictures. Therefore, the color quality parameters (see also Chapter 6) of the illumination of the scene to be recorded, such as color rendering, color gamut, color preference, color harmony, and chromatic lightness, are very important. For this reason, only light sources with high color rendering properties can be applied.

Until recently, film industry has been using tungsten light as a spotlight and tubular fluorescent lamps as floodlights with correlated color temperatures between 2700 and 3200 K. Because the luminous efficacy of tungsten lamps is only in the range of 20–25 lm/W and the lifetime of this lamp types is only about 1000–2000 h, lamp industry encouraged research to develop ceramic discharge lamps with a correlated color temperature of about 3100 K and with a high color rendering index [27]. In Table 4.6, the luminous efficacies of the light sources used to illuminate film scenes are listed.

Figure 4.58 shows the relative spectral power distribution of these lamp types. As can be seen from Figure 4.58, ceramic lamps (like no. 3 in Figure 4.58) contain a continuous spectrum and a remarkable part of their radiation in the red wavelength range between 620 and 780 nm. The tubular fluorescent lamp nos. 1 and 4 have some

**Table 4.6** Luminous efficacy and CCT (correlated color temperature) of the light sources used in indoor TV and cinema production [27].

| Lamp type | CCT (K) | Luminous efficacy (lm/W) |
| --- | --- | --- |
| Halogen tungsten lamps | 2700–3200 | 20–25 |
| Tubular fluorescent lamps | 3200 | 60–90 |
| Ceramic discharge lamps | 3200 | About 90 |

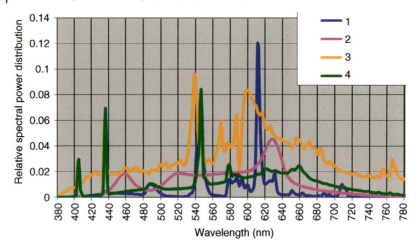

**Figure 4.58** Spectral power distributions of four typical light sources (nos. 1–4) used in indoor TV and cinema production.

strong emission lines at 435.8, 546.1, and 610 nm and a poor spectral distribution in the wavelength range between 440 and 530 nm (blue, cyan, and green) and also for wavelengths longer than 630 nm.

In conventional lighting technology, color rendering is evaluated according to the CIE Publication No. 13.3 [28] (see Section 6.1) by means of 14 test color samples (TCS01–TCS14) illustrated in the upper two rows of Figure 6.9. Their spectral reflectance functions can be seen in Figures 4.59 (samples of low saturation) and 4.60 (samples of high saturation). For lighting engineers and cinematographers, especially the test color sample nos. 9 (red), 12 (blue), 13 (skin tone), and 14 (leaf green) are important.

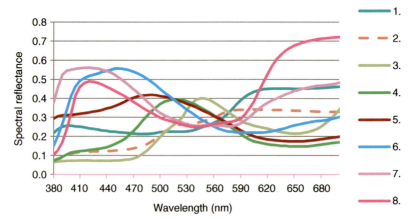

**Figure 4.59** Spectral reflectance functions of the eight desaturated color samples (1–8: TCS1–TCS8) used to characterize the color rendering properties of light sources [28]. Compare with the upper row in Figure 6.9.

## 4.5 Optimum Spectral Power Distributions for Cinematographic Light Sources

**Table 4.7** Special color rendering indices (see Section 6.1) – according to the 14 test color samples of Figures 4.59 and 4.60 for four typical light sources used in indoor TV and cinema productions at a correlated color temperature of about 3200 K.

| Test color sample | Spec. CRI | Tungsten halogen lamp | Ceramic discharge lamp | Tubular lamp (no. 1 in Figure 4.58) | Tubular lamp (no. 4 in Figure 4.58) |
|---|---|---|---|---|---|
| TCS1: old rose | R1 | 99.7 | 97.6 | 96.5 | 90.8 |
| TCS2: Earth | R2 | 99.7 | 97.7 | 96.1 | 90.8 |
| TCS3: yellow–green | R3 | 99.3 | 90.1 | 52.3 | 82.7 |
| TCS4: green | R4 | 99.7 | 96.1 | 87.8 | 89.8 |
| TCS5: cyan | R5 | 99.7 | 95.8 | 90.6 | 90.2 |
| TCS6: light blue | R6 | 99,6 | 97.1 | 84.5 | 85.3 |
| TCS7: light purple | R7 | 99.9 | 91.6 | 87.8 | 90.3 |
| TCS8: purple | R8 | 99.5 | 77.6 | 70.7 | 90.7 |
| TCS9: red | R9 | 98.1 | 42.1 | 78.6 | 78.4 |
| TCS10: yellow | R10 | 99 | 87.3 | 54.9 | 71.8 |
| TCS11: green | R11 | 99.7 | 96.0 | 79.1 | 86.8 |
| TCS12: blue | R12 | 98.8 | 89.6 | 55.8 | 81.9 |
| TCS13: skin tone | R13 | 99.7 | 98.7 | 92.6 | 90.8 |
| TCS14: leaf green | R14 | 99.6 | 93.0 | 66.8 | 89.4 |
| General CRI | $R_a$ | 99.6 | 92.2 | 83.3 | 88.8 |

Table 4.7 shows the special color rendering indices (see Section 6.1) – according to the 14 test color samples of Figures 4.59 and 4.60 – for four typical light sources used in indoor TV and cinema productions.

As can be seen from Table 4.7, the disadvantages of ceramic discharge lamps are their poor color rendering for saturated red (R9) and purple (R8). The weakness of the tubular fluorescent lamp no. 1 for diffuse illuminations can be seen from its yellow–green (R3), purple (R8), red (R9), yellow (R10), blue (R12), and leaf green (R14) color rendering indices (bold numbers in Table 4.7). These colors are very important in daily color applications and in cinematography so that the tubular fluorescent lamp of the type no. 1 had to be replaced by the type no. 4 (see Figure 4.58) to obtain (somewhat) better color rendering.

As mentioned above, for outdoor film productions, light sources of 5400–5900 K are often used as spotlight applications including metal halogen discharge lamps (e.g., HMI or MSR/HR). For floodlight illuminations, two types of tubular fluorescent lamps have been applied. Figure 4.61 shows the spectral distributions of these two lamp types.

As can be seen from Figure 4.61, the spectral distribution curves of these two tubular fluorescent lamps also show the weakness in the wavelength range between 480 and 530 nm, 550 and 570 nm, and longer than 630 nm. Table 4.8 shows the special color rendering indices of these two light sources compared with a HMI discharge lamp.

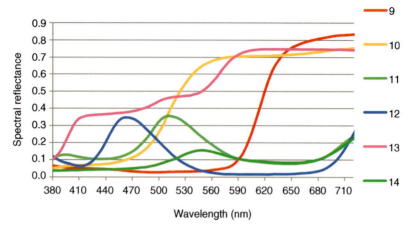

**Figure 4.60** Spectral reflectance functions of the six saturated color samples (9–14: TCS9–TCS14) used to characterize the color rendering properties of light sources [28]. Compare with the middle row in Figure 6.9.

As can be seen from Table 4.8, the tubular fluorescent lamps for 5600 K have similar weaknesses as the lamps for 3200 K (see Table 4.7) for the yellow, red, leaf green, and blue colors, especially no. 1 in Figure 4.61 (see the bold numbers in Table 4.8). The HMI and MSR/HR discharge lamps exhibit color rendering problems with saturated red colors (TCS9).

Since 2006, LED technology has been subject of a very intensive development. Today, LEDs represent a potential light source technology for high-quality TV and cinema applications. At the time of writing this book (2011–2012), daylight white LEDs with a correlated color temperature of about 5700–6500 K exhibited a luminous

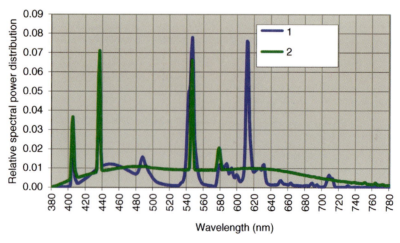

**Figure 4.61** Relative spectral power distributions of two typical light sources used for outdoor TV and cinema production as floodlight illuminations.

**Table 4.8** Special color rendering indices of three light sources used for outdoor TV and cinema production at a correlated color temperature of about 5600 K (see also the caption of Table 4.7).

| Test color | Spec. CRI | HMI discharge lamp | Tubular lamp (no. 1 in Figure 4.61) | Tubular lamp (no. 2 in Figure 4.61) |
|---|---|---|---|---|
| TCS1: old rose | R1 | 88.8 | 90.1 | 90.8 |
| TCS2: Earth | R2 | 94.1 | 95.1 | 90.8 |
| TCS3: yellow–green | R3 | 96.2 | 49.1 | 82.7 |
| TCS4: green | R4 | 89.8 | 86.9 | 89.8 |
| TCS5: cyan | R5 | 90.7 | 97.0 | 90.2 |
| TCS6: light blue | R6 | 92.0 | 82.7 | 85.3 |
| TCS7: light purple | R7 | 93.0 | 86.4 | 90.2 |
| TCS8: purple | R8 | 85.0 | 89.3 | 90.7 |
| TCS9: red | R9 | 57.9 | 64.3 | 78.3 |
| TCS10: yellow | R10 | 85.2 | 56.6 | 71.9 |
| TCS11: green | R11 | 87.3 | 82.5 | 86.8 |
| TCS12: blue | R12 | 90.6 | 69.8 | 81.8 |
| TCS13: skin tone | R13 | 90.2 | 89.5 | 90.8 |
| TCS14: leaf green | R14 | 97.8 | 65.5 | 89.4 |
| General CRI | $R_a$ | 91.2 | 84.6 | 88.8 |

efficacy of about 135 lm/W at the real operating condition with 350 mA and at a temperature at the solder point of 60 °C.

Warm white LEDs with a color temperature of 3200 K have – under similar operating conditions – a luminous efficacy of 105 lm/W. These efficacy values are much better than the luminous efficacy of the discharge lamps, tubular fluorescent lamps, and tungsten halogen lamps.

Further advantages of high-power LEDs are enumerated below:

- A potential to achieve a long life time of 10 000 up to 50 000 h if the thermal management task is optimally solved and the maximal current is not more than 350 mA.
- Dimmable between 0 and 100% (full intensity) without reduction of lifetime.
- Fast response time in the range of a microsecond to a millisecond.
- There is no ultraviolet radiation content and hence no or little damage for the illuminated objects such as human skin, flowers, or textiles.
- There is no infrared radiation content or thermal radiation; hence, there is only a small thermal load for the illuminated objects and no energy is necessary to cool the illuminated area or the room.

At the outset of high-power LED technology – in the time range between 2000 and 2008 – engineers tried to generate white LED light by means of color mixing of white phosphor-converted LEDs of poor color rendering indices with colored LEDs (green, yellow, and red LEDs). The resulting color rendering indices were very good (see Table 4.9) but the combination of the white LEDs with the yellow/red LEDs is very

Table 4.9 Three examples of white high-power LED light sources (see also the caption of Table 4.7).

| Test color | Spec. CRI | Mixed white LED (3150 K) | Warm white phosphor LED (3300 K) | Neutral white phosphor LED (4823 K) |
|---|---|---|---|---|
| TCS1: old rose | R1 | 90.6 | 99.1 | 91.1 |
| TCS2: Earth | R2 | 97.0 | 96.5 | 91.2 |
| TCS3: yellow–green | R3 | 94.0 | 91.3 | **88.7** |
| TCS4: green | R4 | 89.7 | 92.4 | 90.9 |
| TCS5: cyan | R5 | 92.6 | 97.7 | 89.8 |
| TCS6: light blue | R6 | 91.7 | 93.7 | **85.8** |
| TCS7: light purple | R7 | 99.2 | 93.3 | 93.6 |
| TCS8: purple | R8 | 95.6 | 95.1 | 89.2 |
| TCS9: red | R9 | 93.0 | 93.5 | **68.4** |
| TCS10: yellow | R10 | 97.1 | 90.2 | **77.0** |
| TCS11: green | R11 | 83.5 | 92.1 | 89.5 |
| TCS12: blue | R12 | 83.4 | 85.0 | **66.3** |
| TCS13: skin tone | R13 | 91.9 | 98.1 | 90.8 |
| TCS14: leaf green | R14 | 94.4 | 94.1 | 93.4 |
| General CRI | $R_a$ | 93.8 | 94.9 | 90.0 |

complicated regarding keeping the color coordinates of the combined light constant over time and under different environmental conditions, for example, against temperature changes (see also Section 7.1.2).

Between 2009 and 2012, white LED technology based on the mixing principle of efficient and stable blue LED chips and different new phosphor systems (such as YAG phosphor, LuAG phosphor, and Bose phosphor) has been developed and tested. By mixing the blue LED chip with a yellow or green phosphor and a red phosphor, it is possible to have warm white LEDs in the range of 2700–3500 K and neutral and daylight white LEDs up to 5000 K with a very good color rendering index (see also Section 7.1.3).

With the luminous efficiency of 100 lm/W and with high power LEDs of an electrical power of 1 or 2 W, floodlight and spotlight luminaires with an electrical power of maximum 1000 W can be developed for cinematography. Table 4.9 shows three examples for these white high-power LED light sources: a mixed white LED (3150 K, mixed from white and colored LEDs), a warm white phosphor LED (3300 K), and a neutral white phosphor LED (4823 K), together with their special color rendering indices.

As can be seen from Table 4.9, generally, the color rendering indices of the warm white phosphor-converted LED light source at 3300 K are comparable to the mixed white LED light source at 3150 K. These two light sources exhibit no or only little weaknesses in terms of their color rendering properties unlike the neutral white LED (see the bold values in the last column of Table 4.9). The mixed and the warm white LEDs are able to replace the tubular fluorescent lamps for floodlighting applications for indoor productions or for spotlight. However, as of today (2012), there are no white LEDs with a color temperature higher than 5400 K with a very good color rendering index. But it is only a question of time until these daylight white LEDs will be available on the market.

## 4.6
### Visually Evoked Emotions in Color Motion Pictures

This section deals with the visually evoked emotions of motion images. The aim is the modeling of emotional strength starting from a set of technical parameters of the digital video sequence [29]. It will be shown that it is possible to categorize motion image sequences into emotional categories automatically – based on a set of descriptor quantities – and the emotional strength of the input color motion picture can be enhanced. In this section, relevant technical parameters of the video sequences influencing visually evoked emotions will be identified together with a set of relevant psychological factors responsible for these emotions and the numerical correlates of the emotions will be modeled mathematically as a function of the technical parameters.

### 4.6.1
### Technical Parameters, Psychological Factors, and Visually Evoked Emotions

Primarily, it is the story, the sounds, and the actors' play that mediate the emotions of a color motion picture sequence. But this emotional effect is often enhanced or completed by visual effects [30], for example, the color [31], contrast, sharpness, noisiness, or other similar parameters (which will be called "technical" parameters or TPs in this section) of the movie. The aim is the identification of the relevant TPs influencing these "visually evoked" emotions, the identification of some relevant "psychological" factors ($\psi$Fs, for example, *presence* or *interaction*; see Ref. [32]) underlying these emotions, and the modeling of the emotions as a function of the TPs.

In a psychophysical experiment, observers viewed and evaluated 268 test video sequences. For each digital video file of this set of test video clips (containing a wide range of very different scenes and events), a characteristic mathematical descriptor quantity $D$ was computed for each one of the following 12 technical parameters: average luminance, average saturation, color balance, global contrast, global saturation contrast, noisiness, sharpness, sharpness block, contrast of those regions that contain skin colors, speed of camera motion, speed of darkening or lightening, and finally, speed of saturation change.

Then, a $z$ score was computed for each descriptor $D$ and for each clip, by subtracting the mean value of $D$ (for all 268 clips) from $D$ and then dividing the result by the standard deviation of $D$ (for all 268 clips). Five observers (between 20 and 30 years of age) watched each one of the 268 test video clips in a dark room on a large projection wall. After watching, they had to complete two questionnaires. One of the

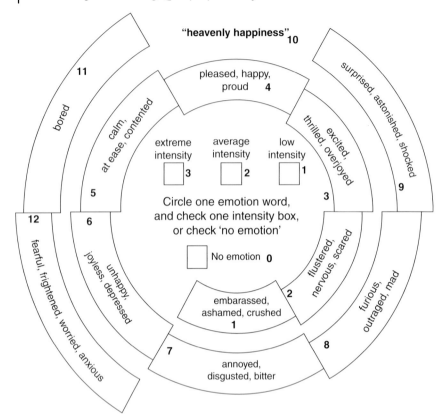

**Figure 4.62** A slightly modified version of the "emotion spiral" [29] proposed by Heise and Calhan [33]. This was used to collect the observer's answers in the psychophysical experiment quantifying visually evoked emotions. Reproduced from Ref. [29] with permission of Scandinavian Colour Institute.

questionnaires concerned the psychological factors. For each clip, subjects had to tell their opinion about the intensity (between −5 and 5) of presence, 3D-ness, persuasion, attention, excitement, and interaction. The mean answer of the five observers constituted a six-dimensional "vector" ($\Psi$) for each clip. The other questionnaire concerned the emotions. Observers completed a slightly modified version of the "emotion spiral" [33] proposed by Heise and Calhan. This modified version can be seen in Figure 4.62.

The experimental setup with an observer watching a test video sequence can be seen in Figure 4.63.

The mean answer of the observers constituted a 12-dimensional "vector" (**e**) for each clip, each emotional category being a dimension of this vector. The mean intensity (between 0 and 3) on the spiral was the corresponding component of this **e**-vector.

**Figure 4.63** An observer watching a video clip and completing the emotional spiral and the questionnaire about the psychological factors. Reproduced from Ref. [29] with permission of Scandinavian Colour Institute.

### 4.6.2
**Emotional Clusters: Modeling Emotional Strength**

The set of 268 test clips could be divided into six emotional clusters (1: Happy; 2: Bored; 3: Surprised; 4: Sad; 5: Fury; 6: Disgusted) based on a cluster analysis of the set of the 268 **e**-vectors. The set of the mean $z$ values $\{z_i\}$ of the descriptors in each emotional cluster showed identifiable features in each one of the six emotional clusters (see Figure 4.64).

In Figure 4.64, the 12 columns of different color for each emotional cluster have the following meaning: 1st (pale purplish blue) average luminance, 2nd (brownish lilac) average saturation, 3rd (drab) color balance, 4th (greenish cyan) global contrast, 5th (dark lilac) global saturation contrast, 6th (yellowish rose) noisiness, 7th (blue) sharpness, 8th (bluish gray) sharpness block, 9th (dark blue) contrast of those regions that contain skin colors, 10th (magenta) speed of camera motion, 11th (yellow) speed of darkening or lightening, and, finally, 12th (cyan) speed of saturation change of the video sequence.

In practice, it is possible to assign any arbitrary video clip (i.e., also beyond the set of 268 test clips of this section) a specific emotional cluster [29] – based on an automatic decision about the clip's similarity to one of the sets of mean $z$ values in the clusters shown in Figure 4.64.

After analyzing the experimental results, a so-called emotional strength (designated by $E$) was computed for every one of the 268 **e**-vectors by summing up the

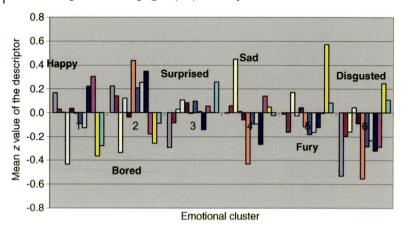

**Figure 4.64** The 12 mean z values depicted by columns of different colors in each emotional cluster. Reproduced from Ref. [29] with permission of Scandinavian Colour Institute.

squared values of the **e**-components except that the square value of "boredom" ($e_{11}^2$) was assigned a negative sign (i.e., it was subtracted from the sum), as can be seen from Equation 4.10.

$$E = e_1^2 + e_2^2 + \cdots + e_{10}^2 - e_{11}^2 + e_{12}^2 \tag{4.10}$$

The mean emotional strength in the different emotional clusters is shown in Figure 4.65. As can be seen from Figure 4.65, emotional strength was maximal in the "Disgust" cluster.

As a next step, the importance of a specific technical parameter was quantified by how effectively it influenced the **e**-vector and the emotional strength $E$. The 12 technical parameters had the following relative importance: noisiness 100, speed of darkening/lightening 96, contrast of skin color regions 59, average luminance 54,

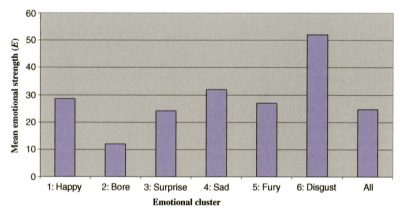

**Figure 4.65** Mean values of emotional strength $E$ in the six emotional clusters and the mean value for all clips. Reproduced from Ref. [29] with permission of Scandinavian Colour Institute.

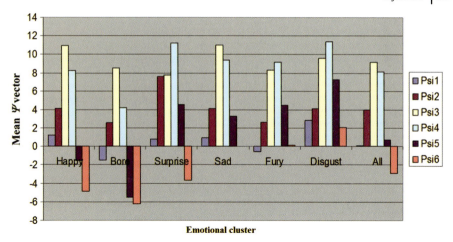

**Figure 4.66** Mean "$\Psi$-vectors" in the six emotional clusters and the mean value for all clips. Reproduced from Ref. [29] with permission of Scandinavian Colour Institute.

color balance 40, global contrast 33, average saturation 31, sharpness 21, sharpness block 21, speed of saturation change 11, speed of camera motion 10, and, finally, global saturation contrast 0. The mean "$\Psi$ vector" (i.e., the "vector" of psychological factors) was also significantly different among the six emotional clusters (see Figure 4.66).

Within each emotional cluster, a multivariate linear regression model was established about the dependence of $E$ on $\{z_i\}$. The performance of model prediction was evaluated. The correlation coefficients ($r^2$) were equal to 0.42, 0.33, 0.19, 0.29, 0.18, and 0.81 for the happy, bored, surprised, sad, fury, and disgusted clusters, respectively.

# References

1 Khanh, T.Q. (2004) Physiologische und Psychophysische Aspekte in der Photometrie, Colorimetrie und in der Farbbildverarbeitung (Physiological and psychophysical aspects in photometry, colorimetry and in color image processing). Habilitationsschrift (Lecture qualification thesis), Technische Universitaet Ilmenau, Ilmenau, Germany.

2 Wunderlich, D. (2003) Analytische und experimentelle Untersuchung zur Verbesserung von farbgemanaged Digitalbildern durch verschiedene Farbwahrnehmungsmodelle (Analytic and experimental investigation to enhance color managed digital images by different color perception models). Master thesis, ARRI, Munich and Technische Universität Ilmenau.

3 Geissler, P., Maier, S., Gottschling, T., Gonschorek, O., Bernt, E., Oehler, A., and Khanh, T.Q. (2004) Farbmanagement zur Angleichung von Monitor und Filmprojektion (Color management to adjust the monitor to film projection). *FKT Fachzeitschrift*, (7).

4 Khanh, T.Q. and Gonschorek, O. (2003) Farbmanagement für die digitale Übertragungskette von einer digitalen

Kamera bis Monitore (Color management for the digital transfer chain from a digital camera to monitors). Annual Meeting of the German Society of Color Science and Application, October 6–10, TAE Esslingen.
5 Fairchild, M.D. (1998) *Color Appearance Models*, Addison Wesley Longman.
6 Hauske, G. (1994) *Systemtheorie der visuellen Wahrnehmung (System Theory of Visual Perception)*, B.G. Teubner Verlag, Stuttgart.
7 Zhu, S.Y., Luo, M.R., Cui, G., and Rigg, B. (2002) Comparing different colour discrimination data sets. Proceedings of the 10th Color Imaging Conference, November 12–15, Scottsdale, AZ, pp. 51–54.
8 Luo, M.R., Cui, G., and Rigg, B. (2001) The development of the CIE 2000 color difference formula. *Color Res. Appl.*, **26**, 340–350.
9 Zhu, S.Y., Cui, G., and Luo, M.R. (2002) New uniform colour spaces. Proceedings of the 10th Color Imaging Conference, November 12–15, Scottsdale, AZ, pp. 61–65.
10 Cui, G.H., Luo, M.R., Rigg, B., Roesler, G., and Witt, K. (2002) Uniform color spaces based on the DIN99 color difference formula. *Color Res. Appl.*, **25**, 282–290.
11 Nadenau, M. (2000) Integration of human color vision models into high quality image compression. Dissertation No. 2296, Ecole Polytechnique Federale de Lausanne.
12 Barten, P.G.J. (1999) *Contrast Sensitivity of the Human Eye and Its Effects on Image Quality*, SPIE Optical Engineering Press, Washington, DC.
13 Röhler, R. (1995) *Sehen und Erkennen (Seeing and Recognition)*, Springer, Berlin.
14 De Lange, H. (1958) Research into the dynamic nature of the human fovea–cortex systems with intermittent and modulated light. I. Attenuation characteristics with white and colored light. *J. Opt. Soc. Am.*, **48**, 777–784.
15 Kelly, D.H. (1972) Adaptation effects on spatiotemporal sine-wave thresholds. *Vis. Res.*, **12**, 89–101.
16 Kelly, D.H. (1979) Motion and vision. II. Stabilized spatiotemporal threshold surface. *J. Opt. Soc. Am.*, **69**, 1340–1349.
17 Kelly, D.H. (1971) Theory of flicker and transient responses. II. Counterphase gratings. *J. Opt. Soc. Am.*, **61**, 632–640.
18 Lee, B.B., Martin, P.R., and Valberg, A. (1989) Sensitivity of macaque retinal ganglion cells to chromatic and luminance flicker. *J. Physiol.*, **414**, 223–243.
19 Lee, B.B., Pokorny, J., Smith, V.C., Martin, P.R., and Valberg, A. (1990) Luminance and chromatic modulation sensitivity of macaque ganglion cells and human observers. *J. Opt. Soc. Am.*, **7** (12), 2223–2236.
20 SMPTE 196M-2003 (2003) *Standard for Motion-Picture-Film Indoor Theater and Review Room Projection Screen Luminance and Viewing Conditions*, Society of Motion Picture and Television Engineers.
21 ITU-R BT.500-11 (2002) *Methodology for the Subjective Assessment of the Quality of Television Pictures*, International Telecommunication Union.
22 ISO/FDIS 12233 (1999) *Photography – Electronic Still Picture Cameras – Resolution Measurements*, International Organization for Standardization.
23 Winkler, S. (2000) Vision models and quality metrics for image processing applications. Thesis no. 2313, Ecole Polytechnique Federale de Lausanne, EPFL.
24 Lubin, J. and Fibush, D. (1997) Sarnoff JND vision model. T1A1.5 Working Group Document #97-612, ANSI T1 Standards Committee.
25 Vollstaedt, A. (2004) Untersuchung wavelet-basierter Bildkompressionsverfahren für die digitale Bildakquisition (Investigation of wavelet based image compression methods for digital image acquisition). M.S. thesis no. 2183/03/D44, ARRI, München and Technische Universität Ilmenau.
26 ITU-R BT.710-3 (1998) *Subjective Assessment for Image Quality in High-Definition Television*, International Telecommunication Union.
27 Khanh, T.Q., Grechana, N., and Möller, K. (2006) Farbwiedergabeeigenschaft in der Film-und Fernsehproduktion: Von den

Lichtquellen bis zur digitalen Bildwiedergabe (Color rendering property in film and TV production: from light sources to digital image rendering). *FKT Fachzeitschrift*, (5), 273–278.

28  CIE 13.3-1995 (1995) *Method of Measuring and Specifying Color Rendering Properties of Light Sources*, Commission Internationale de l'Éclairage.

29  Bodrogi, P., Kwak, Y., Kutas, G., Beke, L., Park, D.S., Czúni, L., and Kim, C.Y. (2008) Modelling visually evoked emotions for color motion images, colour: effects and affects, in *Interim Meeting of the AIC 2008, June 15–18, 2008, Stockholm, Sweden* (eds I. Kortbawi, B. Bergström, and K. Fridell Anter), Scandinavian Colour Institute, Stockholm (CD-ROM).

30  Bíró, Y. (2003) *A hetedik müvészet: a film formanyelve, a film drámaisága (The Seventh Art: Formal Language of the Film, Dramatic Effects of the Film)*, Osiris, Budapest.

31  Gao, X.P., Xin, J.H., Sato, T., Hansuebsai, A., Scalzo, M., Kajiwara, K. Guan, Sh., Valldeperas, J., Lis, M.J., and Billger, M. (2006) Analysis of cross-cultural color emotion. *Color Res. Appl.*, **32** (3), 223–229.

32  Sheridan, T.B. (1992) Musings on telepresence and virtual presence. *Presence-Teleop. Virt.*, **1** (1), 120–125.

33  Heise, D.R. and Calhan, C. (1995) Emotion norms in interpersonal events. *Soc. Psychol. Q.*, **58**, 223–240.

# 5
# Pixel Architectures for Displays of Three- and Multi-Color Primaries

The aim of this chapter is to describe methods to optimize the color gamut of three-primary (RGB) and multi-primary displays (those having more than three primary colors, for example, five: cyan and magenta in addition to red, green, and blue). To obtain a large color gamut, the sets of target colors to be covered have to be carefully chosen to be able to render the colors of the most important natural wide-gamut images.

Besides the target color set, however, there are several further important factors to be considered including color quantization levels, the number of color primaries, the white point, the issues of virtual primaries and technological constraints, and also the visually acceptable luminance ratio between each primary color and the white point of the display. Several sets of optimum wide-gamut color primaries are presented as examples together with the shape of their color gamut in color appearance space. It is shown that the two-dimensional method of color gamut optimization in a chromaticity diagram (shown in Section 2.3) is inferior to three-dimensional optimization techniques in a color space (shown in Sections 5.1 and 5.2) including the brightness information of the target color set.

Non color sequential displays render colors by the fine spatial structure of a subpixel mosaic of their primary colors. There are three-primary mosaics with red, green, and blue subpixels and also so-called multi-primary mosaics consisting of, for example, five colors: saturated red, green, blue, cyan, and magenta. These spatial mosaics are intended to render the spatial color distributions of an original image while the subpixel mosaic itself should not interact with the intended color content; in other words, it should not cause color artifacts.

To avoid such artifacts like the so-called *color fringe* artifact (CFA), a set of principles is described derived from the knowledge of human spatial color vision. These principles can be used to fulfill the requirements of good modulation transfer functions (MTFs), isotropy, good luminance resolution, and high aperture ratio of the optimized display. Examples of optimized multi-primary subpixel architectures are also shown together with the principles of a color image rendering algorithm.

*Illumination, Color and Imaging: Evaluation and Optimization of Visual Displays*, First Edition.
Peter Bodrogi and Tran Quoc Khanh.
© 2012 Wiley-VCH Verlag GmbH & Co. KGaA. Published 2012 by Wiley-VCH Verlag GmbH & Co. KGaA.

## 5.1
### Optimization Principles for Three- and Multi-Primary Color Displays to Obtain a Large Color Gamut

Modern imaging applications demand high color image quality quantified by the classic PSNR metric (see Section 4.4.3.2, Equation 4.9), by S-CIELAB [1], or in the image color appearance framework iCAM [2]. There is an agreement that the appearance of the displayed colors is one of the most important factors of color image quality. Therefore, color enhancement image processing transforms should be implementable on a wide range of realistic color images requiring that the display should be able to render a wide color gamut.

In the absence of a wide-gamut display, the color gamut of the image has to be mapped to the smaller color gamut of the display. There exist several color gamut mapping methods (see, for example, Refs. [3, 4]), but unfortunately, it is inevitable to distort color quality; hence, it is desirable to design wide color gamut displays reducing the need for gamut mapping. Note that if the image is specified, for example, in terms of sRGB (see Section 2.1.2), and the display has a more extended gamut than sRGB, then a reverse algorithm, that is, a color gamut extension method [5], may result in a more pleasing display.

The color gamut of the display depends on the chromaticity and luminance of its color primaries. By increasing the number of primary colors and/or saturating the color primaries (e.g., choosing a narrower spectral bandwidth of the color filters or light sources), color gamut increases. If the introduction of further primary colors (in addition to red, green, and blue) is not possible, then the existing three primaries can also be optimized to minimize the volume of out-of-gamut colors for a set of typical input images. The recent technological development of color displays allows for the usage of wide-gamut multi-primary displays getting in the focus of recent color imaging research.

However, the issue of selecting the number, chromaticity, and luminance of the color primaries needs a firm theoretical basis in order to ensure that a high color image quality is produced by the new display technology. In this section, the principles of such an optimization are described based on the knowledge of human color vision and color appearance. This optimization is carried out in a three-dimensional color appearance space in contrast to the traditional approach of primary color optimization (Section 2.3) using the CIE1931 $x$, $y$ chromaticity diagram (see Figure 5.1).

As can be seen from Figure 5.1, changing the chromaticities of the color primaries, the color gamut (i.e., the area of the triangle in the $x$, $y$ plane) also changes. In this two-dimensional approach, the colors are to be covered only in the chromaticity plane, regardless of their lightness information. By the introduction of more than three primaries, the triangle becomes a polygon spanned by the primaries. In the two-dimensional method, a given set of color primaries is considered to be able to render a given test color (a so-called *target color*, for example, a saturated natural color like a purple flower) if the $x$, $y$ chromaticity coordinates of the color match a certain mixture of the color primaries.

## 5.1 Optimization Principles for Three- and Multi-Primary Color Displays to Obtain a Large Color Gamut

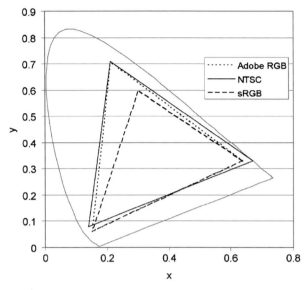

**Figure 5.1** Color primaries in the x, y chromaticity diagram [6]. Comparison of the gamut of the NTSC, Adobe RGB, and sRGB systems. Reproduced with permission from the *Journal of Electronic Imaging*.

But lightness information cannot be neglected as described, for example, by Wood and Sproson [7] focusing on matching the spectral sensitivity of a camera with the display primaries. This approach has become more or less obsolete today because of computational color management imperative in multi-primary color displays due to the complex nature of color calculations. The most widely used example of the two-dimensional color gamut optimization approach is the set of NTSC primaries (see Section 2.3) where the goal was to provide the largest CIE1931 $x$, $y$ chromaticity gamut achievable using the available CRT phosphors at that time.

In this section, it is shown that 2D methods not only are inferior to 3D optimization but can also be misleading. Certain patents and papers on this subject fail to give up the concept of 2D primaries and 2D chromaticity gamut concentrating only on the physical characteristics of the displays and ignoring the human observer's characteristics. For a more advanced gamut optimization, a set of objective metrics was formulated to evaluate the characteristics of color encodings for extended (wide) gamut displays, including gamut volume, color quantization, and complexity of transformation [8].

In this section, a new 3D approach is introduced based on a cost function to optimize a set of color primaries or to compare two sets of color primaries [6]. In this cost function, the result of a psychometric experiment is also included regarding the minimal visually acceptable luminance of the color primaries for the different values of hue. This experiment is described in this section together with a method of minimizing the cost function. This method results in optimal color primaries in the sense of the factors incorporated in the cost function. Computational results of the optimization method (optimum color primaries and optimum color gamuts in color appearance space) will be shown in Section 5.2.

#### 5.1.1
**Target Color Sets**

The most important feature of the display's primary colors is the ability to represent a wide color gamut. As already mentioned above, this is often called *target color gamut* as this gamut is the target of the optimization procedure of the primary colors. The target color gamut shall be transformed into a perceptual color space (e.g., CIE-CAM02) to establish a reasonable visual coverage. Possible target color sets are discussed below.

The first target color set is the set of so-called "legal colors," the widest possible range of colors, that is, a theoretical color gamut with all chromaticities at any luminance level inside the convex boundary of the spectral locus and the purple line inside the $x, y$ chromaticity diagram. Beyond the boundary of "legal colors," there are no physically realizable color stimuli. However, in practice, some perceived colors are more important than others. Therefore, it is desirable to select the target color gamut according to a more realistic criterion.

The second target color set consists of the so-called *optimal surface colors* under a given light source or illuminant. This is the color gamut realized by illuminating a surface of any theoretically possible reflectance curves by a given illuminant. The set of illuminants can, for example, be restricted to CIE D65 and CIE A, related to outdoor and indoor scenes, respectively. At the boundary of this set, the spectral reflectance functions of the colors of highest possible saturation correspond to idealistic step functions between 0 and 1. Of course, natural spectral reflectances (and thus natural chromaticities) are unlikely to be so extreme. The set of optimal surface colors is intended to indicate a theoretical possible limit of highest saturation for object colors.

Due to the limited nature of possible natural pigments or even interference colors, natural spectral reflectances are much more limited than the set of *any* theoretically possible spectral reflectance functions. Hence, for actual color displays, it is more appropriate to define a target color set containing most of the *common natural spectral reflectances* and to concentrate on their coverage by the display primaries, for example, in the CIECAM02 color appearance space. This target color gamut is the gamut of *real-world colors*.

The gamut of real-world colors can be obtained by using a representative set of spectral reflectance functions of natural objects (e.g., faces, leaves, flowers) and artificial objects (Munsell chips, textiles, paints, etc.) under a given light source or illuminant [8]. One of the existing spectral reflectance data sets [9] is based on 4089 samples, the Munsell Limit Color Cascade, the Munsell Matte Atlas, the Royal Horticultural Society Color Charts, colored papers, paint samples, plastics, inks, and textiles. Another data set [8] contains 1793 samples, nonfluorescent Munsell samples, and paints.

Recently, spectral reflectance functions of a set of common indoor objects were measured at the Laboratory of Lighting Technology of the Technische Universität Darmstadt [10], including flowers, leaves, books (covers), food, and textiles. Figures 5.2–5.4 show measured spectral reflectance curves for some selected flowers, books, and food, respectively.

## 5.1 Optimization Principles for Three- and Multi-Primary Color Displays to Obtain a Large Color Gamut | 241

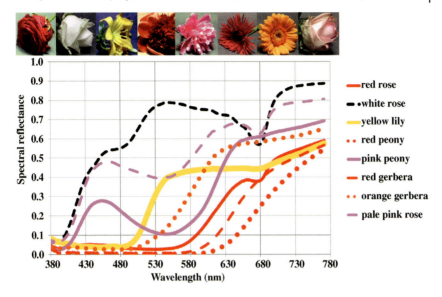

**Figure 5.2** Measured spectral reflectance functions of some selected flowers in an indoor environment [10]. Photos of the same flowers are shown as thumbnails in the same order from left to right as in the legend from top to bottom. *Source*: Technische Universität Darmstadt, Germany.

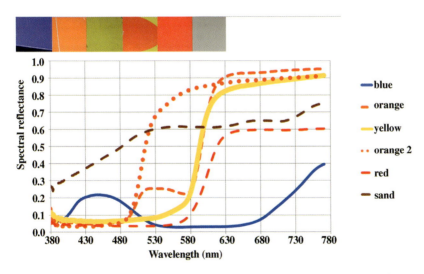

**Figure 5.3** Measured spectral reflectance functions of some selected books (covers) in an indoor environment [10]. Photos of the same books are shown as thumbnails in the same order from left to right as in the legend from top to bottom. *Source*: Technische Universität Darmstadt, Germany.

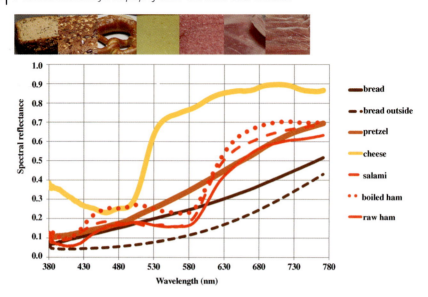

**Figure 5.4** Measured spectral reflectance functions of some selected food items in an indoor environment [10]. Photos of the same food items are shown as thumbnails in the same order from left to right as in the legend from top to bottom. *Source*: Technische Universität Darmstadt, Germany.

As can be seen from Figures 5.2–5.4, the spectral reflectance functions of the important reddish objects extend well beyond 630 nm indicating the importance of illuminating spectral content in the range of 620–660 nm. It is important to keep in mind that natural objects can be metameric to their artificial counterparts or printed reproductions leading to large color shifts if the illuminant is changed.

A comprehensive set of surface reflectance functions is the so-called *SOCS database* [11, 12] with more than 53000 samples including spectral reflectance functions of photographic materials, offset prints, computer color prints, paints, textiles, flowers and leaves, outdoor scenes, and human skin. An alternative target color gamut may consist of the possible color stimuli on a conventional *CRT display* (e.g., sRGB or NTSC).

Such a gamut should perhaps be integrated in a previously mentioned large target color gamut due to certain computer-based standard colors such as operating system colors or the widely used colors of graphs, flowcharts, and diagrams. The inclusion of conventional CRT colors is even more important if the application is a computer's user interface. It should be mentioned that there are also other color sets such as photographic system colors using three-dye sets for common photographic papers under CIE illuminants D50 or D65. Important long-term memory colors (see Section 3.4) can also be added to the target color set.

There are two different aspects of the concept of target color gamut: (1) (more important aspect) percentage of target color gamut covered by the display's gamut and (2) (less important aspect) percentage of the display's color gamut covering the

## 5.1 Optimization Principles for Three- and Multi-Primary Color Displays to Obtain a Large Color Gamut

target color gamut in order not to "waste" parts of the display's gamut outside the target color gamut.

Testing of target color gamut coverage can be accomplished by computing the volume of the intersection of their convex hulls [13] or by sampling the target gamut and testing the display gamut for inclusion of the samples. The latter approach not only is equivalent to the convex hull method but also allows the *weighting* of the different parts of the target gamut according to their importance. Weighting is especially useful for the case of skin colors and grass colors where the color tolerance of the average observer is small; hence, an unweighted sampling would lead to the exclusion of some important target colors. The precision limitations of sampling due to a lower sample density are counterbalanced by the advantage of following some possible *concave* faces of the target gamut.

To illustrate the practice of color primary optimization, a sample optimization procedure [6] is shown in Sections 5.3 and 5.4. At the beginning of this sample procedure, the union of the following target color sets was constructed: the SOCS database [11] under illuminant D65, Pointer's database [9], and the sRGB color gamut using its complete lattice of *rgb* values in a CRT colorimetric characterization model. This united target color set was sampled in the CIECAM02-SCD color space [14] developed specifically for small color difference assessment. The following CIE-CAM02 viewing condition parameters were used (see Section 2.1.9): adapting field luminance of $L_A = 50$ cd/m$^2$, relative background luminance of $Y_b = 20$, and average surround. The white point was set to D65.

A cube was defined in the CIECAM02-SCD $J'$, $a'$, $b'$ color space by the maximum absolute values of $J'$, $a'$, $b'$ for the united target color set. This cube was subdivided into small cubes. The small cubes containing one or more target colors of the united target color set were marked. They were represented by their centers in the sampled target color set. In every step of the optimization procedure, the percentage of target colors covered by the actual value of the color primaries (briefly *coverage*) is computed.

Input data of this *coverage computation* [6] are the actual ("candidate") primaries and the target color set in CIECAM02. First, target color set samples are applied an inverse CIECAM02 transformation by using the white point of the candidate primaries. These sample colors yield a lattice of XYZ tristimulus values. Second, every target color sample is checked if it is inside the actual color gamut; that is, inside the regular polyhedron, the N primaries are spanning in the XYZ tristimulus space. A measure of *coverage* ($p_C^{-1}$) can be defined, for example, by Equation 5.1 [6].

$$p_C = \frac{1}{10^{-7} + 100(c/n)} \tag{5.1}$$

In Equation 5.1, $c$ denotes the number of covered samples of the target color set, $n$ denotes the total number of samples in the target color set, and the constant $10^{-7}$ was included for the improbable case when no samples are covered at all. The reason of defining the measure of coverage by $p_C^{-1}$ and not by $p_C$ is that the optimization procedure *minimizes* (and not maximizes) a weighted sum of the different factors (including $p_C$) (see below).

## 5.1.2
## Factors of Optimization

### 5.1.2.1 Color Gamut Volume

Besides the important criterion of covering a target color set (or, in other words, a target color gamut) for the optimization of the color primaries of the display, there are several other factors that have to be discussed. Parallel to the coverage ratio of a target color set, the *total volume of the color gamut* established by the display's primary colors could, in principle, also be considered, computed in a perceptual color space. The volume of the color gamut in the CIECAM02 $J$, $a_C$, $b_C$ space can be defined, for example, by Equation 5.2:

$$V = \iiint G(J, a_C, b_C) dJ \, da_C \, db_C \tag{5.2}$$

In Equation 5.2, $G(J, a_C, b_C)$ denotes a so-called *characteristic function* that always equals unity within the gamut (as a subset of color space) and zero elsewhere, that is, off the display's color gamut. Nevertheless, the interpretation of an optimal gamut volume is not clear. While a large color gamut is able to represent more colors, it remains a question what percentage of these colors will be actually used on the display. Also, a further factor, *quantization efficiency* (see below), is limited by an unnecessarily large display gamut. According to the above considerations, it is advised [8] not to use this criterion when optimizing the display's color primaries.

### 5.1.2.2 Quantization Efficiency

The next optimization factor is quantization efficiency. Displays driven by digital signals unavoidably quantize the driving signals. Although the criterion of quantization is not related so closely to the optimization of color primaries, for a careful co-optimization, the usage of this criterion is valuable to set up a comprehensive optimization algorithm.

One possible metric of quantization efficiency is the so-called *rgb grid step size*, that is, the average color difference expressed in a modern uniform color space such as CIEDE2000 or CIECAM02-SCD [14]. The advantage of the latter color difference metric is that it is directly based on CIECAM02. This average color difference shall be calculated between two neighboring *rgb* values throughout the color gamut. This can be achieved by random sampling or by systematic sampling, for example, in CIECAM02 and by using the colorimetric characterization model of the display (see Chapter 2).

*Quantization efficiency* can also be computed by assigning a small weight to those samples that are located outside the target color set. This method ensures that, after optimization, the density of *rgb* values will be maximum within the intersection volume of the target color set and the device color gamut. Besides the average color difference, maximum and minimum values can also be used. The total number of *rgb* values outside the target color set may also be of interest but it is quite cumbersome to evaluate. A further useful metric related to quantization efficiency is the so-called *rgb grid uniformity*, that is, the variance of the above-defined color differences representing the uniformity of the *rgb* grid. A final descriptor of quantization efficiency can be defined as the required bit depth per color channel to achieve a given average color difference.

### 5.1.2.3 Number of Color Primaries

The next optimization factor is the number of color primaries to choose: one can use either three primaries or more than three primaries. In the latter case, the number of primaries of a so-called multi-primary display shall be decided. Using a wide-gamut three-primary color display has the advantage of a bijective mapping (i.e., an unambiguous correspondence) between the *XYZ* tristimulus values and the *rgb* values. Three-primary displays also allow backward compatibility with existing hardware and software components including colorimetric characterization models, image formats, and image compression methods.

A multi-primary display offers a large color gamut provided that the color primaries can be made saturated enough although the number of primaries in a multi-primary display is limited by the constraints of state-of-the-art display technology. A multi-primary display also allows for a more flexible correction for interobserver variations of color vision [12] but increasing the number of color primaries introduces some complications.

First, the spatial and/or temporal resolution of the display is compromised by an increased number of primaries. For color displays using a color subpixel mosaic, the number of subpixels increases; hence, the number of pixels capable of displaying any color decreases. In the absence of a bijective mapping between the digital color values (*rgb* and, for example, cyan, magenta, and yellow) and *XYZ* values, the rendering of display colors becomes ambiguous and *metameric* color stimuli appear. Metameric color stimuli have different digital color values and different spectral power distributions although having the same color appearance.

In the absence of an appropriate color image rendering algorithm, gradation and pseudo-contour artifacts may appear. But latter artifacts can be handled with carefully designed rendering algorithms [15]. Observer metamerism, that is, color matching functions deviating from those of the standard observer, is a further factor to take into account in multi-primary color image rendering algorithms [16]. Note that the issue of multi-primary color image rendering on a self-luminous color display is closely related to *multispectral imaging*, that is, the full spectral reconstruction of the objects.

### 5.1.2.4 White Point

The next factor of optimization is the display's *white point* (see also Section 6.5.3). Although it is possible to alter the white point of an existing display by reducing the intensity of one or more primaries, brightness and color gamut decrease inevitably. Therefore, the default white point of the display has to be designed to lie in a certain *white region* in the chromaticity diagram. Results from Section 3.5.3 state that a white point around $u' = 0.19$ and $v' = 0.46$ (corresponding to a correlated color temperature of about 7500 K) was *most preferred* by the observers. The color balance preference of a comprehensive set of pictorial images, however, depended on the image content significantly, ranging between 6000 and 7500 K (Section 3.5.3).

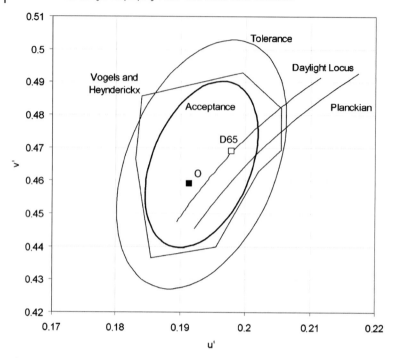

**Figure 5.5** White acceptance ellipse [6], white tolerance ellipse [6], and a white tolerance region [17]. Reproduced with permission from the *Journal of Electronic Imaging*.

To derive a measure of *display white point preference* ($p_W$) for the optimization procedure of the color primaries, two concentric *white tolerance ellipses* were introduced in the following way [6]. The first ellipse (so-called *white acceptance ellipse*, that is, the bold ellipse in Figure 5.5) fits tightly into the tolerance region of Vogels and Heynderickx (polygonal shape in Figure 5.5) [17] while the second ellipse (so-called *white tolerance ellipse*, that is, the thin ellipse in Figure 5.5) was drawn predominantly outside the polygon.

While the optimization procedure of the display's color primaries is being executed, the actual white point falling inside the *acceptance* ellipse is accepted without penalizing the actual set of color primaries ($p_W = 0$). If the actual white point falls into the white *tolerance* ellipse, then the value of $p_W$ becomes proportional to the distance from the "ideal" white point in the $u'$, $v'$ chromaticity diagram (denoted by O, $u' = 0.19$, $v' = 0.46$, see Figure 5.5). If the actual white point falls outside the tolerance ellipse, then the value of $p_W$ becomes – proportional to the distance measured from the D65 illuminant – very high.

### 5.1.2.5 Technological Constraints

The objective of optimization is not only a theoretically optimal color primary set but also the manufacturing of an actual device. Therefore, *technological constraints* shall also be considered, for example, the usage of not impossibly saturated primaries, that is, the distance of the primaries from the spectral locus and their possible maximum luminance values. Also, one may consider only those primary chromaticities that are attainable by filtering a given backlight or projector lamp used in the given display.

In a generic type of optimization intended to be suitable for any display, the following pseudo-technological constraint ($p_T$) was introduced on the color primaries [6]. They had to be further away from the spectral locus – toward the innermost part of the chromaticity diagram, that is, the gamut of "legal" colors – than a given constant value. Otherwise the term $p_T$ became nonzero, specifically, proportional to the distance between the color primary and the spectral locus in the $u'$, $v'$ chromaticity diagram.

Parallel to the above, the color primaries shall also be checked to have all "legal" tristimulus values (all nonnegative and with chromaticities inside the convex boundary of the spectral locus) during the whole procedure of optimization. A measure to avoid virtual primaries ($p_V$) was defined by a suitable constant plus the distance measured from the CIE D65 illuminant in the $x$, $y$ chromaticity diagram for any primary that happens to emerge out of the spectral locus during optimization. The value of $p_V$ equals zero for any primary color inside the spectral locus. If more than one primary is outside the spectral locus, then their distances can be added together [6].

### 5.1.2.6 P/W Ratio

A further complication may arise if the luminance of certain primaries is too low, that is, visually not acceptable. Traditionally, the ratio of the luminance of the primaries (denoted here by $P$) related to the luminance of the peak white (denoted here by $W$), that is, the so-called $P/W$ ratio, was determined so that the desired white point resulted from the mixture of the primaries. The optimization for acceptable $P/W$ ratios represents a further important factor to consider in the design of new wide-gamut color primary sets.

Among the very first problems occurring with the introduction of wide-gamut displays was the *too low brightness* of yellowish colors when the red and green primaries of higher excitation purity were introduced [18]. A good solution seemed to be to introduce a further (fourth) primary, that is, yellow. But the introduction of an additional primary or additional primaries increased the number of those elements of the display that the peak white luminance had to be partitioned to. Therefore, the $P/W$ ratio became less than that for traditional three-primary displays. This partitioning occurs among the subpixels of a pixel on a multi-primary subpixel mosaic, among the differently colored filters of a filter wheel, or among the time slots of the different colors in color sequential displays.

Visually acceptable luminance limits of $P/W$ ratios were examined in a psychophysical experiment [6]. In this experiment, a set of test color stimuli (identified as color primaries) of different hue and saturation (expressed in CIECAM02) were displayed. Six subjects of normal color vision took part in the experiment. Their task was to set that minimal luminance of the test color stimulus that was still acceptable as a display primary color. Test color stimuli of three (high) saturation levels were investigated at 12 different hues (CIECAM02 $H = 389.0$, 374.8, 361.8, 293.1, 267.8,

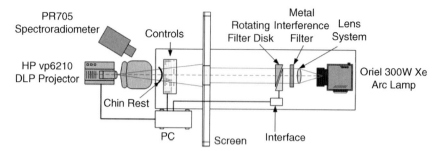

**Figure 5.6** Apparatus of the psychophysical experiment to obtain visually acceptable P/W ratios [6]. Reproduced with permission from the *Journal of Electronic Imaging*.

240.3, 182.9, 155.2, 115.0, 82.7, 56.6, and 24.0, respectively) equally sampling the hue circle.

Test color stimuli were presented on a colorimetrically characterized and calibrated special four-primary display consisting of a three-primary DLP projector illuminating the front of the screen and an exchangeable fourth primary of a quasi-monochromatic light source (a xenon arc lamp filtered by a metal interference filter) illuminating the back of the screen. The subject took place below the DLP projector. The subjects' results of visually acceptable minimum luminance as a display primary were always measured *in situ* by a spectroradiometer. The experimental apparatus is depicted in Figure 5.6.

The test color stimulus was an on-axis disk of 10 cm diameter in steady presentation subtending 4, 30, or 50% of the background while the area of the background was changed. The background was a rectangular white patch ($u' = 0.1822$, $v' = 0.4487$) at three different luminance values: 100, 250, and 400 cd/m².

Results showed that the minimum acceptable lightness or the P/W ratio was only little influenced by saturation (in the high saturation range making sense in case of primary colors) and by background luminance level. Average subjectively acceptable minima of color primary lightness and P/W ratio are shown in Figure 5.7 as functions of hue.

The minimum visually acceptable P/W ratio as a function of hue angle in Figure 5.7 was denoted by $M_{P/W}(h)$. As can be seen from Figure 5.7, yellow and yellow–green primaries ($h = 100-140$) need a high value of CIECAM02 lightness correlate *J* to be visually acceptable as a display color primary. Note that the quantity *J* represents achromatic lightness ignoring the Helmholtz–Kohlrausch effect, that is, the chromatic contribution to lightness (Section 6.5.2). The reason of the maximum of the $M_{P/W}(h)$ function at about 120° is the minimal chromatic contribution to lightness for yellow and yellow–green hues.

Confidence intervals show the interobserver variability. One confidence interval exceeds 100 because in this experiment – unlike real displays – white was not mixed from the color primaries. Visually acceptable minimum P/W ratios have to be kept in mind throughout the optimization procedure to co-optimize this factor with the other factors of optimum color primaries.

## 5.1 Optimization Principles for Three- and Multi-Primary Color Displays to Obtain a Large Color Gamut

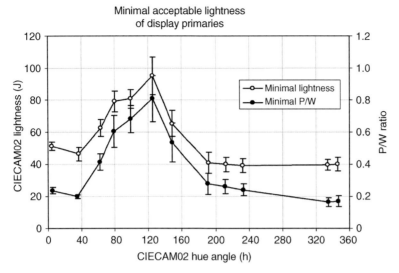

**Figure 5.7** Visually acceptable lightness (left ordinate) and P/W values (right ordinate) as a primary color of a self-luminous display as functions of CIECAM02 hue angle [6]. Reproduced with permission from the *Journal of Electronic Imaging*.

To include the $M_{P/W}(h)$ function of Figure 5.7 in the computational optimization procedure of the color primaries, a measure of the P/W criterion (denoted by $p_{P/W}$) was defined by summing up the differences between the P/W values of the new display's primaries and the interpolated average $M_{P/W}(h)$ curve (see Equations 5.3 and 5.4).

$$p_{P/W} = \sum_{i=1}^{N} \left( o\left( \frac{Y_{P_i}}{Y_W} - M_{P/W}(h_{P_i}) \right) \right)^2 \quad (5.3)$$

$$o(x) = \begin{cases} 0, & x \geq 0 \\ x, & \text{otherwise} \end{cases} \quad (5.4)$$

In Equations 5.3 and 5.4, N represents the number of color primaries, $Y_{P_i}$ is the relative luminance of the ith primary, $Y_W$ is relative white point luminance ($Y_W = 100$), $M_{P/W}(h)$ is the interpolated minimum acceptable P/W function, $h_{P_i}$ is the CIECAM02 hue angle of the ith color primary, and, finally, the function o guarantees that only those primaries are penalized during the optimization procedure that do not reach the value of $M_{P/W}(h)$.

### 5.1.2.7 Roundness

The concept underlying the next factor called *roundness* [19] is based on the hypothesis that, in order not to "overemphasize" any hue region, the ideal color gamut's projection onto the chromaticity plane should be of circular shape. This means that the maximum chroma shall be approximately equal for all hues. Roundness can be measured by the mean value of the eccentricities of maximum chroma for each hue on the CIECAM02 ($a_c$, $b_c$) plane. Ideally, the value of roundness should be zero. Unfortunately, minimizing the above-defined value of the roundness factor places an intolerable restriction on those hues that are easy to display in very high chroma to favor of only a few other hues for which the maximum chroma is low. Latter restriction has no psychophysical justification.

### 5.1.2.8 RGB Tone Scales and Display Black Point

Previously, the factor of *hue constancy by RGB tone scale modifications* was also considered. This was equivalent to defining color primary sets in a way that the hue constancy approximately holds for their tone scale modifications (gamma correction). But today, this is an obsolete criterion because this question is only relevant in the absence of a color characterization model, that is, in the absence of software or hardware color management. *Display black point* seems to be a trivial factor but it is easy to overlook its importance. Color gamut shape, target color set coverage, and many other factors *are* influenced by the magnitude of the radiation emitted by the display at its black point (where the digital values of all color channels are equal to zero).

## 5.2
### Large-Gamut Primary Colors and Their Gamut in Color Appearance Space

In this section, the results of the sample optimization procedure [6] mentioned in Section 5.1 are shown. These results are based on optimizing a *cost function* (*p*) describing every primary set arising when running the optimization procedure. In this cost function, the following factors of Section 5.1.2 were included: coverage of the united test color set of Section 5.1.1 ($p_C^{-1}$), P/W ratio ($p_{P/W}$), white point ($p_W$), the avoiding of virtual primaries ($p_V$), and the pseudo-technological constraint of the display ($p_T$). In the actual optimization procedure mentioned above [6], the cost function (*p*) was computed as the *weighted sum* of the above quantities. A more detailed description of this sample cost function and its components can be found in Ref. [6]. Note that, depending on display technology, various optimization strategies can be developed, including the co-optimization with the subpixel mosaic (see Section 5.4).

The cost function assigns a scalar to the primary set being optimized, while the primary set itself is specified in terms of the *XYZ* tristimulus values of the primaries. The values of the cost function imagined over the 3*N*-dimensional parameter space can be called *cost surface*. The dimensions of parameter space are determined by the number of color primaries (*N*) and the number of *XYZ* tristimulus values (three, that is, *X*, *Y*, and *Z*). The cost surface can be considered as a *scalar field* in 3*N*-dimensional space describing the *suitability* of a point in this *space of display color primary sets*. This cost surface can be reshaped by assigning different weights to the factors of the cost function according to the importance of the different factors.

Sections 5.2.1 and 5.2.2 show the maxima of the cost surface for different optimization conditions [6]. Section 5.2.1 describes several sets of optimum wide-gamut color primaries, while the shapes of the corresponding color gamuts in color appearance space [6] are presented in Section 5.2.2.

### 5.2.1
**Optimum Color Primaries**

In this section, the example of eight possible optimum color primary sets is shown [6]. The "dim" CIECAM02 viewing condition was applied in the output of the optimization because the optimized self-luminous display is used in this viewing condition. The weights of the target function were tuned for the algorithm's best performance in agreement with the *empirical* expectations toward a primary set. From the fact that the cost function components were weighted according to empirical considerations, it follows that the resulting optima are more or less interdependent and none of them can be considered as the ultimate primary set.

Two types of optimizations (Cases I and II) were carried out [6], both types for a three-, a four-, a five-, and a six-primary system, respectively. In Case I, the technological constraint ($p_T$) was included while it was turned off in Case II. Table 5.1 shows the parameters of the eight optimal color primary sets.

As can be seen from Table 5.1, optimum coverage was obtained with four primaries (4P), both with and without the pseudo-technological constraint (Cases I and II). It can also be seen that the white chromaticity differences between 3P and 4P, and 5P and 6P range between 0.004 (well perceptible [20]) and 0.027 (large perceived difference [20]).

The coverage of the optimum primary sets (see Table 5.1) is compared with the coverage of some standard display gamuts in Table 5.2. In Table 5.2, the same unified target color set [6] (as defined in Section 5.1.1) was used.

As can be seen from Table 5.2, all three color primary systems (sRGB, NTSC, and the Adobe RGB) cover significantly less colors of the target color set than the optimized primary systems. Coverage differences range between 4 and 22%.

**Table 5.1** Parameters of the eight optimal color primary sets.

| Opt. prim. set | Case I | | | | Case II | | | |
|---|---|---|---|---|---|---|---|---|
| | TCS cov. (%) | White point | | $\Delta u'v'$ | TCS cov. (%) | White point | | $\Delta u'v'$ |
| | | $u'$ | $v'$ | | | $u'$ | $v'$ | |
| 3P | 92.70 | 0.1982 | 0.4515 | 0.000 | 97.15 | 0.1868 | 0.4426 | 0.000 |
| 4P | 93.45 | 0.1947 | 0.4497 | 0.004 | 99.03 | 0.2014 | 0.4658 | 0.027 |
| 5P | 92.80 | 0.1933 | 0.4555 | 0.006 | 98.96 | 0.1990 | 0.4618 | 0.023 |
| 6P | 93.22 | 0.1877 | 0.4418 | 0.014 | 98.86 | 0.1904 | 0.4670 | 0.025 |

Case I: including the pseudo-technological constraint; Case II: excluding the pseudo-technological constraint. TCS: unified target color set [6] as defined in Section 5.1.1; 3P: three-primary system; 4P: four-primary system; 5P: five-primary system; 6P: six-primary system. $\Delta u'v'$: chromaticity difference from the 3P white point on the $u'$–$v'$ chromaticity diagram. Reproduced with permission from the *Journal of Electronic Imaging*.

**Table 5.2** Coverage of the same unified target color set [6] (TCS) as in Table 5.1 by standard display color gamuts.

|  | TCS coverage (%) |
| --- | --- |
| sRGB | 77.20 |
| NTSC | 89.13 |
| Adobe RGB | 86.44 |

Reproduced with permission from the *Journal of Electronic Imaging*.

### 5.2.2
### Optimum Color Gamuts in Color Appearance Space

Figures 5.8–5.11 depict the color gamuts of the eight optimum color primary sets for the two cases, that is, with (Case I) and without (Case II) the pseudo-technological constraint, as defined in Table 5.1. Figure 5.12 shows the same gamuts in $u'$–$v'$ chromaticity diagrams in "traditional" two-dimensional representations [6].

Comparing the values of target color set coverage with the loci of the primaries in the $u'$–$v'$ chromaticity diagram of Figure 5.12 reveals that the present three-dimensional gamut optimization based on the target function $p$ yields a significantly different result compared to "traditional" two-dimensional gamut optimization in the chromaticity diagram, as shown in Section 2.3.

First, significant differences in the color gamut *area* in the chromaticity plane (i.e., the traditional two-dimensional descriptor of color gamut) may correspond to only small differences of target color set coverage in color appearance space. Second, the polygon line connecting the color primaries in ascending order of hue angle, shown in Figure 5.12, is *concave* for the optimum color primary sets found in the three-dimensional approach of this section.

The latter finding would be a nonsense according to the "traditional" two-dimensional approach. In the three-dimensional color appearance space, however, a number of target color samples of high lightness can be covered by introducing an additional primary only in order to provide the required high lightness (see Figure 5.9, $h = 180°$). This additional primary can be of lower chroma in order not to modify the white point of the device with an additional high-chroma primary.

For "Case II", that is, the optimization *excluding* the pseudo-technological constraint, not all optimum color primary sets exploit the possibility of locating their chromaticity near the spectral locus (in practice, for example, by using lasers). This finding is due to the white point component of the target function. Figure 5.10 depicts the maximum lightness curves at the optimum gamut boundaries, as a function of hue angle. These curves are constructed from the maximum lightness values of the curves representing the gamut boundaries (i.e., for maximum chroma) in Figures 5.8 and 5.9. Figure 5.11 depicts the orthogonal projection of the sampled target color sets [6] onto the CIECAM02 $a_c$–$b_c$ plane together with the projection of the gamuts of the optimum color primary sets.

## 5.2 Large-Gamut Primary Colors and Their Gamut in Color Appearance Space

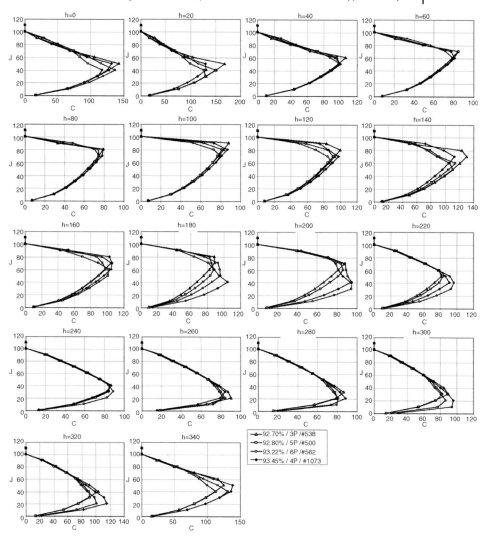

**Figure 5.8** Color gamut boundaries of the optimum color primary sets (3P, 4P, 5P, and 6P) of Table 5.1 in terms of CIECAM02 lightness ($J$), chroma ($C$), and hue ($h$) values. Case I: including the pseudo-technological constraint. The boundary is shown for different hue angles $h$ in the different CIECAM02 $J$, $C$ planes [6]. #: Number of optimization procedure; %: percentage of coverage from Table 5.1. Reproduced with permission from the *Journal of Electronic Imaging*.

The color *primaries* of the 3P color primary systems are generally able to satisfy the visually acceptable $P/W$ ratios, that is, the $p_{P/W}$ component of the target function. But the boundaries of the color *gamut* fail to reach the minimum lightness level corresponding to the visually acceptable $P/W$ ratios (indicated by the bold continuous line in Figure 5.10) between 240° and 310°, that is, in the blue–magenta

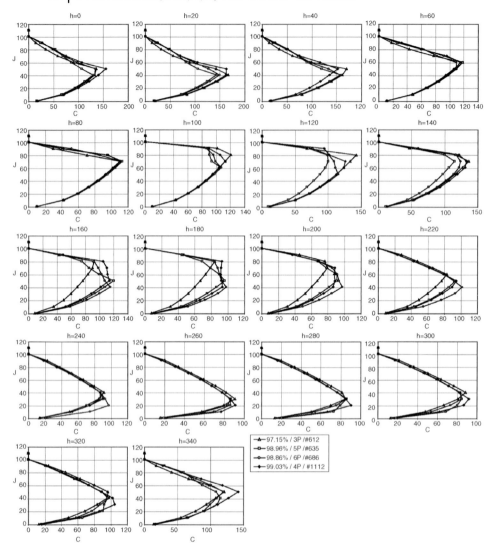

**Figure 5.9** Color gamut boundaries of the optimum color primary sets (3P, 4P, 5P, and 6P) of Table 5.1 in terms of CIECAM02 lightness (J), chroma (C), and hue (h) values. Case II: excluding the pseudo-technological constraint. The boundary is shown for different hue angles h in the different CIECAM02 J, C planes [6]. #: Number of optimization procedure; %: percentage of coverage from Table 5.1. Reproduced with permission from the *Journal of Electronic Imaging*.

region and slightly between 120° and 130°, that is, in the yellow–green region (see Figure 5.10).

In general, the primaries of both the 5P and 6P systems failed to reach the required minimum lightness level [6]. The color gamut exhibited lightness deficiencies in the

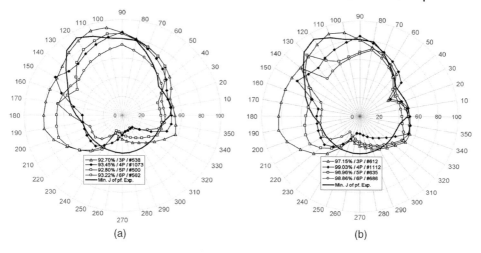

**Figure 5.10** Maximum lightness at the gamut boundary (i.e., for maximum chroma) of the eight optimal color primary sets of Table 5.1 as a function of hue angle, for Case I (a) and Case II (b) [6]. #: Number of optimization procedure; %: percentage of coverage from Table 5.1. Min. J of pf. Exp.: minimum CIECAM02 lightness (J) level corresponding to the visually acceptable P/W ratios. Reproduced with permission from the *Journal of Electronic Imaging*.

same regions as the 3P system plus in the red region for the 5P system and almost everywhere for the 6P system. The reason is that, in the case of N primaries, the entire luminance of the white point is distributed among the N primaries because the white point is the sum of the primaries.

Increasing the weight of the P/W ratio ($p_{P/W}$) in the cost function relative to the weight of target color set coverage ($p_C$) for the 5P system, the optimization

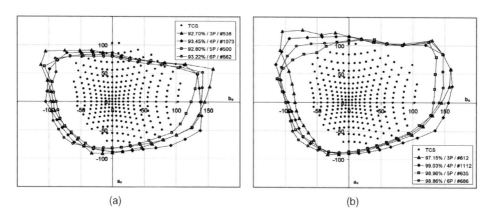

**Figure 5.11** Orthogonal color gamut projections of the eight optimal color primary sets to the CIECAM02 $a_c$–$b_c$ plane. (a) Case I and (b) Case II. TCS: target color set [6] as defined in Section 5.1.1. #: Number of optimization procedure; %: percentage of coverage from Table 5.1. Reproduced with permission from the *Journal of Electronic Imaging*.

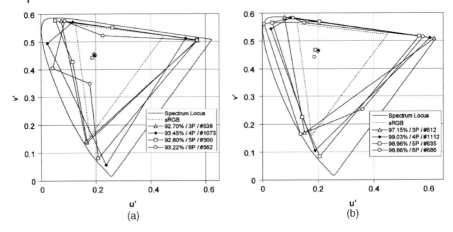

**Figure 5.12** Chromaticity diagram-based, that is, "traditional two-dimensional," representation of the color gamuts of the eight optimal color primary sets [6]. (a) Case I and (b) Case II. #: Number of optimization procedure; %: percentage of coverage from Table 5.1. Reproduced with permission from the *Journal of Electronic Imaging*.

procedure yielded coverage values between only 82 and 88% (compare with Table 5.1). This means that there is a trade-off between target color set coverage and satisfying $P/W$ ratios to be decided by considering the aim of the actual color display development.

Taking a closer look at the color gamut shapes reveals that the 5P and 6P gamuts are able to render low lightness–high chroma samples. 3P gamuts are better in representing high chroma–high lightness samples while 4P systems represent a trade-off between 3P and 5P–6P systems. This is especially apparent from Figure 5.8 ($h = 200°$ and $h = 140°$) or from Figure 5.9 ($h = 180°$ and $h = 120°$).

The above considerations also explain why the introduction of additional primaries *fails to increase coverage* compared to an *optimum 4P gamut*. Significant increase of gamut volume happens only in the green–blue ($h = 120$–$220°$) hue range in Case I. Among the four tested primary sets (3P, 4P, 5P, and 6P), the 4P system gives the highest target color set coverage in both cases (I and II).

The reason is that a four-primary system joins the increased coverage of a multi-primary system with the fact that the luminance of the white point is distributed among only four primaries. If an additional fifth or sixth primary is introduced, then the lightness of the white point still equals 100 but, on average, the lightness of the primaries will be necessarily lower. Independent of whether the luminance of the white point is distributed among subpixel areas of a color mosaic display or among color time slot lengths of a color sequential display, the average lightness of the primaries is inversely related to the number of color primaries. Introducing more and more primaries makes the gamut more and more flat at the edges in CIECAM02, thus being less and less able to render high lightness–high saturation colors.

## 5.3
### Optimization Principles of Subpixel Architectures for Multi-Primary Color Displays

Recently, due to the improvement of direct-view displays, especially flat panel displays, a plethora of new colored image forming mosaics became technologically feasible. The colored mosaic consists of small homogeneous differently colored *subpixels* of different shapes. A repeating spatial pattern of subpixels builds up the so-called *pixels*, that is, elementary combinations of subpixels that are able to display any color including the white point of the display.

A pixel may consist of red, green, and blue subpixels for a common three-color primary display. A pixel may also consist of more than three colored subpixels such as red, green, blue, cyan, magenta, and yellow to yield a multi-primary subpixel architecture. The aim of this section is to formulate a usable set of criteria, that is, a set of *visual optimization principles* for multi-primary subpixel architectures. The easiest way to arrange subpixels is the classic RGB stripe architecture (see Figure 5.13).

As can be seen from Figure 5.13, the RGB stripe architecture consists of a pattern of square-shaped pixels with three vertical red, green, and blue subpixels of the same size. Using the color image rendering method of *pixel rendering*, the unit of displaying spatial information is a pixel consisting of the three subpixels and the resolution of the display equals the number of pixels. A pixel is able to display any desired color inside the color gamut.

A better spatial resolution can be attained by using the color image rendering method of *subpixel rendering*. The extra spatial information of the image may come from a special storage method of graphical objects such as vector graphic shapes, for example, true type letters or curves or functions described by numbers. Using subpixel rendering, a spatially fine resolution of the silhouettes of these graphical objects is achievable.

As learned from Sections 2.3 and 5.2, the color gamut of the RGB stripe architecture (as a three-primary display) is less than the color gamut of, for example, a four-primary display (see, for example, Figure 5.11). Examples of extended color gamut *projectors* (without colored subpixel mosaics) were presented [18, 21, 22]. A

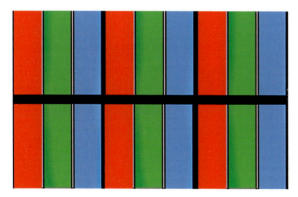

**Figure 5.13** RGB stripe subpixel architecture.

four-primary display using the stripe layout was also described [23]. However, a stripe layout cannot avoid the color fringe artifact (see the visual optimization principles below). New subpixel architectures incorporating an optimized arrangement of the "multi-primary" subpixels can yield an extended color gamut together with optimized spatial color appearance [24].

### 5.3.1
### The Color Fringe Artifact

In the method of subpixel color image rendering, the driving value of every single subpixel of a location in the subpixel architecture of the display equals the driving value as a primary element of a desired color in the corresponding location of the stored input image. By addressing the single subpixels, the horizontal resolution of the display is increased three times. But the display of achromatic information (e.g., black letters on a white background) leads to the so-called *color fringe artifact* [24]. The color fringe artifact arises due to the violation of some optimization principles to be discussed later.

The color fringe artifact becomes acute on the RGB stripe architecture if vertical lines, tilt lines, or some rotational symmetric figures are to be displayed. In the case of vertical white lines, the improper color of the subpixels at the border of the subpixel rendered image becomes visible. Equation 5.5 defines a measure ($M_{CFA}$) for the extent of the color fringe artifact [24].

$$M_{CFA} = \sum_{a \in A} \min(d_{fg}, d_{bg}) \qquad (5.5)$$

In Equation 5.5, $A$ is the subpixel rendered object, $a$ is a unit area in the image, $d_{fg}$ is the chromaticity distance (in the $u'$–$v'$ chromaticity diagram) between the possible chromaticities of the unit area $a$ and the foreground color, and $d_{bg}$ is a similar distance for the background color. The size of the unit area $a$ can be, for example, one pixel. The unit area is moving over the object area $A$ by scanning the whole object area $A$. Due to the minimum criterion in Equation 5.5, only the region of the contour of the object will be of importance.

Figure 5.14 illustrates the appearance of a bold Times New Roman letter "m" on an RGB stripe display with (a) and without (b) subpixel rendering.

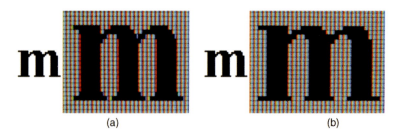

(a)                                               (b)

**Figure 5.14** Appearance of a bold Times New Roman letter "m" (small letters) on an RGB stripe display with subpixel rendering (a) and without subpixel rendering (b) [24]. The large letters are magnified photos taken on a real LCD. Reproduced with permission from the *Journal of Electronic Imaging*.

As can be seen from Figure 5.14, in the subpixel rendered letter, the three legs of the letter "m" seem to have red strokes on the left. Without subpixel rendering, the color fringe artifact is not visible. In Figure 5.14, real magnified photos taken on an LCD can also be seen.

### 5.3.2
### Optimization Principles

In this section, the term "logical pixel" is used in the sense of a "group of related subpixels" not definitely belonging to the same pixel. The summation of the single colors of the subpixels of a logical pixel shall be able to yield *achromatic* output (gray or white). The number of the contributing subpixels of a given primary color is generally one within a logical pixel but logical pixels of certain multi-primary subpixel architectures may have more than one subpixels of the same chromaticity.

To improve the RGB stripe architecture, a software algorithm was developed enhancing the horizontal resolution of the fonts and hence increasing the readability of small letters at the cost of some remaining extent of the color fringe artifact [25]. In this algorithm, the continuous input, for example, a mathematical description of the font, is mapped on the level of subpixels instead of whole pixels.

Low-pass prefiltering (so-called *anti-aliasing*) was incorporated [25] to minimize the color fringe artifact caused by rendering a spatial *luminance* distribution at the subpixel level. The method is called *displaced box filter RGB decimation* and the system became known as *ClearType*™. In another research direction, advanced three-primary subpixel architectures and image rendering algorithms were developed [26] (see Section 5.4.1).

Concerning the general visual optimization principles for the multi-primary subpixel architectures, the most fundamental general statement is that the digital image rendered on the display shall be as close as possible to the ideal continuous original image as it is seen by the display user's visual system. This can be checked by applying, for example, the image color appearance (iCAM) framework (including the spatial color vision properties of the human visual system) to both the original image and the rendered image and computing the color differences in every pixel. Keeping the above general concept in mind, the specific optimization principles described in the following subsections can be formulated [24].

#### 5.3.2.1 Minimum Color Fringe Artifact

One principle is the requirement of *minimum CFA*, that is, minimum visible chromatic error at the edge of spatial patterns. The color fringe artifact is most disturbing in the achromatic case, that is, at the border of black letters on a white background. As seen from Figure 5.14, RGB stripe architectures fail to avoid CFA, especially without spatial low-pass prefiltering. It is easy to imagine that, on a black background, a "non-rgb" sequence (e.g., g-b-r or b-r-g) will not be perceived as white, either. What the user perceives is the additive color mixture of two neighboring color pixels (cyan–red for g-b-r and blue–yellow for b-r-g).

ClearType™ technology was developed to prevent this artifact in text rendering but it enables only the horizontal smoothing of letters or similar shapes [25]. To facilitate the perfect blending of single-colored light from subpixels, the so-called "checkerboard" arrangement of the subpixels seems to yield a good result. This means that the subpixels are positioned in an *alternating* way so that *no neighboring* of the same primary colors can arise. This kind of subpixel arrangement reduces the color fringe artifact since a better color blending is possible. The checkerboard arrangement is different from the RGB stripe architecture where identical subpixels are positioned in the neighborhood of each other.

#### 5.3.2.2 Modulation Transfer Function

A high value of the modulation transfer function means fine spatial resolution. Originally, the MTF concept was used to describe the *contrast transmission* capabilities of optical devices, for example, lens systems in photography. MTF can be easily adopted for self-luminous displays since it is a curve representing the ratio of the output (rendered image) and input (original image) contrast as a function of line frequency: $MTF(\nu) = M_o/M_i$, where $M_o$ and $M_i$ indicate the modulation (or contrast) of the output and input images containing a square wave grating, that is, alternating black and white lines of line frequency $\nu$.

Modulation (M) is defined as $M = (L_{max} - L_{min})/(L_{max} + L_{min})$, where $L_{max}$ is the maximum and $L_{min}$ is the minimum luminance of the grating. A simplified MTF definition for displays is the *maximum* number of representable alternating vertical black and white *line pairs* over a given area. It is essential for a display to be able to present the *largest possible* number of line pairs over a given area. In the conventional RGB stripe subpixel architecture, it is impossible to generate very thin lines (e.g., 1 or 2 subpixels width) of any color (inside the gamut) including white or gray. The minimum width of such lines is the width of the whole pixel of the architecture.

#### 5.3.2.3 Isotropy

Besides the possibility of displaying thin white or gray lines, a more severe criterion is the possibility of displaying separate *dots* of any color. If MTF is high for both the vertical and the horizontal directions, then this requirement can be fulfilled. This consideration leads us to the next important principle: *isotropy*. Isotropy is the directional independence of the subpixel architecture. For the conventional RGB stripe architecture, simple horizontal *line rendering* can be accomplished easily but subpixel rendering will not work in vertical directions. Tilt lines and calligraphic letters cannot be displayed optimally on a conventional RGB stripe architecture.

A synopsis of the first three principles (Sections 5.3.2.1–5.3.2.3) leads to a further principle, that is, it should be possible to draw thin and long rectangular stripes not only vertically and horizontally but also in any direction. Also, within these thin stripes, the summation of the color of the crossed subpixels should be able to yield any color within the gamut, including gray and white.

The contrast sensitivity of the human visual system is poorer for oblique than for vertical and horizontal orientations for middle and high spatial frequencies. This is the so-called *oblique effect* [27]. Anyway, it is worth ensuring the complete isotropy, that is, directional independence of the subpixel architecture, because the elimination of the color fringe artifact is very important in any direction.

The color fringe error emerging from the incomplete summation of the subpixel colors is even more visible if some adjacent subpixels of the same color merge and create single-color blocks such as at the legs of the letter "m" in Figure 5.14. The color fringe artifact is thought to be more robust than the oblique effect that is only of minor importance. Also, it is not a disadvantage if the display outperforms the human visual system in terms of oblique spatial resolution.

#### 5.3.2.4 Luminance Resolution

In the human visual system, the relatively low spatial resolution of short-wavelength ("S") sensitive receptors (so-called S-cones) in the retina would allow the placing of blue (or at least bluish) subpixels in a sparser manner. In order to reach the proper color balance (i.e., an appropriate display white point), the area of these blue subpixels should be larger than the area of the other subpixels.

But, unfortunately, the blue subpixels cannot exhibit high enough luminance. Therefore, their large size would visually divide the architecture and block the visually uniform blending of the light output from all subpixels. Hence, these large-sized blue subpixels would evoke visible textures in those areas where the subpixel architecture is intended to look homogeneous.

To overcome this, instead of using a smaller number of large blue subpixels, the use of a higher number of normal-sized blue subpixels at the same display driving address seems to be a better solution. This may allow for the above "less spatial resolution in blue" criterion. It is well known that in the human visual system, it is the *luminance* channel (the sum of middle-wavelength sensitive and long-wavelength sensitive photoreceptors, that is, the sum of the M- and L-cones) that carries the *finest spatial details* of the image.

According to the spectral sensitivity of the luminance (L + M) channel of the human visual system, the luminance of the green (or greenish) subpixels is highest among all subpixel colors. Therefore, referring to the above considerations, *green spatial resolution* shall be greater than the resolution of the other display color primaries. In the three-primary Bayer pattern [28], subpixel arrangement used recently in CCD and CMOS filtering mosaic arrays of digital cameras, green spatial resolution is twofold. Nevertheless, it should be kept in mind that, for self-luminous displays, white balance (the white point criterion) should also be kept in mind (see Figure 5.15).

As can be seen from Figure 5.15, it is not possible to achieve white balance for a self-luminous display by the aid of the Bayer pattern that is designed for image sensor arrays.

#### 5.3.2.5 High Aperture Ratio

The visually useless area of the display (i.e., the area used for wiring or other electronics components where no light is emitted) should be *minimized* to achieve a good luminous efficiency. To achieve this, the surface of the display should be covered completely with simple (regular) two-dimensional figures that are easy to manufac-

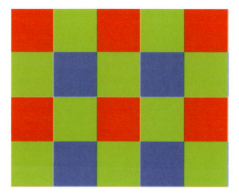

**Figure 5.15** Illustration of a Bayer pattern [28], that is, a three-primary subpixel arrangement used in CCD filtering mosaic arrays of digital cameras, not for self-luminous displays. Green spatial resolution is twofold; hence, self-luminous displays may not achieve a white balance.

ture, for example, as parts of a filter mosaic. It is well known from geometrical considerations that only a limited set of such regular shapes can be used: triangles, rectangles, and hexagons.

A possible spatial arrangement for subpixel color image rendering would be a stochastic pattern in which the human visual system cannot discover any systematic order, similar to the solutions of inkjet printing. But such a pattern would probably imply undesirable manufacturing costs, not mentioning that this arrangement would work only at relatively high spatial resolutions contradicting the basic idea of subpixel color image rendering.

## 5.4
### Three- and Multi-Primary Subpixel Architectures and Color Image Rendering Methods

In this section, examples of optimized three-primary and multi-primary subpixel architectures are described designed according to the design principles of Section 5.3. They yield color images of better visual quality than previous three-primary color architectures including the RGB stripe architecture.

### 5.4.1
**Three-Primary Architectures**

Optimization principles point toward a set of different advanced RGB subpixel architectures reducing the number of blue subpixels and their switches and drivers in a self-luminous color display and enhancing visual display quality. This is the group of PenTile Matrix™ architectures [29]. An example is shown in Figure 5.16.

As can be seen from Figure 5.16, the PenTile Matrix™ subpixel architecture fulfills the optimization principles in the following way. It increases MTF in both

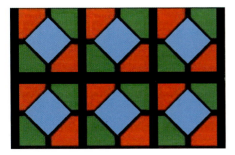

**Figure 5.16** Schematic drawing to illustrate an example from the family of PenTile Matrix™ subpixel architectures [29].

horizontal and vertical directions. Every pixel is divided into five subpixels: two red, two green, and one blue. The data driver for the blue subpixel is shared with a neighboring pixel's blue subpixel reducing the number of data drivers on the display [30]. A prototype AMLCD panel demonstrated improved color image quality [26].

In a computation, it was shown that the PenTile™ panel with half the number of subpixels could achieve near-equivalent MTF to an RGB stripe display of the same overall spatial resolution [31]. The PenTile™ method also exhibits good isotropy (rotational symmetry) due to the symmetrical arrangement of the red and green subpixel checkerboard layout [32]. Hexagonal RGB architectures [33] further increase isotropy. Figure 5.17 shows an example.

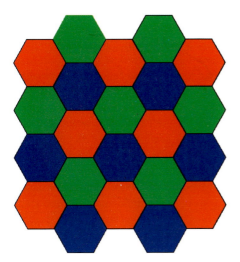

**Figure 5.17** Schematic drawing to illustrate an example of a hexagonal RGB subpixel architecture [33].

## 5.4.2
### Multi-Primary Architectures

According to the optimization principles of Section 5.3, two multi-primary subpixel architectures will be shown as examples [34]. The aim of these architectures is not only to enlarge the color gamut of the display but also to comply with the optimization principles of their spatial layout. This means a trade-off between the color gamut principles (Section 5.1) and the principles derived from spatial color vision (Section 5.3). The first example consists of hexagonal logical pixels while each logical pixel is divided into six equal triangles as subpixels, with six primary colors (see Figure 5.18).

In the first sample architecture shown in Figure 5.18, three of the primary colors may be the common red, green, and blue primaries ($P_1$, $P_3$, $P_5$), while the additional ones can be chosen as other chromaticities ($P_2$, $P_4$, $P_6$), for example, yellow, cyan, and magenta. The arrangement on the right-hand side of Figure 5.18 is the rotated version of the arrangement on the left-hand side. It was rotated by 30° counterclockwise.

In the second sample architecture shown in Figure 5.19, a pixel consists of seven subpixels of hexagon shape with seven color primaries.

As can be seen from Figure 5.19, the shape of the whole pixel looks like a flower in which the chromaticities and the order of the six primaries ($P_1$, $P_2$, $P_3$, $P_4$, $P_5$, $P_6$) may be identical to the previous architecture, that is, red, green, blue, yellow, cyan, and magenta. In addition, a white subpixel can be placed in the middle ($P_7$) to enhance the overall brightness of the display. In this case, the chromaticity of white should be carefully chosen to match the chromaticity of white generated by the additive mixture of the other six subpixels. Including an additional white color primary is often used in recent DLP (digital light projection) displays to increase

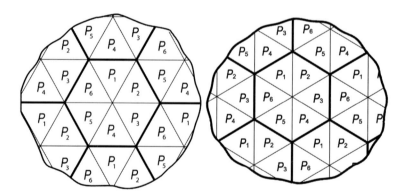

**Figure 5.18** Six-primary architecture for multi-primary subpixel image rendering [24]. Left: original arrangement; right: original arrangement rotated by 30° counterclockwise. If both $P_1$ and $P_4$ are red, both $P_2$ and $P_5$ are green, $P_3$ blue, and $P_6$ cyan, then a 4P (four-primary) color gamut arises. Reproduced with permission from the *Journal of Electronic Imaging*.

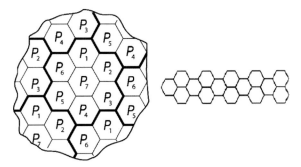

**Figure 5.19** Seven-primary architecture for multi-primary subpixel image rendering [24]. Left: arrangement of the primary colors; right: a sequence of subpixels taken from the architecture. This is the thinnest possible horizontal line containing all primary colors. Reproduced with permission from the *Journal of Electronic Imaging*.

output luminance. The disadvantage of a white pixel in the middle is that – in the absence of an intelligent color image rendering software – the $P/W$ ratios of the primaries $P_1$–$P_6$ are reduced.

In Figures 5.18 and 5.19, the surface of the display is covered so that no neighboring subpixels of the same color emerge – according to the design principle of minimizing the color fringe artifact described in Section 5.3.2.1. Accordingly, these architectures are predicted to exhibit a reduced extent of the color fringe artifact due to the possibility of a more uniform placement of the primary colors on the hexagonal grid (a sample CFA calculation is presented in Table 5.4).

These hexagonal architectures have an increased rotational symmetry – this is why it is so easy to avoid adjacent subpixels of the same chromaticity on the subpixel mosaic. It is also possible to display thin gray or white lines in many directions according to the design principle described in Section 5.3.2.3, that is, the principle of isotropy. The value of their MTF is also higher than that of the RGB stripe architecture for all directions in the plane of the display.

Unlike for the case of the RGB stripe, addressing neighboring subpixels in case of the hexagon structures along an imaginary line will not result in color blocks of the same primaries. Consequently, displaying, for example, vertical black and white lines, subpixels can be used instead of whole pixels, such that half of the subpixels are lit in a pixel, for example, $P_4P_3P_2$ from one pixel and $P_5P_6P_1$ from the upper right neighboring pixel in Figure 5.18.

Table 5.3 shows a comparison of the vertical and horizontal MTFs and the addressability of the hexagon structures of Figures 5.18 and 5.19 together with the classic RGB stripe for the example of two letters "n" of different font types rendered by them as shown in Figure 5.20.

As can be seen from the renderings of Figure 5.20, the different polygons, that is, triangles, rectangles, and hexagons, exhibit dissimilar geometric covering properties; hence, they are not easy to compare among themselves. The first row of Table 5.3 contains the number of subpixels in the same rectangular area for all layouts shown in Figure 5.20. For a reasonable comparison, approximately the same number of

**Table 5.3** Comparison of the hexagonal architectures of Figures 5.18 and 5.19 with the RGB stripe layout in terms of MTFs and addressability [24].

|  | RGB stripe | Hexagon (six-prim.) | Hexagon (seven-prim.) | Hexagon (six-prim. rotated) |
|---|---|---|---|---|
| Figure number | 5.13 | 5.18 (left) | 5.19 | 5.18 (right) |
| Number of subpixels | 324 | 300 | 304 | 312 |
| MTF vertical | 6 | 7.5 | 6 | 5 |
| MTF horizontal | 4.5 | 3.5 | 8 | 6 |
| Vertical addressability | 12 | 15 | 38 | 25 |
| Horizontal addressability | 9 | 18 | 16 | 12 |

Reproduced with permission from the *Journal of Electronic Imaging*.

subpixels (between 300 and 324) were included for each of the four subpixel layouts shown in Figure 5.20.

The second row of Table 5.3 shows the vertical MTF of the four subpixel layouts. According to the simplified definition of MTF in Section 5.3.2.2, this is the maximum number of vertical black and white line pairs within the rectangular area shown in Figure 5.20. The third row of Table 5.3 shows the horizontal MTF of the four subpixel layouts, similar to the vertical one. The fourth row of Table 5.3 shows vertical addressability. This is the maximum number of displayable (thin) horizontal lines of any color (including black and white) in different positions in the rectangular area

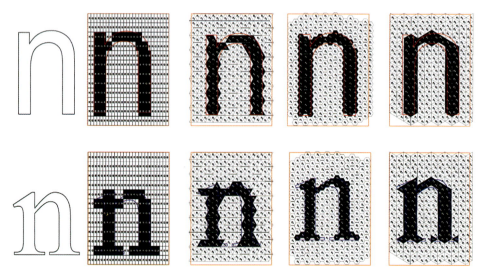

**Figure 5.20** Subpixel rendered black Arial and Times New Roman "n" letters on white background [24]. Arial letters are in the first row and Times New Roman letters are in the second row. Architectures: first column – original continuous image; second column – RGB stripe (Figure 5.13); third column – six-primary (Figure 5.18, left); fourth column – seven-primary (Figure 5.19); fifth column – six-primary rotated (Figure 5.18, right). Reproduced with permission from the *Journal of Electronic Imaging*.

shown in Figure 5.20. The fifth row of Table 5.3 shows horizontal addressability. This is the maximum number of displayable (thin) vertical lines, similar to the previous one.

As can be seen from Table 5.3, the addressability of the RGB stripe is exactly double its MTF for a given direction (vertical or horizontal). But the addressability of the hexagon architectures is usually higher than twice the MTF of the same direction. This feature of the hexagon patterns originates in the nature of the texture of triangles and hexagons. Unlike the grid of squares, triangles and hexagons cover the plane so that it is possible to represent thin lines not only at the center of the logical pixels (surrounded by thick black lines in Figures 5.18 and 5.19) but also between two such pixels – for example, by using the rightmost subpixels of the left pixel and the leftmost subpixels of the right pixel.

In the arrangements depicted in Figure 5.18, shifting the hexagonal pixel grid (designated by the thick black lines) by one subpixel, other hexagonal pixels can be formed, each of them comprising all of the six color primaries. As pointed out above, this does not increase the MTF of the structure but it enables a more accurate spatial positioning of the thin subpixel rendered lines.

It can also be seen from Table 5.3 that, for the case of the seven-primary hexagon architecture, not only the whole "flower shape" can be used to represent rows but also the altering sequence of one and two subpixels involving all seven primaries (see the right-hand side of Figure 5.19). This in turn increases vertical addressability.

In Figure 5.20, black subpixel rendered Arial and Times New Roman type lowercase letters "n" are shown on white background. According to the basic idea of subpixel rendering, these letters "n" are composed of those subpixels only of which the major part is inside the outline of the letter "n" – if the geometric centers of the subpixels are considered – regardless of their primary colors. To quantify the color fringe artifact for the four different layouts shown in Figure 5.20, a calculation method can be derived for the special case of black letters on white background using the general formula of Equation 5.5.

Since the stroke width of the letters "n" is similar to the width of the logical pixels in all four layouts in Figure 5.20, the color fringe artifact can be quantified by considering the number of subpixels of each of the different color primaries inside the letters "n". For minimum CFA, the number of subpixels of the different color primaries should be *equal* inside the letters "n" to be able to produce "white" in the neighborhood of the letters "n". If these numbers are different from each other, then there will be a chromatic difference between "white" and the actual chromaticity of the background of the letter, corresponding to the concept of Equation 5.5.

The percentage values of these numbers of subpixels of different color primaries inside the letters "n" (related to the total number of subpixels inside the letters "n") are listed in Table 5.4 together with the mean percentage value, its standard deviation (STD), and the maximum deviation from the mean.

Assuming an ideal case (thus ignoring subpixel rendering), the percentage values in any column (1–7) should be equal and the value of the STD should be zero. Thus, these STD values can be considered as a measure of CFA for the special case of black letters on white background as shown in Figure 5.20. Considering the rotational

**Table 5.4** A sample calculation of the color fringe artifact for black letters on a white background based on a method derived from Equation 5.5.

| Arch. | Font | Total number of subpixels | Percentage of the number of subpixels of the color primaries inside the letters "n" | | | | | | | Mean (CFA) | STD | Max. dev. |
|---|---|---|---|---|---|---|---|---|---|---|---|---|
| | | | 1 r | 2 g | 3 b | 4 c | 5 m | 6 y | 7 w | | | |
| RGB stripe | Arial | 80 | 26.3 | 27.5 | **46.3** | — | — | — | — | 33.3 | 11.2 | 12.9 |
| | Times | 77 | 28.6 | 27.3 | **44.2** | — | — | — | — | 33.3 | 9.4 | 10.8 |
| Six-prim. | Arial | 73 | 19.2 | 19.2 | 15.1 | 15.1 | 15.1 | 16.4 | — | 16.7 | 2.0 | 2.5 |
| | Times | 63 | 17.5 | 15.9 | 17.5 | 15.9 | 15.9 | 17.5 | — | 16.7 | 0.9 | 0.8 |
| Seven-prim. | Arial | 88 | 14.8 | 13.6 | 14.8 | 13.6 | 14.8 | 13.6 | 14.8 | 14.3 | 0.6 | 0.6 |
| | Times | 53 | 15.1 | 11.3 | 15.1 | 15.1 | 15.1 | 17.0 | 11.3 | 14.3 | 2.1 | 3.0 |
| Six-prim. rotated | Arial | 83 | 15.7 | 16.9 | 15.7 | 18.1 | 18.1 | 15.7 | — | 16.7 | 1.2 | 1.4 |
| | Times | 67 | 14.9 | 13.4 | 16.4 | 19.4 | 19.4 | 16.4 | — | 16.7 | 2.4 | 3.2 |

Number of subpixels of each of the different color primaries inside the letters "n" (shown in Figure 5.20) related to the total number of subpixels inside the letters "n" as percentage values. STD values can be considered as a measure of CFA. Numbers 1–7 in the second row represent the color primaries and letters "r, g, b, c, m, y, and w" along with these numbers are examples for the primaries, red, green, blue, cyan, magenta, yellow, and white [24]. Reproduced with permission from the *Journal of Electronic Imaging*.

symmetry (isotropy) of the hexagon architectures, it is not astonishing that they perform better in terms of this measure than the RGB stripe.

In the RGB stripe architecture, the number of blue subpixels is almost twofold inside the letters "n" compared to the red and green ones. This is indicated by bold numbers in Table 5.4. This causes a visible color fringe artifact on an RGB stripe. For all other architectures in Table 5.4, the value of STD is less implying a more uniform participation of the subpixels of different color primaries inside the letters "n".

As can be seen from Table 5.4, the seven-primary architecture (see Figure 5.19) has overall the smallest STD value (0.6) occurring for the case of rendering the Arial type letter. For the Times letter, however, the six-primary architecture of the original arrangement performed best (STD = 0.9).

In the examples shown in Figure 5.20 and Table 5.4, subpixel rendering was performed without low-pass filtering. But in practice, low-pass filtering is always applied to the RGB stripe architecture [25] to reduce the color fringe artifact. By applying low-pass filtering to the hexagon architectures in more than one direction, it is expected that the visual performance of the hexagon architectures would be further improved, especially for text rendering.

### 5.4.3
**Color Image Rendering Methods**

In this section, an image rendering method is shown for multi-primary subpixel architectures. The algorithm computes the driving values of the color subpixels to render an original image on a multi-primary subpixel architecture. This method can also be used without subpixel rendering. The method is based on an error function enabling a proper chromaticity reproduction together with an enhanced luminance resolution [24, 35]. Further methods can be found in Ref. [36].

The color image rendering method [24, 35] of this section is intended for the multi-primary subpixel architectures of Section 5.4.2 (Figures 5.18 and 5.19) with six or seven primaries but the method can also be adapted to any multi-primary system featuring $n$ primaries ($n > 3$). The method is described in terms of device-independent XYZ tristimulus values.

The input for a given area of the subpixel architecture is the luminance distribution of the original image at the level of subpixels and the whole color distribution of the original image at the level of the whole (logical) pixels. This means that it is the luminance information that is given at a higher spatial resolution corresponding to the high luminance resolution of the human visual system according to Section 5.3.2.4. The algorithm itself is described below.

Let $C_i$ denote the relative driving value or *weight* of subpixel $P_i$ in a pixel. $C_i$ is in the interval of [0, 1] and $i$ is indexing the primaries, $i = 1, \ldots, n$. The problem of multi-primary color image rendering is that the decomposition of a three-dimensional input vector XYZ is not straightforward since the $(3 \times n)$ matrix containing the XYZ tristimulus values of the *color primaries* is – unlike matrix **P** in Section 2.1.4 – not a square matrix ($n > 3$); hence, it is not invertible.

For $n = 6$, to render the color of any arbitrary $XYZ_{orig}$ original input tristimulus value, the following algorithm can be applied. Let us divide the six color primaries of the pixel into two groups, for example, $\{P_1, P_2, P_3\}$ and $\{P_4, P_5, P_6\}$. Change the driving values of the first group ($C_i$, $i = 1, 2, 3$) and calculate the corresponding XYZ output in every optimization step by summing up the output of the *first* group. A so-called *remainder* $XYZ_{rem}$ is also calculated as the difference vector between the input color ($XYZ_{orig}$) and the current value of XYZ.

This remainder can be rendered unambiguously in the three-dimensional space of relative driving values of the second group of primaries $\{P_4, P_5, P_6\}$ by using a linear matrix transform. The solution is valid only if $C_i$ is in the interval [0, 1] for $i = 4, 5,$ and 6. Therefore, the so-called *out-of-color-gamut* error term $E_{col}$ is calculated by Equation 5.6.

$$E_{col} = \Sigma \Delta_i \tag{5.6}$$

In Equation 5.6, the symbol $\Delta_i$ is defined by Equation 5.7 for $i = 4, 5,$ and 6.

$$\begin{aligned} \Delta_i &= 0 & \text{if } C_i \in [0, 1] \\ \Delta_i &= -C_i & \text{if } C_i < 0 \\ \Delta_i &= C_i - 1 & \text{if } C_i > 1 \end{aligned} \tag{5.7}$$

In the optimization, each variable of the first set $\{P_1, P_2, P_3\}$ is changed in the interval [0, 1] to minimize $E_{col}$ until $E_{col}$ becomes less than a color tolerance value or equals zero. Any general $n$-dimensional optimization technique can be used. A more detailed description of the algorithm can be found in Ref. [24].

There may be identical color primaries among the values in the first group but the second set of primaries $\{P_4, P_5, P_6\}$ has to span a real three-dimensional space in the XYZ space of tristimulus values. The above method yields the weight factors of each color primary to represent a desired color output within a single pixel without any subpixel level rendering.

To include subpixel rendering, further considerations are required. In this case, luminance information is represented at a higher spatial resolution than color information, namely, at the spatially highly resolved subpixel level. A method can be derived at the cost of some chromaticity error. The method is based on the finding that the decomposition of $XYZ_{orig}$ is mathematically not unique for more than three primary colors. There are an infinite number of linear combinations of the weights of the six primary colors and it is possible to choose one single combination of weights with such a luminance distribution among the subpixels that yields the best approximation of the luminance distribution of the original image at the subpixel resolution.

Subpixel rendering is incorporated by adding a second error function term to the algorithm. This is the so-called *subpixel luminance error term* ($E_{lum}$) supplementing the out-of-color-gamut error term $E_{col}$. The subpixel luminance error term ($E_{lum}$) is the sum of the absolute differences of the luminance ratios of the subpixels within a logical pixel (denoted by $Y_i/Y_1$) and the same luminance ratios in the original image (denoted by $Y_{i0}/Y_{10}$). Equation 5.8 contains the computational formula of $E_{lum}$ for the example of a six-primary system.

$$E_{lum} = \sum_{i=2}^{6} (|Y_i/Y_1 - Y_{i0}/Y_{10}|) \tag{5.8}$$

The total error function to be minimized by the optimization procedure is written in Equation 5.9.

$$E_{tot} = \alpha E_{col} + \beta E_{lum} \tag{5.9}$$

In Equation 5.9, $\alpha$ and $\beta$ are suitable weight parameters. If $\alpha = 0$, then no color information is taken into account, only luminance information. During the optimization procedure, each weight of the set $\{P_1, P_2, P_3\}$ is changed in the interval $[0, 1]$ to minimize $E_{tot}$ until $E_{tot}$ becomes less than a suitable tolerance value. Any general three-dimensional optimization technique can be used again. For the case of the seven-primary architecture, a similar method can be used. The only modification is that four primaries have to be selected to optimize in the first group. The second group must contain three elements again that should be linearly independent in a three-dimensional space.

### Acknowledgment

Authors would like to acknowledge Ms. Nathalie Krause (Laboratory of Lighting Technology of the Technische Universität Darmstadt) for measuring the spectral reflectance functions of the indoor objects and making their photos in Figures 5.2–5.4.

# References

1. Zhang, X. and Wandell, B.A. (1996) A spatial extension of CIELAB for digital color reproduction. SID'96 Digest, pp. 731–735.
2. Fairchild, M.D., Johnson, G.M., Kuang, J., and Yamagutchi, H. (2004) Image appearance modeling and high-dynamic-range image rendering. Proceedings of ACM-SIGGRAPH, First Symposium on Applied Perception in Graphics and Visualization.
3. Green, Ph. and MacDonald, L. (2002) *Color Engineering*, John Wiley & Sons, Ltd., pp. 297–314.
4. CIE 156-2004 (2004) *Guidelines for the Evaluation of Gamut Mapping Algorithms*, Commission Internationale de l'Éclairage.
5. Horiuchi, T., Uno, M., and Tominaga, Sh. (2010) Color gamut extension by projector-camera system. *Lecture Notes Comput. Sci.*, **6453**, 181–189.
6. Beke, L., Kwak, Y., Bodrogi, P., Lee, S.D., Park, D.S., and Kim, C.Y. (2008) Optimal color primaries for three- and multi-primary wide gamut displays. *J. Electron. Imaging*, **17**, 023012.
7. Wood, C.B.B. and Sproson, W.N. (1977) The choice of primary colors for color television. *BBC Eng.*, (January), 19–35.
8. CIE 168 (2005) *Criteria for the Evaluation of Extended-Gamut Color Encodings*, Commission Internationale de l'Éclairage.
9. Pointer, M.R. (1980) The gamut of real surface colors. *Color Res. Appl.*, **5** (3), 145–155.
10. Krause, N., Bodrogi, P., and Khanh, T.Q. (2011) Spectral reflectance functions of indoor natural products and materials. Annual Meeting 2011 of the German Society of Color Science and Application, October 4–6, Braunschweig, Germany.
11. ISO/TR 16066:2003 (2003) *Graphic Technology – Standard Object Color Spectra Database for Color Reproduction Evaluation (SOCS)*, International Organization for Standardization.
12. Yamaguchi, M., Teraji, T., Ohsawa, K., Uchiyama, T., Motomura, H., Murakami, Y., and Ohyama, N. (2002) Color image reproduction based on the multispectral and multiprimary imaging: experimental evaluation, in *Color Imaging: Device Independent Color, Color Hardcopy and Applications VII (Proc. SPIE 4663)*, SPIE, pp. 15–26.
13. O'Rourke, J. (1998) *Computational Geometry in C*, Cambridge University Press.
14. Luo, M.R., Cui, G., and Li, C.J. (2006) Uniform color space based on CIECAM02 color appearance model. *Color Res. Appl.*, **31** (4), 320–330.
15. Brill, M.H. and Larimer, J. (2005) Avoiding on-screen metamerism in N-primary displays. *J. SID*, **13** (6), 509–516.
16. Murakami, Y., Ishii, J.I., Obi, T., Yamaguchi, M., and Ohyama, N. (2004) Color conversion method for multi-primary display for spectral color reproduction. *J. Electron. Imaging*, **13** (4), 701–708.
17. Vogels, I.M.L.C. and Heynderickx, I.E.J. (2004) Optimal and acceptable white-point settings of a display. Proceedings of IS&T/SID 12th Color Imaging Conference, pp. 233–238.
18. Roth, S., Ben-David, I., Ben-Chorin, M., Eliav, D., and Ben-David, O. (2003) Wide gamut, high brightness multiple primaries single panel projection displays. SID Symposium Digest, vol. 34, pp. 118–121.
19. Kwak, Y., Lee, S.D., Choe, W., and Kim, C.Y. (2004) Optimal chromaticities of the primaries for wide gamut 3-channel display. *Proc. SPIE*, **5667**, 319–327.
20. Bieske, K. (2009) Investigation of the perception of changes of the color of light for the development of dynamic illumination systems. Ph.D. thesis, Technische Universität Ilmenau, Ilmenau, Germany (in German).
21. Ajito, T., Obi, T., Yamaguchi, M., and Ohyama, N. (2000) Expanded color gamut reproduced by six-primary projection display. *Proc. SPIE*, **3954**, 130–137.
22. Ajito, T., Obi, T., Yamaguchi, M., and Ohyama, N. (1999) Multiprimary color display for liquid crystal display projectors using diffraction grating. *Opt. Eng.*, **38**, 1883–1888.

23 Hiyama, I., Ohyama, N., Yamaguchi, M., Haneishi, H., Inuzuka, T., and Tsumura, M. (2002) Four-primary color 15-in. XGA TFT-LCD with wide color gamut. Proceedings of EuroDisplay 2002, pp. 827–830.

24 Kutas, G., Choh, H.K., Kwak, Y., Bodrogi, P., and Czúni, L. (2006.) Subpixel arrangements and color image rendering methods for multi-primary displays. *J. Electron. Imaging*, **15**, 023002.

25 Betrisey, C., Blinn, J.F., Dresevic, B., Hill, B., Hitchcock, G., Keely, B., Mitchell, D.P., Platt, J.C., and Whitted, T. (2000) Displaced filtering for patterned displays. SID Symposium Digest, pp. 296–299.

26 Elliott, C.H.B., Han, S., Im, M.H., Higgins, M., Higgins, P., Hong, M.P., Roh, N.S., Park, C., and Chung, K. (2002) Co-optimization of color AMLCD subpixel architecture and rendering algorithms. SID Symposium Digest, pp. 172–175.

27 Long, G.M. and Tuck, J.P. (1991) Comparison of contrast sensitivity functions across three orientations: implications for theory and testing. *Perception*, **20** (3), 373–380.

28 Bayer, B.E. (1976) Color imaging array. U.S. Patent No. 3,971,065.

29 Elliott, C.H.B. (1999) Reducing pixel count without reducing image quality. *Inform. Display*, **15** (12), 22–25.

30 Elliott, C.H.B. (2000) Active matrix display layout optimization for sub-pixel image rendering. Proceedings of 1st International Display Manufacturing Conference, pp. 185–187.

31 Credelle, T. *et al*. (2002) MTF of high resolution PenTile™ matrix displays. Proceedings of EuroDisplay 2002, pp. 159–162.

32 Elliott, C.H.B. and Hellen, C. (2002) Rotatable display with sub-pixel rendering. U.S. Patent Application 20020186229.

33 McCartney, R.I., Jr. (1994) Color mosaic matrix display having expanded or reduced hexagonal dot pattern. U.S. Patent No. 5,311,337.

34 Kwak, Y., Choh, H.K., Bodrogi, P., Czuni, L., and Kranicz, B. (2006) Pixel structure for flat panel display apparatus. USPTO Application No. 20060290870.

35 Kwak, Y., Choh, H.K., Bodrogi, P., Schanda, J., and Kutas, G. (2006) Apparatus and method for rendering image, and computer-readable recording media for storing computer program controlling the apparatus. USPTO Application No. 20060017745.

36 Berbecel, G. (2003) *Digital Image Display: Algorithms and Implementation*, John Wiley & Sons, Ltd.

# 6
# Improving the Color Quality of Indoor Light Sources

## 6.1
### Introduction to Color Rendering and Color Quality

An important task of modern lighting design concerns the selection of new light sources with a spectral power distribution that achieves sustained high-level user acceptance of the color perception of typical colored objects illuminated by the new light source. The level of acceptance is usually based on a comparison of actual object colors with the colors of the same objects under a so-called reference light source or reference illuminant. This property of the light source is called color rendering.

Color rendering is defined by the International Commission on Illumination as the "effect of an illuminant on the color appearance of objects by conscious or subconscious comparison with their color appearance under a reference illuminant" [1]. The reference light source is either daylight or tungsten light. The users usually refer to their long-term memory to recall how the objects should look like under the artificial (test) light source (see Figure 6.1). In practice, color rendering is one of the most important properties when a light source is to be selected.

There are numerous artificial light sources whose light color (i.e., the color stimulus of the light source itself and not the light reflected from colored objects) is very similar to a certain phase of daylight (at a given time of the day) or to tungsten light. This means that these light sources do not "disturb" the color perception of a white surface or the color perception of a spectrally neutral gray object as long as the light source itself is perceived to exhibit an acceptable white tone. However, even in the latter case, problems may occur with the color perception of spectrally selective (i.e., colored) object surfaces illuminated by the light source. The reason is that some spectral ranges may be missing or insufficiently represented in the spectral power distribution of the light source distorting the color perception of certain objects. In this case, the color rendering of the light source is said to be low.

The numerical description of the color rendering property of a light source is a complicated task playing an important role in today's practice of lighting engineering [3]. The color rendering of a test light source is large when the color appearance of a set of important reflecting surface colors matches their appearance under the reference illuminant. This matching is carried out by the user of the test light source

*Illumination, Color and Imaging: Evaluation and Optimization of Visual Displays*, First Edition.
Peter Bodrogi and Tran Quoc Khanh.
© 2012 Wiley-VCH Verlag GmbH & Co. KGaA. Published 2012 by Wiley-VCH Verlag GmbH & Co. KGaA.

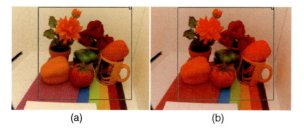

(a)                                 (b)

**Figure 6.1** To assess the color rendering property of a test light source, the user of the test light source compares the color appearance of the colored objects under the test light source (b) with their color appearance under a reference light source (a) [2]. In practice, the reference side (a) is not present. Hence, long-term color memory constitutes the basis of the user's color rendering assessment.

visually by comparing the two appearances under the test and reference conditions (in most cases from long-term memory) because the reference condition is usually absent in real-world situations either consciously or very often unintentionally. The color rendering property of light sources is described by the general color rendering index $R_a$ of the CIE [1]. A detailed description of the computation method of the general color rendering index can be found in a CIE publication [1]. Here only the most important steps of this method (1–7) are summarized (see Figure 6.2).

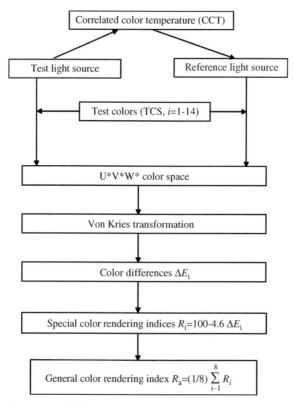

**Figure 6.2** Block diagram to demonstrate the steps of computing the CIE color rendering index [1].

**Step 1.** A reference illuminant should be selected having the same correlated color temperature (CCT) as the test light source. If the CCT of the test light source is less than 5000 K, then the reference illuminant will be a blackbody radiation of the same color temperature. If the CCT of the test light source is equal to 5000 K or more than 5000 K, then a phase of daylight of the same CCT will be selected as reference. The color difference $\Delta E$ shall be less than $5.4 \times 10^{-3}$ on the $u, v$ color diagram.

**Step 2.** Fourteen color samples of the Munsell color atlas are selected as test color samples (TCS). The first eight TCS serve as the basis to compute the general color rendering index $R_a$. For the last six TCS, only so-called special color rendering indices are computed.

**Step 3.** The CIE 1931 tristimulus values X, Y, Z are computed for the 14 TCS under the test light source and the reference illuminant. Then, the CIE 1960 UCS coordinates $u$, $v$ and the CIE 1964 $U^*$, $V^*$, $W^*$ values are calculated.

**Step 4.** Test light source chromaticity is transformed into reference illuminant chromaticity by the *von Kries* transformation.

**Step 5.** The 14 CIE 1964 color differences are calculated for the 14 TCS ($\Delta E_i$, $i = 1$–14) between the $U^*$, $V^*$, $W^*$ values under test light source and reference illuminant.

**Step 6.** For every TCS ($i = 1$–14), a so-called special color rendering index is computed in the following way: $R_i = (100 - 4.6 \Delta E_i)$.

**Step 7.** The general color rendering index ($R_a$) is defined as the arithmetic mean of the first eight special color rendering indices.

Color rendering (or, in other words, color fidelity), however, is only one aspect of the more general concept of light source color quality and other aspects such as color discrimination capability are equally important (see Section 6.4.1). Yet any one of these aspects needs an individual experimental method and a separate numerical prediction method, that is, an index to predict its magnitude for the case of a given test light source. If a reliable computation method of a particular color quality index exists, then the user of the light source can apply the most appropriate index for a given application. For example, an electrician working with wires of different colors can use a color discrimination index (CDI). Similarly, a lamp designer can use a color preference index for textile applications to describe the color preference of textiles for a given user group. If a light source is to be used for several different purposes, then the corresponding color quality indices can be weighted according to their relevance.

Note that color quality is only an issue of indoor lighting. For indoor lighting, the users' color expectations can be rather demanding, for example, for offices or households. Just the opposite is true for the case of outdoor lighting such as street lighting. Certain outdoor light sources such as low-pressure sodium lamps are of very poor color quality. The reason is that for outdoor applications, energy efficiency and visual performance (e.g., twilight visual performance or twilight brightness efficiency) are more relevant than color quality.

## 6.2
## Optimization for Indoor Light Sources to Provide a Visual Environment of High Color Rendering

Visual color rendering experiments have shown that the value of the general color rendering index ($R_a$) introduced in Section 6.1 generally does not agree well with the perceived color rendering property of test light sources. The predicted rank order of color rendering among a set of test light sources is often wrong especially if this set contains red–green–blue light emitting diode light sources (RGB LEDs) or phosphor-converted white LEDs (pcLEDs) together with other light sources [3]. While a comparison among classical fluorescent light sources shows a better agreement. The optimization of the color rendering of modern indoor light sources needs a visually more appropriate target quantity (other than $R_a$). To define this quantity, several visual experiments have been conducted.

Section 6.2.1 presents the methods and the results of these visual color rendering experiments and related experiments on the more general concept of light source color quality (see Section 6.1). Section 6.2.2 is devoted to more recent color rendering prediction methods different from the CIE color rendering index. Section 6.2.2.1 describes the deficits of the computation method of the CIE color rendering index, while Section 6.2.2.2 deals with some recent proposals to redefine the color rendering computation method, including the computation methods for the other aspects of light source color quality.

### 6.2.1
### Visual Color Fidelity Experiments

In the classic visual color rendering (or color fidelity) experiment, the task of the observer concerns the comparison of the color appearance of two identically reflecting homogeneous color samples under the test light source and the reference light source and the rating of their color difference visually. To illustrate this, a typical experiment is described below. In this experiment [4], observations were carried out in a double-chamber viewing booth like the one shown in Figure 6.3.

The walls of the chambers were painted white with a luminance well within the photopic range of vision (240 cd/m$^2$) for all light sources illuminating the booth. The test light source illuminated the left chamber and the reference light source illuminated the right chamber. The light passed through a diffuser plate. The spectral power distributions (SPDs) of the light sources were measured on a white standard placed on the bottom of the chambers. Colorimetric properties of the light sources (correlated color temperature CCT; chromaticity coordinates $x$, $y$, and CIE color rendering indices $R_a$, see Table 6.1) were computed from these SPDs.

Two groups of test and reference light sources were investigated, one at approximately 2700 K and the other one at approximately 4500 K. The 2700 K group consisted of two types of warm white phosphor LED (HC3L and C3L), an RGB LED cluster (RGB27), and two warm white fluorescent lamps (FL627 and FL927) as test light sources and a tungsten halogen reference light source (TUN). The 4500 K group consisted of two

## 6.2 Optimization for Indoor Light Sources to Provide a Visual Environment of High Color Rendering

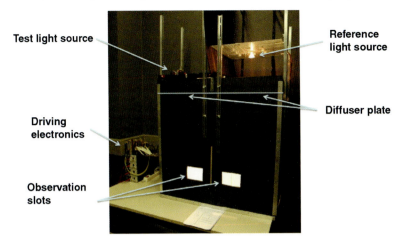

**Figure 6.3** Double-chamber viewing booth [4]. Reproduced with permission from *Color Research and Application*.

**Table 6.1** Correlated color temperatures, chromaticity coordinates (x, y), and CIE color rendering indices ($R_a$) of the light sources of the visual color fidelity experiment [4].

| Light source | Description | CCT (K) | x | y | $R_a$ |
|---|---|---|---|---|---|
| HC3L | T; white phosphor LED | 2798 | 0.448 | 0.401 | 97 |
| C3L | T; white phosphor LED | 2640 | 0.476 | 0.432 | 67 |
| RGB27 | T; RGB LED | 2690 | 0.462 | 0.414 | 17 |
| FL627 | T; a 4000 K fluorescent lamp filtered by a color filter plus the diffuser plate to get close to 2700 K | 2786 | 0.456 | 0.415 | 64 |
| FL927 | T; a 3000 K fluorescent lamp filtered by the diffuser plate to get close to 2700 K | 2641 | 0.466 | 0.414 | 90 |
| TUN | R; tungsten halogen | 2762 | 0.460 | 0.419 | 97 |
| HC3N | T; white phosphor LED | 4869 | 0.349 | 0.355 | 95 |
| C3N | T; white phosphor LED | 4579 | 0.363 | 0.393 | 69 |
| RGB45 | T; RGB LED | 4438 | 0.361 | 0.355 | 22 |
| FL645 | T; the same 4000 K fluorescent lamp as FL627 filtered by another Rosco color filter and the diffuser plate to get close to 4500 K | 4423 | 0.365 | 0.371 | 69 |
| FL945 | T; a 5400 K fluorescent lamp filtered by the diffuser plate to get close to 4500 K | 4391 | 0.366 | 0.372 | 92 |
| HMI | R; a gas discharge light source filtered by the diffuser plate to get close to 4500 K | 4390 | 0.362 | 0.353 | 92 |

T: test light source; R: reference light source.

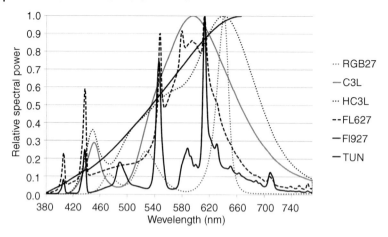

**Figure 6.4** Relative spectral power distributions of the light sources of the 2700 K group (see Table 6.1) [4]. Reproduced with permission from *Color Research and Application*.

types of white phosphor LED (HC3N and C3N), an RGB LED cluster (RGB45), and two cool white fluorescent lamps (FL645 and Fl945) as test light sources and a HMI reference light source. Figure 6.4 shows the relative SPDs of the 2700 K light sources, while Figure 6.5 shows the relative SPDs of the 4500 K light sources of Table 6.1.

Two identical copies of matte color papers (pairs of identical uniform stand-alone test color samples subtending about $4° \times 3°$ viewed from the observation slot, one under the test light source and the other one under the reference light source) were observed on a gray background ($L = 59 \, \text{cd/m}^2$). Seventeen pairs were observed altogether, one after the other. Each pair had a different color. A grayscale color difference anchor (see also Ref. [5]) helped scale the perceived color differences

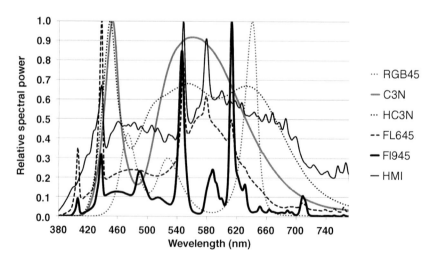

**Figure 6.5** Relative spectral power distributions of the light sources of the 4500 K group (see Table 6.1) [4]. Reproduced with permission from *Color Research and Application*.

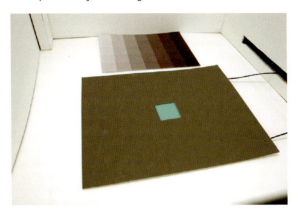

**Figure 6.6** Reference side of the viewing booth [4] with a 4° × 3° square test color sample (matte paper) on a gray background and the color difference grayscale anchor to scale color differences. One grayscale step was equal to $\Delta L^* = 2$. Observers saw this scene when looking into the reference (right) chamber through the observation slot (compare with Figure 6.3). Reproduced with permission from *Color Research and Application*.

between the test and the reference chamber of the viewing booth (see Figure 6.6). The unit of visual color difference ($\Delta E_{vis}$) was equal to *one step* on the gray scale.

Twelve of the 17 test color samples were taken from the Macbeth ColorChecker® chart (nos. 1–12) and five from the NIST color set [6] (nos. 13–17) (see Figure 6.7). The spectral power distribution of the light reflected from each test color sample under each light source was measured *in situ* by a well-calibrated spectroradiometer.

Eight observers of normal color vision completed the experiment by evaluating each color difference once. Before the observation, subjects were trained to get familiar with assessing color attributes (hue, chroma, lightness) by using well-known

**Figure 6.7** Test color samples (1–17) [4]. Top: Macbeth ColorChecker® chart (1–12); bottom: NIST CQS ver. 7.1 color set [6] (13–17, that is, VS1, VS4, VS7, VS10, and VS13). Reproduced with permission from *Color Research and Application*.

color training material. After this training phase, each of the color differences was evaluated visually in the following way. First, observers had to rate the color difference on a five-step ordinal rating scale: (1) excellent; (2) good; (3) acceptable; (4) not acceptable; and (5) very bad. This constituted the variable $R$ for the evaluation. Second, observers had to assess the degree of similarity between the test and the reference side by putting a cross on a graphical rating scale (variable $P$). This scale was used to check the consistency of the observers' judgments and to help observers scale the total visual difference ($\Delta E_{vis}$) in their third task by the aid of the grayscale color difference anchor of Figure 6.6.

Observers were taught to look into one of the chambers through one of the two observation slots binocularly by placing their head always directly to the slots for total immersion. The observation through the slots was used to achieve complete immersion and adaptation. At the beginning, observers had to look into the reference chamber for 10 min to adapt. In the observation phase, they were taught to look into the reference chamber and – after at least 2 s – change their line of sight to look into the test chamber. Then, in turn, after at least 2 s, subjects were looking into the reference chamber again. Observers had to repeat this procedure until they could assess the color difference and fill the questionnaire.

For comparison with theoretical predictions, color differences were computed for each of the 10 test light sources in Table 6.1 and each of the 17 test colors of Figure 6.7 by using six color difference formulas: CIELAB ($\Delta E^*_{ab}$), CIEDE2000 ($\Delta E_{2000}$, $k_L = k_C = k_H = 1$) [7], the Euclidean difference in CIECAM02 $J$–$a_C$–$b_C$ space [8] ($\Delta E_{02}$), CIECAM02-LCD, CIECAM02-SCD, and CIECAM02-UCS [9]. In these formulas, the viewing parameter values shown in Table 6.2 were used. These values depended slightly on the light source (except $F$, $c$, and $N_C$). Table 6.2 shows only their average values.

The z-scores of the variables $P$ ($z_P$, visual similarity of the test color samples under the test and the reference light source) and $\Delta E_{vis}$ ($z_{\Delta Evis}$, visual color difference compared to the grayscale anchor) were computed from the mean and standard deviation of all data of each observer. Visual color difference ($z_{\Delta Evis}$) decreased according to a quadratic function ($r^2 = 0.75$) as perceived similarity ($z_P$) increased (see Figure 6.8).

The inverse tendency in Figure 6.8 corroborates the consistency of the observers' judgments and the suitability of the above-described experimental method. The values of $\Delta E_{vis}$ and $z_{\Delta Evis}$ were compared with the calculated color differences ($\Delta E_{calc}$)

Table 6.2 Average viewing parameters used to compute color differences.

| Viewing parameter | Value |
| --- | --- |
| CIELAB $Y_n$ [a] | 323 cd/m² [b] |
| CIECAM02 $Y_b$; $L_A$ | 19; 62 cd/m² [b] |
| CIECAM02 $F$; $c$; $N_C$ | 1.0; 0.69; 1.0 |

[a] Reference white luminance measured at a white standard placed on the bottom of the viewing booth.
[b] Average value. The actual value depends on the light source causing about ±2% variation.

## 6.2 Optimization for Indoor Light Sources to Provide a Visual Environment of High Color Rendering

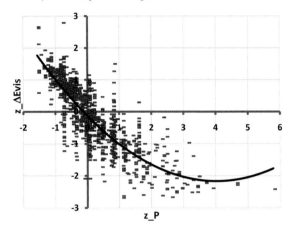

**Figure 6.8** Visual color difference ($z_{\Delta Evis}$) as a function of perceived similarity ($z_P$) [4]. The whole data set is shown (1318 points). Fit line: quadratic, $r^2 = 0.75$. Reproduced with permission from Color Research and Application.

according to the six color difference formulas. The correlation coefficients between $\Delta E_{vis}$ and $\Delta E_{calc}$ were computed for the whole visual data set (see Table 6.3).

As can be seen from Table 6.3, the CIECAM02-UCS metric yielded the best correlation. Correlation coefficients are higher for $z_{\Delta Evis}$ than for $\Delta E_{vis}$ indicating the effect of interindividual differences (see Section 6.6).

The above-described advantages of the CIECAM02-UCS were recently corroborated in a similar visual color rendering experiment [10]. It was found that the best correlation with the visual color differences was achieved when the color differences were computed in the CIECAM02-UCS uniform color space [9] instead of the $U^*$, $V^*$, $W^*$ color space of the CIE color rendering index (see Step 5 in Section 6.1). The CIECAM02-UCS uniform color space performed also better than CIELAB, CIEDE2000, or the Euclidean difference in the $J$, $a_C$, $b_C$ space of the CIECAM02 color appearance model. In a further similar visual color rendering study [11], the color space of the CIECAM02 color appearance model yielded significantly better correlation with visual color differences than the $U^*$, $V^*$, $W^*$ color space of the CIE color rendering index for all three investigated color temperatures, 2700, 4000, and 6500 K.

**Table 6.3** Pearson's correlation coefficients ($r$) between visual color differences $\Delta E_{vis}$ (and their z-scores $z_{\Delta Evis}$) and $\Delta E_{calc}$ for the visual color difference data set.

|  | CIELAB | CIEDE2000 | CIECAM02 | CAM02-LCD | CAM02-SCD | CAM02-UCS |
|---|---|---|---|---|---|---|
| $\Delta E_{vis}$ | 0.596 | 0.609 | 0.645 | 0.651 | 0.650 | 0.654 |
| $z_{\Delta Evis}$ | 0.647 | 0.660 | 0.698 | 0.704 | 0.702 | 0.706 |

All correlations were significant at $p = 0.01$ (two-sided).

## 6.2.2
### Color Rendering Prediction Methods

Section 6.1 described the current CIE method of predicting the color rendering property of light sources. In this section, some more recent color rendering prediction methods are presented trying to solve the deficits of the CIE color rendering index. First, these deficits will be summarized in Section 6.2.2.1. Then, in Section 6.2.2.2, some recent proposals are shown to redefine the color rendering computation method including the computation methods for the other aspects of light source color quality.

#### 6.2.2.1 Deficits of the Current Color Rendering Index

Reasons of the deficits of the value of the current color rendering index $R_a$ have already been identified [12]. One serious problem concerns the choice of eight test color samples used in the computation method of $R_a$. These test color samples are unsaturated and the color appearance of more saturated colors cannot be described correctly, especially if the spectral reflectance functions of the saturated colors interact with certain light source spectra such as RGB or white LED spectra as illustrated by Figure 6.9.

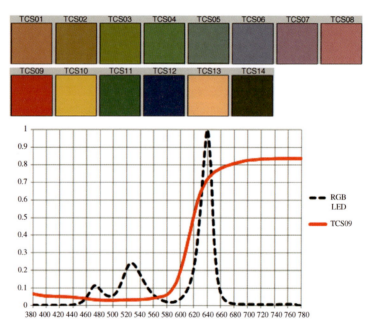

**Figure 6.9** Effect of the choice of eight unsaturated test color samples in the CIE color rendering method. Top: test color samples TCS01–TCS08 used to compute the value of the general color rendering index $R_a$; middle: saturated test color samples TCS09–TCS14; bottom: interaction of an RGB LED light source spectrum with the spectral reflectance function of the saturated red TCS09.

The example of Figure 6.9 shows that the red maximum (at 640 nm) of the RGB LED light source (CCT = 2690 K) enhances the saturated red of the test color sample TCS09 significantly. Therefore, the value of the special color rendering index of this TCS is equal to −180 in this example. This effect cannot be described by the general color rendering index $R_a$ (which is $R_a = 17$ in this example) because $R_a$ is based on the eight unsaturated TCS (TCS01–TCS08) only.

The choice of the reference light source constitutes the next deficiency. The continuous transition among the reference light sources is not easy to interpret visually because the concept of correlated color temperature turned out to be uncertain in a visual experiment. A seven-value scale of perceived white tones, however, seemed more adequate for the subjects: T27, T30, T35, T42, D50, D65, and D95 [13]. A computational problem also arises because the current CRI method is discontinuous at 5000 K (a blackbody radiation below 5000 K and a phase of daylight above 5000 K, see Section 6.1). In addition to this, illuminants of low color temperatures (e.g., 2000 K) cause the perception of a yellow tint in the lit scene as human color constancy breaks down (see Figure 6.10).

The use of an outdated chromatic adaptation formula (von Kries transformation) to compute the current color rendering index causes a further problem. The von Kries transformation is inaccurate especially for large differences in the state of chromatic adaptation between the test light source and its reference light source (e.g., 6500 K versus 3000 K). New color rendering methods often require bridging such differences. An inaccurate chromatic adaptation formula leads to inaccurate color appearance predictions under the test and reference light sources. Consequently, the magnitude of predicted color difference between the test and reference conditions will be inaccurate and this causes a false value of the color rendering index.

Color difference formulas represent a further deficit. The correlation between Euclidean color differences in the $U^* V^* W^*$ color space and visual color differences is low. Such a color space is called visually nonuniform. In a uniform color space,

**Figure 6.10** Illustration of imperfect color constancy. Color appearance of a still life of colored objects illuminated by a light source of low color temperature.

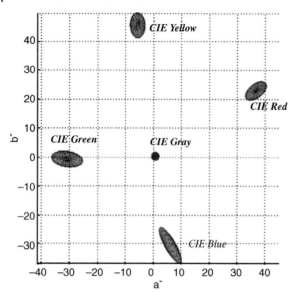

**Figure 6.11** Tolerance ellipsoids projected onto the CIELAB $a^*$–$b^*$ plane [14].

however, all test colors (T) with a constant perceived color difference ($\Delta E_{vis}$) between T and a fixed reference color R should lie on a sphere surface around R and the diameter of this sphere should be constant for all reference colors in the whole color space. Instead of these spheres, constant perceived color difference can be represented by ellipsoids of varying axes depending on the reference color. To visualize a nonuniform color space, Figure 6.11 shows the example of so-called tolerance ellipsoids projected to the CIELAB $a^*$–$b^*$ plane. Colors lying on the surface of these ellipsoids represent on average a *tolerable* color difference compared to the color center [14].

It can be seen from Figure 6.11 that visual tolerance ellipsoids depend on the central color. Therefore, CIELAB is said to be visually nonuniform for the evaluation of tolerable color differences.

The next deficit is associated with the way the general color rendering index ($R_a$) is computed from the eight special color rendering indices corresponding to eight unsaturated test color samples. $R_a$ is an arithmetic mean value that is unable to describe the color shifts of all possible object colors (including saturated colors) between the test and reference light sources correctly. This effect is especially significant for the case of modern pigments in combination with novel light sources, for example, LEDs. A further problem arises if a nonexpert user tries to interpret any value of the general color rendering index. It is not easy to understand what, for example, $R_a = 83$ means and especially what a color rendering index *difference*, for example, $\Delta R_a = 3$ means as a visual quality criterion of a test light source. Finally, as mentioned earlier, there are several other important aspects of color quality in addition to color rendering such as color preference, gamut, or harmony not correlating with the value of $R_a$ in the visual experiments [2].

### 6.2.2.2 Proposals to Redefine the Color Rendering Index

Possibilities to improve the CIE color rendering computation method were considered among others in the framework of CIE Technical Committee 1–33 from 1991 onward. In 1999, a new method (R96$_a$) was proposed [15]. In this method (not endorsed internationally), the following solutions were suggested to the deficits mentioned in Section 6.2.2.1.

1) Test color samples were selected from the Macbeth ColorChecker® chart containing saturated and unsaturated colors as well as important memory colors.
2) Instead of the continuous choice of reference light sources (Planckian radiators and phases of daylight), only six fixed reference illuminants are used: D65, D50, P4200, P3450, P2950, and P2700 (P represents a Planckian radiator of the subsequently indicated color temperature [13].
3) The CIE chromatic adaptation formula [16] is recommended (Note that this is outdated today).
4) The color stimulus of the test color samples is transformed into D65 both for the test and for the reference light sources and then color differences are computed in CIELAB.

Recently, numerous new color rendering indices were introduced in addition to the above-described R96$_a$ method. To illustrate the current state of ongoing international effort, some typical methods are selected and described below.

**CQS.** The method of the National Institute of Standards and Technology (NIST, Gaithersburg, MD, USA) assigns every test source a so-called color quality scale (CQS) value [6]. Color differences are computed in CIELAB color space. In recent versions of the method, more uniform color spaces are used. Instead of the CIE test colors, the CQS method uses 15 saturated colors from the Munsell color atlas (VS1–VS15), distributed around the hue circle. The CIE color rendering index penalizes those test light sources that increase the saturation of the test color sample – compared to the reference light source. This, however, ignores the observers' general preference of more saturated colors. To take this preference into account, the CQS method does not penalize any increase of saturation caused by the test light source but it does penalize sources with smaller rendered color gamut areas. In addition to these features, test light sources of extreme correlated color temperatures obtain lower CQS values. Saturation corrected color differences of the 15 test color samples VS1–VS15 are combined by computing a root mean square (RMS) value and not arithmetic averaging.

**SBI.** The method of the Technische Universität Ilmenau (Germany) is a subjective evaluation index. It needs input from subjects to compute the value of SBI summarizing their impressions about the different aspects of perceived color quality [17]. Subjects rate their impressions about the change of brightness, brilliancy, purity, hue, attractivity, and naturalness between the test and reference light sources.

**CRI-CAM02UCS.** The CRI-CAM02UCS method of the University of Leeds (UK) [10] computes color differences ($\Delta E_{UCS}$) in the CIECAM02-UCS uniform color space [9], computes a special color rendering index $R_{UCS,i} = 100 - 8.0 \Delta E_{UCS,i}$

for every test color sample ($i$), and then a general index $R_{UCS,i}$ as the average of the special indices. The same reference light source and the same test color samples are used as in the CIE method. Note that any spectral reflectance function required by a specific application can be used as a test color sample. As a recent chromatic adaptation transformation method is a basic component of the CIECAM02 color appearance model [18], CIECAM02-based color difference formulas or uniform color spaces such as CIECAM02-UCS are suitable to quantify perceived color differences under the different adaptation luminance levels and chromaticities.

**RCRI**. The method of the Technische Universität Darmstadt (Germany) [4] computes 17 CIECAM02-UCS color differences for the 17 dedicated test color samples seen in Figure 6.7. In the RCRI method, these color differences are classified into five categories predicting the observer's judgments (ordinal ratings) about the similarity of the color appearance of the test color sample under the test and reference light sources (1: excellent; 2: good; 3: acceptable; 4: not acceptable; and 5: very bad). The classification is based on the similarity of a computed color difference to the values of the mean CIECAM02-UCS color differences corresponding to the five categories (1: 2.01; 2: 2.37; 3: 3.75; 4: 6.53; 5: 11.28). For example, if the computed color difference is equal to 2.1 (11.1), then the corresponding category will be excellent (very bad). The mean CIECAM02-UCS color differences corresponding to the five categories were determined experimentally with a reasonable interpersonal agreement [4]. The RCRI method uses the same reference light source as the CIE color rendering method described in Section 6.1. The value of RCRI is computed from the number of excellent and good ratings ($N$) among the 17 test color samples in the following way: $RCRI = 100(N/17)^{1/3}$. This value informs the user of the test light source about how many test color samples match the reference color appearance well (rating 2) or excellently (rating 1).

**nCRI.**. This recent method was developed in CIE TC 1–69. It was not endorsed by the CIE. The method will be published in the next future. It contains very significant improvements compared to Fig. 6.2.

## 6.3
## Optimization of Indoor Light Sources to Provide Color Harmony in the Visual Environment

As seen in Section 6.1, color harmony is an important psychological dimension of light source color quality [2]. Color harmony expresses an esthetic judgment about the relationship among the colors of some selected objects of (at least approximately) homogeneous color in a lit scene. The number of selected objects can be two (two-color combinations), three (three-color combinations), or more (multi-color combinations). According to the general aim of this chapter, that is, optimizing indoor light source spectral power distributions for best color quality, this section describes a mathematical method (Szabó et al.'s method [19, 20]) to optimize a light source so as to obtain best color harmony for a set of test color combinations illuminated by this optimized light source. Test color combinations in the color harmony rendering index method are analogous to the stand-alone test color samples in the method of the color rendering index (see Section 6.1). However, the color harmony rendering index computes differences between the values of predicted color harmony between the test and reference light sources instead of the color differences computed in the color rendering method.

### 6.3.1
### Visual Color Harmony Experiments

Subjects are known to be able to assess the color harmony of color combinations consistently on ordinal rating scales [19–21]. In these visual experiments, either test color samples are simulated on a well-characterized and well-calibrated color monitor [19, 21] or paper samples are illuminated in a viewing booth [20]. In the experiment serving as the basis of Szabó et al.'s model (Section 6.3.2 [19]), nine observers of normal color vision scaled their color harmony impression on a −5 to +5 scale (−5: disharmonious; +5: harmonious) for two- and three-color combinations on the monitor. A gray background was shown for 2 s between the subsequent combinations to eliminate any afterimage effect. Figure 6.12 shows an example.

Test samples were chosen to sample the whole CIECAM02 $J$, $a_C$, $b_C$ space. Test samples of three different lightness values at three different levels of chroma were used as well as the most saturated colors for each examined hue, 2346 two- and 14 280 three-color combinations altogether.

### 6.3.2
### Szabó et al.'s Mathematical Model to Predict Color Harmony

The principle of color harmony is completeness according to Goethe [22], order according to Chevreul [23], and balance according to Munsell [24]. Classical harmony principles are described by Nemcsics [25]. Judd and Wyszecki [26] defined harmony

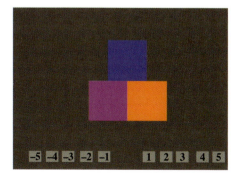

**Figure 6.12** Visual scaling of color harmony on a −5 to +5 scale (−5: disharmonious; +5: harmonious) for two- and three-color combinations [19]. Example: illustration of a three-color combination. Subjects had to click on the appropriate number corresponding to their color harmony impression.

as "when two or more colors seen in neighboring areas produce a pleasing effect, they are said to produce a color harmony." For computational light source optimization, it is very important to devise *mathematical formulas* quantifying color harmony impression. These formulas should be preferably based on the recently internationally widely used CIECAM02 color appearance model due to the built-in chromatic adaptation transformation. In this section, Szabó et al.'s two-color harmony formulas (CHF) are presented [19].

These CHF formulas include two-color combinations of fixed hue and also with variable hue among the constituent colors. It should be mentioned that independent CHF formulas were also described for three-color combinations [19]. CHF formulas were confirmed by an independent color harmony experiment based on reflecting color samples and a significant correlation was found between the model and the visual data ($r^2 = 0.81$ [20]). An alternative formula, Ou and Luo's so-called CH value, also exists. This CH formula predicts the perceived color harmony impression of two-color combinations [21].

From the results of the color monitor experiment underlying the model ([20], see Section 6.3.1), visually scaled color harmony was analyzed mathematically, as a function of the CIECAM02 chroma, lightness, and hue differences and sums among the constituent colors of the two- and three-color combinations. An adequate interobserver and intraobserver agreement was found. The classical harmony principles did not always result in a good visual harmony impression. The principle of "equal lightness" yielded negative harmony scores and the data set of nonequal lightness produced better results. For two-color combinations, the principle of "complementary hue" was the main principle of classical color harmony theories but this principle caused sometimes negative visual harmony values. However, the "equal hue" and "equal chroma" principles corresponded to the best visual color harmony scores.

For three-color combinations, the "equal hue" property had a high visual color harmony score. Three-color combinations of equal lightness had negative values and the "neighboring hue" principle generated high visual harmony scores. Experimental results also showed that different formulas should be used for two- and three-color combinations. In addition to this, it turned out that different formulas should be used for combinations of the same hue and for combinations of different hues.

Equation 6.1 shows the formula predicting the color harmony impression of two-color combinations of the same hue. Factors of Equation 6.1 are defined in Table 6.4.

$$CHF = 0.283 \cdot (3.275 JDIFF - 0.643 JSUM + 2.749 CDIFF + 4.773 HP) - 5.305 \tag{6.1}$$

Equation 6.2 contains the formula predicting the color harmony impression of two-color combinations of different hues. Factors of Equation 6.2 are defined in Table 6.5.

$$CHF = 0.47 \cdot (0.515 JDIFF + 0.391 JSUM + 0.205 CDIFF + 1.736 CSUM + 2.187 HDIFF + 5.104 HP) - 2.283 \tag{6.2}$$

**Table 6.4** Factors of the CHF formula predicting the color harmony of two-color combinations of the same hue [19].

| Factor | Equation |
| --- | --- |
| Lightness difference (JDIFF) | $2.33 \times 10^{-5}|\Delta J|^3 - 0.004|\Delta J|^2 + 0.211|\Delta J| + 0.246$ |
| Lightness sum (JSUM) | $0.0268 J_{sum} - 0.656$ |
| Chroma difference (CDIFF) | $3.87 - 0.066|\Delta C|$ |
| Hue preference (HP) | $0.361 \sin(1.511 h) + 2.512$ |

$|\Delta J|$: absolute value of lightness difference between the two constituent colors; $J_{sum}$: sum of the lightness of the two colors; $|\Delta C|$: absolute value of chroma difference between the two constituent colors; $h$: hue angle of the color combination. All quantities are computed in the CIECAM02 color appearance model.

### 6.3.3
### A Computational Method to Predict Color Harmony Rendering

Perceived color harmony among the illuminated objects is subject to significant changes when the light source illuminating the color samples changes from a reference light source to a test light source. A distortion of color harmony can be observed if there are nonsystematic color shifts (concerning both magnitudes and directions) among the samples in CIECAM02 color space [20]. To quantify the change of color harmony, a *color harmony rendering index* was suggested [20] to be equal to a linear transformation of the predicted color harmony (CHF) differences of a set of test color sample combinations between the test light source and the reference light source.

These differences are intended to characterize the *color harmony rendering* property of the test light source by quantifying the extent of distortion (or amelioration) of

**Table 6.5** Factors of the CHF formula predicting the color harmony of two-color combinations of different hues [19].

| Factor | Equation |
| --- | --- |
| Lightness difference (JDIFF) | $2.5 \times 10^{-5}|\Delta J|^3 + 3 \times 10^{-3}|\Delta J|^2 - 2.2 \times 10^{-2}|\Delta J| + 0.158$ |
| Lightness sum (JSUM) | $0.027 J_{sum} - 0.656$ |
| Chroma difference (CDIFF) | $-0.053|\Delta C| + 1.172$ |
| Chroma sum (CSUM) | $-0.051 C_{sum} + 2.36$ |
| Hue difference (HDIFF) | $8 \times 10^{-5}|\Delta h|^2 - 0.0279|\Delta h| + 2.3428$ |
| Hue preference (HP) | $\frac{1}{2}\left[4 \times 10^{-5}(h_1)^2 - 0.0127 h_1 + 1.4035 + 4 \times 10^{-5}(h_2)^2 - 0.0127 h_2 + 1.4035\right]$ |

$|\Delta J|$: absolute value of lightness difference between the two constituent colors; $J_{sum}$: sum of the lightness of the two colors; $|\Delta C|$: absolute value of chroma difference between the two constituent colors; $C_{sum}$: sum of the chroma of the two colors; $|\Delta h|$: absolute value of hue angle difference between the two constituent colors; $h_1, h_2$: hue angles of the two colors. All quantities are computed in the CIECAM02 color appearance model.

perceived color harmony. Equation 6.3 shows the special color harmony rendering index ($R_{hr,i}$) for the $i$th test color combination.

$$R_{hr,i} = 100 + k(\text{CHF}_{i,\text{ref}} - \text{CHF}_{i,\text{test}}) \tag{6.3}$$

where $\text{CHF}_{i,\text{ref}}$ is the value of color harmony calculated under the reference light source for the $i$th combination of test samples, $\text{CHF}_{i,\text{test}}$ is the value of color harmony calculated under the test light source for the $i$th combination of test samples, and $k$ is a constant establishing a usable color harmony rendering scale. A general index ($R_{hr}$) can also be computed as the arithmetic average of the special indices of all combinations of test samples.

The value of $R_{hr}$ can be used to evaluate and optimize the color harmonic property of the different light sources computationally. As an example, all possible two-color combinations of a set of 34 colored test objects were considered. The number of such combinations is equal to $34 \cdot 33 \div 2 = 561$. Predicted color harmony values ($\text{CHF}_{i,\text{test}}$) were computed for this set of 561 two-color combinations under a set of 42 current test light sources and their reference light sources ($\text{CHF}_{i,\text{ref}}$). Reference light sources were determined by using the CIE method (see Step 1 in Section 6.1). Five hundred sixty-one special color harmony rendering indices ($R_{hr,i}$) were computed for each test light source and then the mean value of $R_{hr}$ by using the constant value of $k = -100$ in this example. In addition, special CRI-CAM02UCS color rendering indices ($R_{UCS,i} = 100 - 8.0 \Delta E_{UCS,i}$) [10] were computed for the 34 test objects and the mean index ($R_{UCS}$) of these 34 values ($R_{UCS,i}$; $i = 1$–34) was calculated.

The 34 test objects are characteristic objects of modern indoor scenes covering a wide range of hues and saturations. Their spectral reflectance functions can be seen in Figure 6.13.

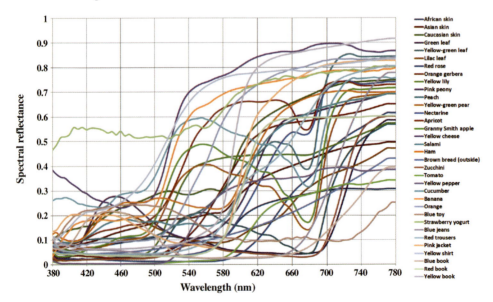

Figure 6.13 Spectral reflectance functions of the 34 test objects selected to evaluate the color harmony rendering property of light sources.

## 6.3 Optimization of Indoor Light Sources to Provide Color Harmony in the Visual Environment

**Table 6.6** A set of 42 light sources (LS), their correlated color temperatures (CCT), mean CRI-CAM02UCS special color rendering indices ($R_{UCS}$) [10] of the 34 color samples, and the mean color harmony rendering indices ($R_{hr}$) of the 561 color combinations [20].

| LS | CCT | $R_{UCS}$ | $R_{hr}$ | LS | CCT | $R_{UCS}$ | $R_{hr}$ |
|---|---|---|---|---|---|---|---|
| pcLED1 | 6344 | 69 | 123 | pcLED6 | 3098 | 74 | 111 |
| pcLED2 | 4579 | 67 | 128 | pcLED7 | 3099 | 76 | 115 |
| pcLED3 | 4870 | 93 | 97 | pcLED8 | 2974 | 76 | 103 |
| FL1 | 4374 | 59 | 129 | pcLED9 | 3348 | 87 | 88 |
| FL2 | 4445 | 88 | 99 | pcLED10 | 4234 | 61 | 112 |
| HMI | 4370 | 86 | 104 | pcLED11 | 4309 | 66 | 111 |
| retrLED | 2683 | 87 | 99 | pcLED12 | 4333 | 67 | 113 |
| pcLED4 | 2640 | 65 | 127 | pcLED13 | 5046 | 56 | 113 |
| pcLED5 | 2797 | 95 | 91 | pcLED14 | 3982 | 61 | 118 |
| FL3 | 2773 | 56 | 131 | pcLED15 | 4061 | 76 | 106 |
| FL4 | 2637 | 88 | 102 | pcLED16 | 4821 | 86 | 100 |
| TUN | 2762 | 96 | 108 | pcLED17 | 5650 | 61 | 114 |
| MLED1 | 2775 | 93 | 92 | pcLED18 | 5575 | 66 | 117 |
| MLED2 | 3042 | 92 | 99 | pcLED19 | 5046 | 56 | 113 |
| MLED3 | 3032 | 94 | 96 | pcLED20 | 5225 | 58 | 111 |
| MLED4 | 4520 | 82 | 98 | pcLED21 | 5043 | 63 | 118 |
| MLED5 | 4541 | 82 | 98 | pcLED22 | 6369 | 59 | 113 |
| MLED6 | 4947 | 88 | 83 | pcLED23 | 4966 | 64 | 121 |
| MLED7 | 6476 | 85 | 112 | pcLED24 | 6540 | 66 | 120 |
| MLED8 | 6451 | 85 | 85 | pcLED25 | 6369 | 59 | 113 |
| MLED9 | 6219 | 96 | 96 | pcLED26 | 5018 | 62 | 120 |

pcLED: phosphor-converted white LED; TUN: tungsten halogen; FL: fluorescent; HMI: a filtered gas discharge lamp; retrLED: a white retrofit LED lamp; MLED: theoretical multicomponent LED light source.

The set of 42 test light sources covers a set of phosphor-converted white LED light sources, fluorescent lamps, a filtered gas discharge lamp, a white retrofit LED lamp, a tungsten halogen lamp, and a set of nine theoretical multicomponent LED light sources optimized for high color rendering. These light sources are listed in Table 6.6.

Figure 6.14 illustrates the relationship between the mean color rendering property ($R_{UCS}$) and the mean harmony rendering property ($R_{hr}$) for the example of Table 6.6.

As can be seen from Figure 6.14, there is a significant *negative* correlation ($r^2 = 0.62$) between the mean color rendering property ($R_{UCS}$) and the mean harmony rendering property ($R_{hr}$). Another harmony rendering index computation [20] using the CIE color rendering index ($R_a$) instead of $R_{UCS}$, different test color combinations, and different test light sources also yielded *negative* correlation with $r^2 = 0.35$. These results indicate that a trade-off is required between the two properties (rendering and harmony) when optimizing the relative spectral power distribution of a light source.

Furthermore, to assess the color harmonic property of a light source, it is not enough to work with a mean harmony rendering index. To elucidate this, let us take a

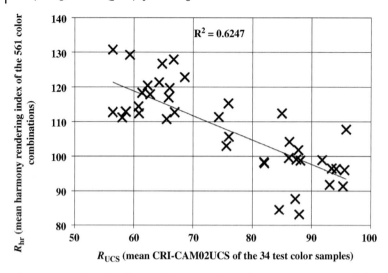

**Figure 6.14** Relationship between the mean color rendering property ($R_{UCS}$) and the mean harmony rendering property ($R_{hr}$) for the example of Table 6.6.

closer look at the special harmony rendering indices of two selected light sources in the above example, FL1 ($R_{hr} = 129$) and TUN ($R_{hr} = 108$). The two histograms of the 561 special harmony rendering indices of these two light sources are compared in Figure 6.15.

As can be seen from Figure 6.15, although FL1's mean harmony rendering index is greater than TUN's index, FL1 has a much broader histogram than TUN exhibiting numerous less harmonically rendered color combinations that tend to break down

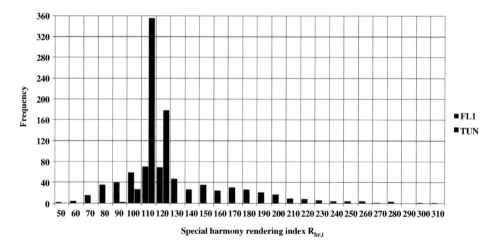

**Figure 6.15** Histograms of the 561 special harmony rendering indices of FL1 ($R_{hr} = 129$) and TUN ($R_{hr} = 108$) (see Table 6.6). Abscissa: harmony rendering index categories ($j = 1$–$27$; $C_j = 50, 60, \ldots, 300, 310$). Ordinate: frequency, that is, the number of test color samples for which $C_{j-1} < R_{hr,i} \leq C_j$.

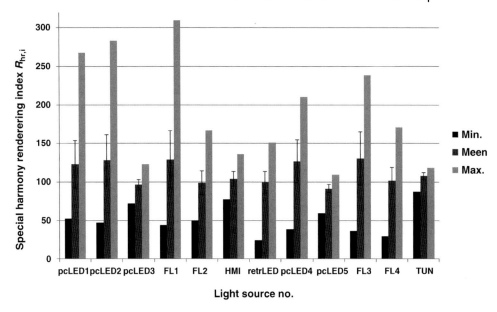

**Figure 6.16** Harmony rendering properties of the first 12 test light sources of Table 6.6 including the minimum, mean, and maximum $R_{hr,i}$ values for the 561 combinations as well as 10 times the 95% confidence intervals of the mean values.

the impression of harmony of the illuminated scene if attention is paid to these combinations or if some of the more harmonic combinations are missing in an actual scene. Figure 6.16 shows a graphical representation of the harmony rendering properties of the first 12 test light sources of Table 6.6 including the minimum, mean, and maximum $R_{hr,i}$ values for the 561 combinations as well as the 95% confidence intervals of the mean values.

As can be seen from Figure 6.16, the tungsten halogen light source (TUN) exhibits a balanced harmony rendering property comparable to HMI, pcLED3, and pcLED5. The other light sources (including all fluorescent lamps), however, turn out to distort the harmony of several color combinations. Hence, the optimization of the relative spectral power distribution of the light source should take the entire distribution of special color harmony rendering indices into account.

## 6.4
## Principal Components of Light Source Color Quality

### 6.4.1
### Factors Influencing Color Quality

As already mentioned in Section 6.2, light source color quality includes different aspects of the observer's color perception and judgment about the colored objects in the environment lit by the light source [2, 3, 27]. *Color fidelity* (defined in Section 6.1) represents a classic property. An equivalent term to color fidelity is *color rendering* [1]. Some visual tasks require that the light source should alleviate the perception of small color differences of reflecting color samples. This is the *color discrimination* property of the light source. *Visual clarity* is related to the general brightness sensation of the lit environment and also to the feeling of global contrasts, that is, the presence of large color differences among the objects [28].

It is also believed that the perception of fine shadings with continuous *color transitions* on the surface of the object is a further important aspect. *Color preference* is related to an esthetic judgment about the object colors usually paying special attention to their vividness and naturalness. *Color harmony*, in turn, expresses an esthetic judgment about the *relationship* among the colors of some selected objects in the scene. The concept of memory color rendering was also introduced [29, 30] based on the similarity of an object color under the test light source to long-term memory colors. This aspect can be called *memory* property because it corresponds to the observer's judgment about whether the color distribution of the object under a test light source is congruent with its long-term memory color distribution.

For a comprehensive optimization of the relative spectral power distribution of light sources, it is important to use different mathematical formulas for the different properties of color quality. Such formulas (e.g., the color rendering index) enable the user to predict the perceived magnitude of a property of color quality (e.g., color fidelity) from the instrumentally measured spectral power distribution. For a comprehensive design of color quality, one important question concerns the interdependence (correlations) among these predictors. The aims are to find out the number of noncorrelating indices, to find a possible trade-off among the properties of color quality, and to obtain a reasonable target function for the optimization of light source spectral power distributions.

To this end, Guo and Houser [31] compared several existing predictors computationally. Their computation included the CIE color rendering index ($R_a$) [1], Judd's "flattery" index ($R_f$) [32], Thornton's CPI [33], Thornton's CDI [34], Xu's color rendering capacity (CRC) [35], Fotios' cone surface area (CSA) [36], and Pointer's color rendering index ($R_P$) [37]. Significant correlations were found between $R_a$ and $R_f$, CPI, CRC, $R_P$; $R_f$ and CPI, CRC, $R_P$; CPI and CDI, CRC, $R_P$; CDI and CRC, CSA, $R_P$; CRC and CSA, $R_P$; and CSA and $R_P$.

From the analysis of these correlations, two principal components were extracted: (1) a color gamut (color area or color volume)-based component (explaining CSA, CRC, and $R_P$); and (2) a reference light source-based component (explaining $R_a$, $R_f$, CPI, and CDI). These two principal components of color quality were corroborated by Rea and Freyssinier-Nova [38] who stated that a gamut area-based index together with $R_a$ was suitable to predict visual judgments about color discrimination, vividness, and naturalness.

In accordance with this finding, Hashimoto and Nayatani [39] proposed a gamut area-based index (FCI) together with $R_a$ to account for color quality. A further approach to describe color rendering proposed by Yaguchi et al. [40] uses categorical color names. Combined fidelity–preference indices were also introduced, for example, the NIST Color Quality Scale (see Section 6.2.2.2) [41] and Schanda's combined preference–rendering index [42]. As already mentioned in Section 6.3, Szabó et al. [20] compared their color harmony rendering index $R_{hr}$ with $R_a$ for a set of test light sources computationally and found negative correlation between $R_a$ and $R_{hr}$. They also included CQS [41] (see Section 6.2) in their computation and $R_a$ correlated significantly with CQS.

Concerning the visual experiments of color quality (other than the visual color rendering task described in Section 6.2.1), the assessment of the observer's color preference plays an important role where it is crucial to discern color preference-related tasks from the color fidelity task, that is, the evaluation of color differences between a test and a reference condition. In a typical color preference study, subjects evaluated the attractivity, the naturalness, and the so-called *suitability* of different fruit and vegetable colors (suitability refers to an easy visual decision when choosing preferred fruits and vegetables) under different mixtures of LED light sources compared with a tungsten halogen and a fluorescent light source [43]. The CIE color rendering index ($R_a$) did not correlate well with the subjective judgments of the observers. But attractiveness correlated well with the gamut area index (see also Ref. [38]).

In another experiment, color discrimination ability was investigated among small color differences by using the so-called *Circle 32 test* [44] in which subjects had to arrange 32 desaturated color swatches equally distributed along the CIELAB hue circle. Color discrimination ability diminished significantly only in case of spectrally very imbalanced light sources (RGB and RGBA LED clusters) and it correlated well with the CIE color rendering index $R_a$.

In a further study [45], 18 observers evaluated the vividness and naturalness of an arrangement of colored objects (real fruit, vegetables, and the Macbeth ColorChecker® chart) under six light sources (three warm white at 3000 K and three cool white light sources at 4600 K) with high color rendering and low size of the color gamut [38], low color rendering and high gamut, and high rendering *and* high gamut. The highest amount of visual naturalness exhibited the light sources of both high rendering and high gamut property. However, the high color rendering property alone did not yield high visual naturalness. Best vividness was achieved by the light sources of lower color rendering and high gamut.

An important study of color quality validating the so-called color quality scale (see Section 6.2.2.2) [6] was carried out in NIST's Spectrally Tunable Lighting Facility [46]. This is a room at the NIST in which every relative light source spectral power distribution can be produced with good accuracy by the aid of a complex light source. Different white LED spectra ($70 < R_a < 95$ at 3000 and 4000 K) were simulated. The spectra were chosen to either increase or decrease the saturation of different object colors, for example, real fruit, sushi dish, artificial flowers, or the own skin. Pairs of lights were presented alternately and subjects had to choose which of the two lights made the objects look better. The CQS scale [6] worked significantly better than the CIE color rendering index by describing the subjects' choice.

So-called color acceptability studies were also carried out about the *color appearance of entire* (three-dimensional) *objects* (containing fine shadings and textures) or their two-dimensional projections on a display [47]. But up to now, no descriptor of color acceptability (in the above sense) has been proposed due to the complexities of modeling image color appearance and image color differences. Color acceptability corresponds to the observer's judgment about whether the entire color distribution of the object under the test light source is congruent with its color distribution stored in long-term color memory.

In a comprehensive visual color quality experiment, Nakano et al. [48] evaluated five different scenes with colored objects illuminated by 10 different light sources on 18 semantic differential scales (high fidelity–low fidelity, easy to tell colors–difficult to tell colors, warm–cool, saturated–desaturated, merry–sober, good–bad, clear–faint, strong–weak, comfortable–unpleasant, beautiful–ugly, cheerful–gloomy, soft–hard, preferable–dislike, new–old, vivid–inert, yellow–blue, red–green, and bright–dark) by four observers. The five scenes were simulated on a well-calibrated computer screen by using hyperspectral images. Two principal components with contribution rates of 70% (first component) and 16% (second component), respectively, were found. The second component correlated well with $R_a$ ($r^2 = 0.74$).

In the next section (Section 6.4.2), the correlations among selected visual scales of color quality (in some respect similar to the Nakano et al.'s [48] experiment) will be analyzed without making numerical predictions from physically measured properties. The comprehensive psychophysical method of obtaining these visual scales will be shown using a detailed questionnaire of color quality. In Section 6.4.3, visual results will be interpreted by a four-factor model of color quality resulting from a principal component analysis of the visual scales.

### 6.4.2
**Experimental Method to Assess the Properties of Color Quality**

In this section, a comprehensive method of the visual assessment of color quality is described. Observers had to answer a detailed color quality questionnaire by regarding a complex visual stimulus, a so-called *tabletop* arrangement. This tabletop arrangement was a *still life* consisting of various colored objects representing real indoor scenes in a viewing booth (see Figure 6.17).

The viewing booth was illuminated from above diffusely by three different light sources. Only one light source illuminated the still life at a time; the other two light sources were obstructed by a movable mechanical checkerboard cover so that their light did not reach the viewing booth. The experimenter could change the light source by moving the cover. Each light source had a correlated color temperature of about 2600 K. A tungsten incandescent lamp (INC, reference), a compact fluorescent lamp (CFL), and a white phosphor LED lamp (LED) were used.

Observers were not aware of the type of the light source. Light sources were denoted by numbers in the experiment while the light source itself was hidden. The still life included a white standard in the middle to ensure good chromatic adaptation. The mean luminance of the white standard was $107 \pm 6 \, \text{cd/m}^2$ including the variation among the three light sources. Light source colorimetric data are summarized in Table 6.7, while Figure 6.18 shows their relative spectral power distributions (both measured at the white standard).

6.4 Principal Components of Light Source Color Quality | 297

**Figure 6.17** Still life arrangement (150 cm × 80 cm × 80 cm) to assess the different properties of color quality visually in binocular viewing. The distance between the 10 cm × 15 cm PTFE white standard and the observer's eye was 70 cm.

**Table 6.7** Colorimetric data of the light sources measured at the white standard.

| Light source | INC | CFL | LED |
|---|---|---|---|
| $L$ (cd/m$^2$) | 106 | 111 | 110 |
| $x$ | 0.469 | 0.484 | 0.467 |
| $y$ | 0.412 | 0.423 | 0.422 |
| CCT (K) | 2589 | 2480 | 2684 |
| $R_a$ | 99.7 | 83.9 | 89.4 |

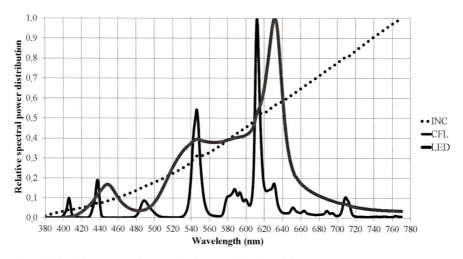

**Figure 6.18** Relative spectral power distributions of the three light sources.

Thirty observers of normal color vision participated in the experiment. Their color vision was tested by Farnsworth's D-15 test. Observers were looking at the still life covering their whole visual field. They were asked to assess the different properties of color quality visually by putting crosses on the different rating scales of the questionnaire concerning one specific property of color quality.

Each question consisted of the assessment of one or more items. Observers had to assess only one given item at a time. This item was either a single object or a combination of objects or the whole still life. They compared the appearance of the current item under the three light sources. One observer rated a given item under each light source only once. But during this rating process, the experimenter was allowed to uncover any light source any time according to the observer's request before moving to the next object or to the next question but it was required to view the still life under a given light source for at least 3 s.

This time was supposed to be enough to readapt to the slight changes of chromaticity among the light sources, especially due to the white standard that was always visible in the middle of the still life. Observers did not require a longer observation time under each light source possibly because then their short-term memory of object color appearance would have faded out and made the comparison difficult.

Before filling the questionnaire, every observer was trained to distinguish among the different properties of color quality very thoroughly. A synoptic overview was given about all questions of the questionnaire. The definition of each property was written on the questionnaire and the differences among these properties were explained. The following nine questions (Q1–Q9) were asked about the color quality of the still life.

- **About brightness (Q1).** Assess the brightness of the following items (see Figure 6.17): white standard, roses, a specific green leaf, the blue wool, and finally, the whole still life. The fixed value of 1.00 was assigned to the brightness under the tungsten incandescent lamp (INC, reference). Observers had to put two crosses on the questionnaire corresponding to the brightness they perceived under the CFL and the LED, compared to INC ($\equiv 1.00$). For CFL and LED, values greater than 1 were allowed. For the evaluation of the experimental results, the average rating of these five items for each light source and each observer constituted variable $V_1$ called "brightness" for 90 cases: each of the 30 observers and each of the 3 light sources (for INC, $V_1 \equiv 1.00$).
- **About attractiveness (Q2).** Assess attractiveness according to an esthetic judgment of the colors of the following items: pullover, the right doll's skin tone, red dahlia, rose dahlia, orange dahlia, green pepper, blue wool, and the water lily, related to INC ($\equiv 1.00$). For CFL and LED, values greater than 1 were allowed. Assess whether you like the color appearance of the *individual* colors and not the combination of colors. The average rating of these eight items for each light source and each observer constituted variable $V_2$ called "preference" (for INC, $V_2 \equiv 1.00$).
- **About color harmony (Q3).** Assess the color harmony of each of 19 combinations of watercolors (12 two-color and 7 three-color combinations, 3 cm diameter each,

**Figure 6.19** Nineteen two-color and three-color combinations of watercolors used to assess color harmony. The diameter of each circular watercolor was 3 cm.

see Figure 6.19). Only one color combination was seen at a time. It was put in the middle of the still life on the gray cover, 60 cm from the observer's eye. It was explained that an esthetic judgment about the *relationship* among the constituent colors was required. The average rating of these 19 items for each light source and each observer constituted variable $V_3$ called "harmony" (for INC, $V_3 \equiv 1.00$).

- **About color categories (Q4).** Assess whether the color categories within adjacent pairs of circular watercolors ($d = 3$ cm) are visually clearly distinct and recognizable in the watercolor box in the middle of the still life below the white standard (see Figure 6.17). Following seven pairs were assessed: yellow/orange, orange/red, red/purple, purple/lilac, and lilac/blue in the upper row, and blue/dark green and dark green/grass green in the lower row. The observer was instructed that – in an ideal case – it should be easy to distinguish the color categories of the adjacent watercolors (e.g., orange from yellow) visually. The average rating of these seven items for each light source and each observer constituted variable $V_4$ called "categories" (for INC, $V_4 \equiv 1.00$).
- **About color gamut (Q5).** Assess the extent of the color gamut of the whole scene whether all possible colors are present including saturated colors, especially with respect to the Macbeth ColorChecker® chart in the still life (see Figure 6.17). This rating constituted variable $V_5$ called "gamut" (for INC, $V_5 \equiv 1.00$).
- **About the continuity of color transitions (Q6).** Assess the presence of fine color transitions (shadings) within the following objects: the roses on the right, the green leaf, and the blue wool. The observer was instructed that there should be as many delicate color shades on the object as possible. The average rating of these three items for each light source and each observer constituted variable $V_6$ called "transition" (for INC, $V_6 \equiv 1.00$).

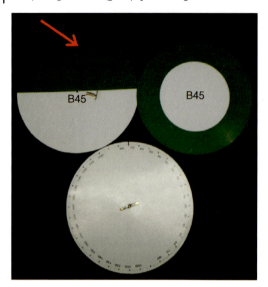

**Figure 6.20** Color circles ($d = 21$ cm) to assess small color differences. Upper left circle: cover plate with the color center that was a homogeneous half circle containing a 2 cm × 2 cm slot indicated by a red arrow in which a part of the color transition of the lower circle was visible; upper right circle: two symmetric printed continuous color transitions (2 × 180°, from top to bottom, one on the left and one on the right) between the color center and a second different color; bottom: two symmetric scales between 0° and 180° to read the observer's setting of just noticeable color difference in degrees.

- **About small color differences (Q7).** Observers had to turn a paper circle of a diameter of $d = 21$ cm viewed from a distance of 60 cm. The circle contained two copies of the same printed continuous color transition, each covering 180 degrees from the top to the bottom. One color transition was on the left-hand side and the other one on the right-hand side. The color transition was distributed along a line on the CIELAB $a^*$–$b^*$ plane between a color center ($C_1$) and a second different color ($C_2$). $\Delta E_{ab}^*$ values between $C_1$ and $C_2$ ranged between 7.0 and 20.0 corresponding to 0.04–0.11 CIELAB color difference units per degree. CIELAB $L^*$ was held constant for all colors at $L^* = 50$. Observers had to turn this (lower) paper circle under another (cover) paper circle (see Figure 6.20). The upper half of the cover circle was printed homogeneously (color center $C_1$) and contained a 2 cm × 2 cm slot in which a part of the color transition of the lower circle was visible. The lower half of the cover circle contained the denotations of the eight color centers (A0–D45) (see Figure 6.21). The eight pairs of color centers ($C_1$) and second colors ($C_2$) are depicted in the CIELAB $a^*$–$b^*$ diagram of Figure 6.22.

The observer had to turn the lower circle until the color difference to the upper circle (cover plate) became just noticeable. Then, the scale on the back of the lower circle ($\alpha$) was read. $\alpha = 0$ corresponds to complete physical agreement with the color center and $\alpha = 180°$ corresponds to the second color (i.e., the maximum color difference). The observer's answer was transformed in the following way:

### 6.4 Principal Components of Light Source Color Quality | 301

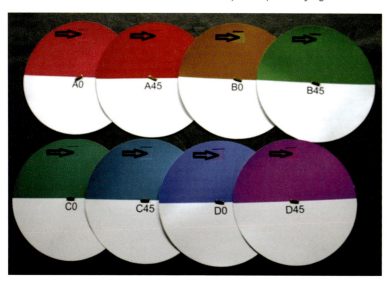

**Figure 6.21** Eight color circles (cover plates, $d = 21$ cm) to assess small color differences. The upper half circle was the homogeneous color center with a 2 cm × 2 cm slot where a part of the continuous color transitions printed on the lower circle can be seen (indicated by black arrows). The white half circle contains the denotations of the eight color centers.

$a' = (180° - a)/180°$. Then, the mean value of $a'$ was computed for the eight circles and this mean value constituted variable $V_7$ called "difference". There was no reference light source in this question.

- **About the similarity to memory colors (Q8).** Assess the similarity of the color appearance of four objects (a specific green leaf of the still life, the jeans below the dolls, the right doll's skin tone, and the roses) compared to your long-term *memory*

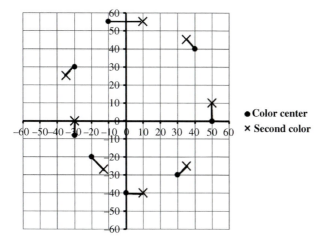

**Figure 6.22** Eight pairs of color center ($C_1$) and second color ($C_2$) in a CIELAB $a^*$–$b^*$ diagram.

colors of the same objects as you remember them, that is, by using an internal reference. The average ratings of the above four items constituted variable $V_8$ called "*memory*". There was no reference light source in this case. Observers had to put three crosses on the questionnaire for every object, corresponding to the similarity to the internal reference they perceived under the tungsten light source, the compact fluorescent lamp, and the white phosphor LED lamp. The maximum value of each scale was equal to 1.00 corresponding to the subject's long-term memory color.

- **About color fidelity (Q9).** Assess the similarity of the color appearance of six objects (carrot, green pepper, roses, the specific green leaf, jeans, and the right doll's skin tone) under CFL and LED to the color appearance of the same object under the reference light source (INC). This is the conventional task of visual color rendering. The average rating of these six items constituted variable $V_9$ called "fidelity" (for INC, $V_9 \equiv 1.00$).

### 6.4.3
### Modeling Color Quality: Four-Factor Model

The data set of 810 values of the variables $V_1$–$V_9$ (9 questions × 30 observers × 3 light sources) described in Section 6.4.2 constituted the input of a factor analysis. The method of principal component analysis was chosen. It was based on the $\{V_1$–$V_9\}$ correlation matrix. Four factors were extracted and rotated. Table 6.8 shows the factor loadings.

Factor loadings greater than 0.6 (indicated by bold letters in Table 6.8) were used to impute a label to every factor in the following way: "preference" to $F_1$, "brightness" to $F_2$, "gamut" to $F_3$, and "difference" to $F_4$. This represents a so-called four-factor model of color quality. In this sense, the preference factor ($F_1$) is responsible for five properties, preference, harmony, memory, fidelity, and transitions. The brightness

**Table 6.8** Factor loadings of the variables $V_1$ – $V_9$.

| Variable | Question | Factor loadings | | | |
|---|---|---|---|---|---|
| | | $F_1$ | $F_2$ | $F_3$ | $F_4$ |
| | Factor labels | Preference | Brightness | Gamut | Difference |
| $V_2$ | Preference | **0.860** | 0.247 | −0.065 | 0.134 |
| $V_3$ | Harmony | **0.802** | −0.069 | −0.088 | 0.121 |
| $V_8$ | Memory | **0.676** | 0.122 | 0.027 | −0.485 |
| $V_9$ | Fidelity | **0.655** | −0.257 | 0.424 | 0.002 |
| $V_6$ | Transition | **0.636** | −0.134 | 0.257 | −0.369 |
| $V_1$ | Brightness | −0.143 | **0.927** | −0.115 | 0.038 |
| $V_4$ | Categories | 0.403 | **0.645** | 0.462 | −0.038 |
| $V_5$ | Gamut | −0.052 | 0.012 | **0.926** | 0.034 |
| $V_7$ | Difference | 0.076 | 0.030 | 0.050 | **0.887** |

Bold values were used to impute labels to the factors.

factor ($F_2$) seems to be in charge of $V_1$ (brightness) and $V_4$ (categories). The gamut factor ($F_3$) is responsible for gamut and the difference factor ($F_4$) for color differences.

It can also be seen from Table 6.8 that five properties (preference, harmony, memory, fidelity, and transition) could be described by the first (preference) factor ($F_1$). This finding is advantageous for light source design because it suggests that there is a common underlying color quality property. Increasing that common property ($F_1$), for example, at a first phase of the optimization procedure of the relative spectral power distribution of the light source, all five properties will increase to obtain a light source satisfying all five color quality criteria. But increasing any of the five properties does not necessarily mean an increase in the other property because only part of their variance is explained by $F_1$ (see Table 6.8). Thus, optionally, at a second phase of the optimization procedure, either property can be fine-tuned according to the expected field of application of the light source.

The association among certain properties of color quality does not mean that they do not require separate numerical indices. Just the opposite is true: it is important to develop an appropriate correlate (or index) for every property that can be the target of a specific optimization for a special lighting task where, in the meantime, a satisfactory result can be obtained for the associated properties. A good example is color preference and color fidelity: optimizing for color fidelity does not contradict to preferred colors but the *most* preferred colors cannot be reached. For *general* lighting purposes, it is not appropriate to optimize color preference because the deviations from fidelity (especially chroma enhancing deviations) may misrepresent some important objects and lead to invalid decisions of the light source user.

### 6.4.4
### Principal Components of Color Quality for Three Indoor Light Sources

In this section, the color quality factors $F_1$–$F_4$ of Section 6.4.3 are analyzed for the example of the three indoor light sources in Figure 6.18. Figure 6.23 compares the

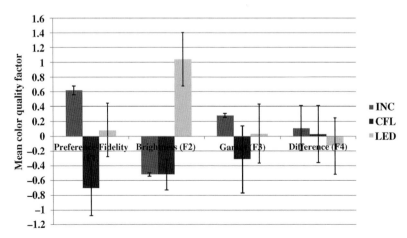

**Figure 6.23** Mean color quality factors ($F_1$–$F_4$) for the test light sources. Mean values of 30 observers and their 95% confidence intervals are shown.

values of these color quality factors ($F_1$–$F_4$) for the three light sources in the four-factor model. Mean factor values of 30 observers and their 95% confidence intervals are shown. Factor values were computed by a regression method.

As can be seen from Figure 6.23, the test light sources exhibit different rank orders according to the different factors. The tungsten light source (INC) has the highest ranking for $F_1$ (preference) followed by LED and CFL. Nonoverlapping confidence intervals showed that this tendency was significant. LED has the highest ranking for $F_2$ (brightness) followed by INC and CFL (latter two were not significantly different).

The effect of light source on $F_2$ was significant (ANOVA, $F=53.3$, $p<0.001$). CFL had a smaller gamut factor ($F_3$) than INC but the overall effect of light source on gamut was not significant at the 5% level (ANOVA, $F=2.8$, $p=0.068$). The type of light source did not affect factor $F_4$ ("difference") significantly. Finally, note that the tungsten lamp (INC) was a reference light source with fixed scale values (1.00) for all variables but $V_7$ and $V_8$. This is the reason of the INC's small confidence intervals except for $F_4$ that accounts mainly for $V_7$.

The color difference factor ($F_4$) did not exhibit significant changes among the present test light sources indicating that this factor may not be relevant for optimizing the color quality of high-grade light sources. Indeed, color discrimination ability diminished significantly only in case of spectrally very imbalanced light sources (RGB and RGBA LED clusters) in a previous study [36].

## 6.5
### Assessment of Complex Indoor Scenes Under Different Light Sources

Complex indoor scenes consist of numerous illuminated objects of different colors whose color appearance changes when different light sources are used to illuminate them. The perceived color quality of the environment may be distorted or enhanced by installing a new light source in the room. The choice of the light source should enhance the observer's *judgment* about all aspects of perceived color quality. Judgment scales may exhibit a nonlinear relationship with the underlying perceptual quantity. Section 6.5.1 deals with such a relationship between the perceived color differences of colored test objects and the observer's corresponding similarity judgment where differences and similarity are meant in the context of color fidelity, between the test and the reference light sources.

Another important question is how to quantify the brightness perception of colored objects in a complex indoor scene in association with color fidelity, color harmony, and color gamut. This will be elucidated in Section 6.5.2 by the example of the 34 test color samples and 42 test light sources already described in Section 6.3.3. It will be pointed out that – after ensuring a high mean luminance level – luminous efficiency is not a reasonable target of optimizing an indoor lit environment. Finally, the issue of whiteness perception and light source chromaticity is discussed in Section 6.5.3 by the example of the above 42 test light sources.

## 6.5.1
### Psychological Relationship between Color Difference Scales and Color Rendering Scales

Deficits of the current color rendering index were described in Section 6.2.2.1 including the complication arising when interpreting the scale of the general color rendering index. This section deals with the final step of any color rendering computation method, that is, the conversion (linear or nonlinear) of a color difference value to a color rendering index. The prediction of observer judgments about the color appearance of typical objects under a test light source in a built indoor environment is crucial in order to characterize the color rendering of that test light source.

In Step 6 of Section 6.1, the color rendering index ($R$) of a test light source is computed from a predicted color difference ($\Delta E$) of a test color sample between the test light source and the reference light source: $R = 100 - 4.6 \Delta E_{U^*V^*W^*}$. The latter formula has been defined on a technological basis [1], where the quantity $R$ represents a technical term to quantify the color rendering property of the test light source. But Step 6 of the computation method of the color rendering index should carry a psychological meaning; that is, it should be a prediction of the observer's *judgment* $R_p$ about the *similarity* of the colors of the objects in a scene illuminated by the test light source to the colors of the objects in the same scene but illuminated by a reference light source.

In this sense, the aim of this section is to describe the result of a psychophysical experiment to establish this psychological relationship $R_p(\Delta E_{UCS})$ between instrumentally measured color differences (expressed in the CIECAM02-UCS units) and the observers' color rendering judgments. Advantages of using such a psychological relationship include the easy interpretation of such an index for nonexpert users. One possible relationship (RCRI) was introduced in Section 6.2.2.2. The value of RCRI [4] informs the user of the test light source about how many test color samples yield a good or excellent matching to the reference color appearance.

In the RCRI experiment, however, only homogeneous test color samples were included and no real test objects of characteristic color distributions were used. For practice, it is important to assess the color differences of real objects. Real objects (natural or artificial, for example, flowers, fruits, vegetables, or toys) have typical color distributions [49] and these color distributions change characteristically if a different light source illuminates the same objects, for example, a white LED lamp instead of, for example, a tungsten lamp. It is also important to find out the corresponding color similarity rating on a more continuous scale (unlike the RCRI scale) allowing intermediate values like 1.7 or 2.3 and not only the ordinal ratings 1 (excellent), 2 (good), 3 (acceptable), 4 (not acceptable), and 5 (very bad) of the RCRI method. The apparatus of the visual experiment to establish the continuous psychological relationship $R_p(\Delta E_{UCS})$ is shown in Figure 6.24.

Light sources of the experiment included 1 incandescent reference light source and 10 test light sources of different color rendering properties with $R_a$ values ranging between 18 and 93 (see Figure 6.25). White points (at about 2800 K) were

## 306 | 6 Improving the Color Quality of Indoor Light Sources

**Figure 6.24** Apparatus of the visual experiment to establish the continuous psychological relationship $R_p(\Delta E_{UCS})$. (1) Lighting board with white LEDs, RGB LEDs, and incandescent lamps to produce the 10 test light sources and the incandescent reference light source; (2) diffuser plate (the space between the board and the plate is surrounded by white walls to provide diffuse illumination dismounted by taking this photo); (3) white standard; (4) the artificial test objects of the still life, during the experiment fresh objects and the observer's hand were also assessed; (5) switch for the observer with two buttons, one for the actual test light source and one for the reference light source; (6) computer with a dedicated software user interface to control the light sources and to input the values of the two scales (color difference perception scale and similarity judgment scale) by the aid of two sliders; (7) electronic circuitry to drive the lighting board by the same software.

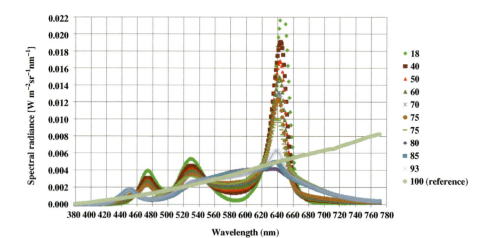

**Figure 6.25** Spectral power distributions of the 11 light sources in the visual experiment to establish the continuous psychological relationship $R_p(\Delta E_{UCS})$. The values of the CIE general color rendering index ($R_a$) values are shown in the legend.

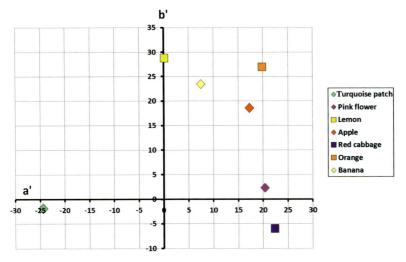

**Figure 6.26** Test objects in the CIECAM02-UCS $a'-b'$ diagram under the tungsten reference light source. Compare with Table 6.9. Every subject tested her or his own hand. The diagram shows the example of one specific observer.

matched visually at the white standard (see Figure 6.24). Illuminance values measured at the white standard ranged between 701 and 710 lx.

The 10 test light sources were mixtures of white LEDs of high and low color rendering properties as well as RGB LEDs. All light sources were mounted on a lighting board (no. 1 in Figure 6.24) to provide homogeneous illumination via the diffuser plate (no. 2) on the table where a still life was arranged consisting of a white standard (no. 3) and several artificial test objects (no. 4). During the visual experiment, homogeneous color patches, fresh fruits, and flowers were also assessed (seven objects altogether) (see Table 6.9 and Figure 6.26).

All test objects were visible at the same time in the still life arrangement (no. 4 in Figure 6.24) together with some further objects. All 15 observers had normal color vision confirmed by Farnsworth's D-15 test. Observers were assessing one test object at a time by switching between the test light source (min. 2 s) and the reference light source (min. 2 s) as many times as required by using the two buttons (no. 5 in Figure 6.24). The observer's task was to judge the *similarity* of each test object between their test and reference color appearance to establish the psychological color rendering scale. The observer had to make a continuous judgment by setting a slider on the similarity judgment scale at the user interface of a dedicated psychophysical computer program (no. 6 in Figure 6.24). Observers had to readapt to the reference white for at least 10 s every time after looking at the computer monitor (no. 6 in Figure 6.24) to enter the similarity judgment.

The similarity judgment scale ranged between 1 and 6. Whole numbers were assigned semantic categories (1: very good; 2: good; 3: moderate; 4: low; 5: bad; 6: very bad). But the observer was required to make continuous judgments similar to the

**Table 6.9** Seven test objects and their CIECAM02-UCS $J'$, $a'$, $b'$ values computed under the tungsten reference light source (compare with Figure 6.26).

| Test object | Type | Photo | CIECAM02-UCS | | |
|---|---|---|---|---|---|
| | | | $J'$ | $a'$ | $b'$ |
| Turquoise | Homogeneous patch | | 70.4 | −24.4 | −1.9 |
| Pink flower | Real | | 53.6 | 20.4 | 2.3 |
| Orange | Real | | 58.2 | 19.9 | 26.9 |
| Lemon | Artificial | | 76.2 | 0.1 | 28.8 |
| Apple | Artificial | | 69.1 | 17.3 | 18.6 |
| Red cabbage | Artificial | | 37.5 | 22.4 | −6.0 |
| Banana | Real | | 76.4 | 7.5 | 23.4 |

school grade system (e.g., 1.7) in certain countries. The driver electronics of the lighting board (no. 7 in Figure 6.24) was controlled by the same computer program. A mean $\Delta E_{UCS}$ value was assigned to the change of the color difference distribution of each natural and artificial object between each test light source and the reference light source.

Similarity ratings (visual scaling of $R_p$) of 15 observers are described for the seven objects listed in Table 6.9 as a function of their instrumentally measured mean color differences ($\Delta E_{UCS}$) between the test light source and the reference light source. Figure 6.27 shows the mean similarity ratings ($R_p$) of 15 observers and their 95% confidence intervals for each of the seven objects individually as a function of $\Delta E_{UCS}$.

As can be seen from Figure 6.27, the $R_p(\Delta E_{UCS})$ of all seven objects exhibit similar nonlinear monotonically decreasing tendencies modeled by third-order polynomial connecting curves. However, similarity ratings were object dependent, observers being more critical about certain objects (apple, banana, pink flower, and lemon) and less serious about the other objects (turquoise color patch, red cabbage, and orange). The latter tendency depended on the magnitude of $\Delta E_{UCS}$. Mean similarity ratings ($R_p$) of 15 observers and all 7 objects are depicted in Figure 6.28. Intervals show the 95% confidence intervals of these overall mean similarity ratings.

## 6.5 Assessment of Complex Indoor Scenes Under Different Light Sources

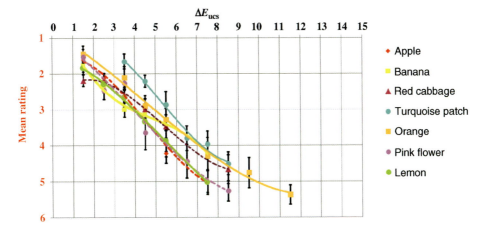

**Figure 6.27** Mean similarity ratings ($R_p$) of 15 observers and their 95% confidence intervals for each of the seven objects as a function of $\Delta E_{UCS}$. Connecting curves for each object: third-order polynomials fitted to the experimental data. Red curve: fit curve to the mean tendency of all color objects and all observers (see Figure 6.28 and Equation 6.4).

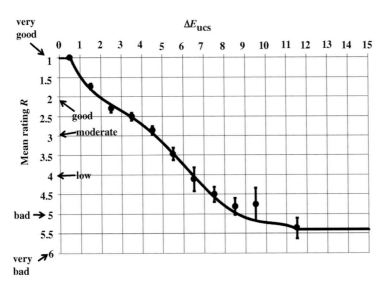

**Figure 6.28** Dots: mean similarity ratings ($R_p$) of 15 observers × 7 test objects. Intervals: their 95% confidence intervals. Abscissa: instrumentally measured color differences ($\Delta E_{UCS}$). Continuous curve: fit curve to the mean tendency of all objects (see Equation 6.4). Labels to the left ordinate (very good, good, moderate, low, bad, and very bad) represent semantic rating categories assigned to the whole numbers of the continuous similarity rating scale. Compare with Table 6.11.

**Table 6.10** Values of the polynomial coefficients of Equation 6.4.

| $i$ | 0 | 1 | 2 | 3 | 4 | 5 |
|---|---|---|---|---|---|---|
| $a_i$ | 0.24389926 | 1.78724426 | −0.68216738 | 0.14116555 | −0.01249592 | 0.00039117 |

As can be seen from Figure 6.28, mean similarity ratings (dots) reveal a characteristic nonlinear decreasing tendency. The following function was fitted to the experimental results running within the confidence intervals:

$$R = 1 \quad \text{if } \Delta E_{UCS} < 0.5$$
$$R = \sum_{i=0}^{5} a_i (\Delta E_{UCS})^i \quad \text{if } 0.5 \leq \Delta E_{UCS} \leq 11.5 \quad (6.4)$$
$$R = 5.39952461 \quad \text{if } \Delta E_{UCS} > 11.5$$

Polynomial coefficients of Equation 6.1 are listed in Table 6.10.

Equation 6.4 was fitted to the psychophysical similarity rating data set of 15 observers × 7 objects. Hence, it can be considered as a psychological relationship between the instrumentally measured color differences and the final step of the computational method of the color rendering index. Equation 6.4 adds an important component to any color rendering method, that is, the prediction of the observer's expected similarity rating (judgment) if the magnitude of the color difference between the test light source and the reference light source is known. Equation 6.4 helps the nonexpert user interpret the scale of any color rendering index in the following way. Any value of any color rendering (or fidelity) index can be associated with a corresponding $\Delta E_{UCS}$ color difference value. By using Equation 6.4, a similarity rating can be computed. Table 6.11 elucidates this by the example of the CRI-CAM02UCS ($R_{UCS}$) color rendering metric.

As can be seen from the example of Table 6.11, the $R_{UCS}$ scale [10] can be interpreted by the nonexpert user in terms of common expressions such as "very good" or "moderate". This interpretation represents an added value to the color rendering index. Note that instead of Equation 6.4, other mathematical functions can also be used to fit the experimental data shown in Figure 6.28.

**Table 6.11** Interpretation of the values of the CRI-CAM02UCS ($R_{UCS}$) color rendering metric [10] by the aid of Equation 6.4.

| $R_{UCS}$ | $\Delta E_{UCS}$ | R (Equation 6.4) | Semantic rating |
|---|---|---|---|
| 96.8 | 0.4 | 1.00 | Very good |
| 92.0 | 1.0 | 1.48 | Good–very good |
| 84.8 | 1.9 | 1.99 | Good |
| 76.8 | 2.9 | 2.3 | Good–moderate |
| 68.8 | 3.9 | 2.7 | Moderate |

Compare with the labels at the ordinate of Figure 6.28.

## 6.5.2
### Brightness in Complex Indoor Scenes in Association with Color Gamut, Rendering, and Harmony: A Computational Example

In this section, perceived lightness of object colors is analyzed in relation to other aspects of color quality: color gamut, color rendering, and color harmony. In an indoor environment, a high enough overall luminance level should be maintained to ensure good visual acuity for good visual performance. This is important, for example, to read black letters on a white sheet of paper. Corresponding illuminance levels are defined in ISO standards, for example, ISO 9241-6. For example, a luminance level of $200 \, cd/m^2$ represents an acceptable value for good visual acuity on white surfaces for working in an office environment. Provided that an appropriate luminance level can be ensured, attention should be focused on the color quality aspects of the indoor scene and the relative spectral power distribution of the light source should be optimized for the relevant aspects of color quality in a given application.

The first aspect is the lightness of the colored objects in the scene, that is, their brightness relative to the brightness of a white surface illuminated by the same light source. When evaluating the color quality of an illuminated indoor scene visually, observers usually assess the lightness of the objects. To describe lightness impression correctly, it has to be kept in mind that more chromatic objects are perceived to be lighter than less chromatic objects of the same luminance. This is the so-called Helmholtz–Kohlrausch effect [50]. The effect depends on hue significantly. In this section, a predictor formula ($L^{**}$) is presented [51]. Alternative models of the Helmholtz–Kohlrausch effect include the more complex In-CAM (CIELUV) model [52]. The $L^{**}$ formula applies a CIELAB hue ($h$)- and chroma ($C^*$)-dependent correction to the CIELAB lightness ($L^*$) of the object color in the following way:

$$L^{**} = L^* + [2.5 - 0.025 \, L^*][0.116|\sin((h-90°)/2)| + 0.085]C^* \tag{6.5}$$

In Equation 6.5, $L^{**}$ represents a numerical correlate for the lightness of the object color including the chromatic component according to the Helmholtz–Kohlrausch effect. It is interesting to plot the value of $L^{**}$ against the value of relative luminance ($Y_{rel}$), that is, the luminance of the object divided by the luminance of a perfect white in the scene. Figure 6.29 shows this relationship by the example of the 34 test colors and the first 12 test light sources of Table 6.6, that is, from the light source pcLED1 up to the light source TUN.

Besides compression of the relative luminance scale for higher values – which is well known from CIELAB $L^*$ – the hue and chroma dependence of $L^{**}$ for the different object colors can also be seen from Figure 6.29. Some objects belong to the same range of relative luminance (e.g., $Y_{rel} = 20-40$) but they exhibit different relative lightness dependence on relative luminance, for example, ham, pear, and the yellow book. Therefore, to improve the lightness of certain (characteristic) colored objects in a complex illuminated indoor scene, it is desirable to use an (average) chromatic lightness metric such as $L^{**}$ as a target of light source optimization.

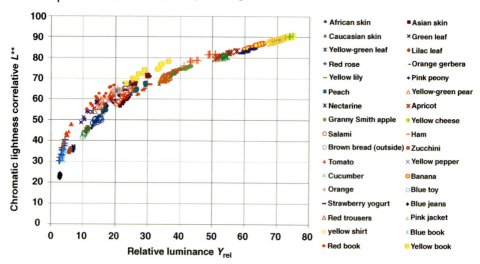

**Figure 6.29** Lightness correlate ($L^{**}$) [51] as a function of relative luminance ($Y_{rel}$). Example: 34 test color samples and the first 12 test light sources in Section 6.3.3. The relationship $L^{**}(Y_{rel})$ is shown for each test color separately.

The next question is how lightness perception is related to the other aspects of color quality. To elucidate this, the relationship among the correlate of mean lightness ($L^{**}$), mean special color fidelity index ($R_{UCS}$), mean special color harmony rendering index ($R_{hr}$, see Equation 6.3), and color gamut ($G$) was analyzed for the example of the 34 test color samples (or 561 color combinations in case of $R_{hr}$) and the 42 test light sources of Section 6.3.3. Mean values were computed by considering all 34 color samples. Color gamut ($G$) was estimated in CIECAM02 $J$, $a_C$, $b_C$ color space by summing up the distances in this space between the members of all possible two-color combinations among the 34 color samples and dividing by 10 000. The number of such combinations is equal to $34 \cdot 33 \div 2 = 561$, equivalent to Section 6.3.3. Figure 6.30 shows the relationship between $L^{**}$ and $G$. The 42 test light sources were classified into three CCT groups, warm (CCT < 3500 K), neutral (3500 K ≤ CCT < 5500 K), and cool (CCT ≥ 5500 K).

As can be seen from Figure 6.30, mean lightness ($L^{**}$) exhibits only a modest (about 6%) variability among the 42 light sources. The percentage value was calculated as the difference between the maximum and the minimum $L^{**}$ value divided by the mean $L^{**}$ value. Individual objects, however, may vary much more in lightness (see Figure 6.29). For example, the lightness of the red rose varies as much as 29% among the first 12 light sources of Table 6.6. Therefore, the choice of typical reflecting objects in the given application (e.g., flowers) should be considered by improving the *lightness* aspect of the light source.

It can also be seen that the warm CCT group provides on average more mean lightness for the particular set of the present 34 test objects than the neutral and the cool group. Note that cool white light sources tend to enhance the lightness of bluish test objects. Gamut (with a variability of about 18%) correlates well with lightness (with $r^2 = 0.53$, 0.66, and 0.87 for warm, neutral, and cool, respectively) at least within

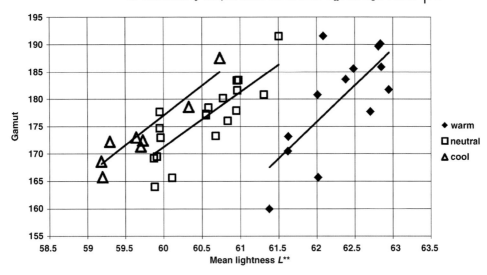

**Figure 6.30** Relationship between the lightness correlate $L^{**}$ [51] and color gamut G. Each symbol represents the average of 34 test color samples (see Section 6.3.3). Example of 42 light sources from Section 6.3.3. Light sources are sorted into three groups, warm (CCT < 3500 K), neutral (3500 K ≤ CCT < 5500 K), and cool (CCT ≥ 5500 K).

each CCT group. Figure 6.31 shows the relationship between the mean lightness correlate $L^{**}$ and $R_{UCS}$.

As can be seen from Figure 6.31, the color rendering index $R_{UCS}$ (with a variability of about 53% among the 42 light sources) correlates well with mean lightness (with $r^2 = 0.68$, 0.72, and 0.59 for warm, neutral, and cool, respectively) within each CCT

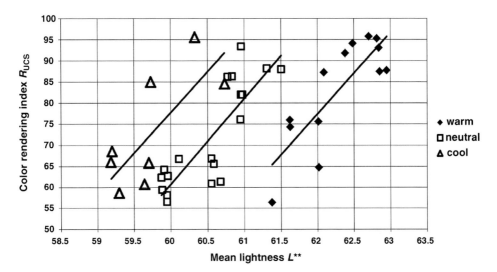

**Figure 6.31** Relationship between the lightness correlate $L^{**}$ [51] and the color rendering index $R_{UCS}$. See also the caption of Figure 6.30.

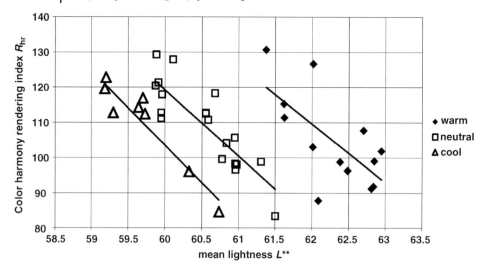

**Figure 6.32** Relationship between the lightness correlate $L^{**}$ [51] and the color harmony rendering index $R_{hr}$. See also the caption of Figure 6.30.

group. The absolute maximum of mean lightness *and* color rendering can be obtained in the warm CCT group. Figure 6.32 shows the relationship between mean lightness $L^{**}$ and the harmony rendering index $R_{hr}$.

As can be seen from Figure 6.32, there is a negative correlation between the color harmony rendering index $R_{hr}$ (with a variability of about 44% among the 42 light sources) and mean lightness (with $r^2 = 0.45$, 0.72, and 0.87 for warm, neutral, and cool, respectively) within each CCT group. Figure 6.33 shows the relationship between color gamut ($G$) and the color rendering index $R_{UCS}$.

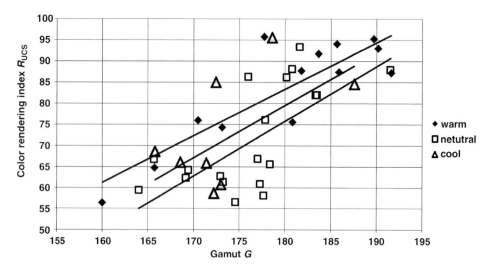

**Figure 6.33** Relationship between color gamut $G$ and the color rendering index $R_{UCS}$. See also the caption of Figure 6.30.

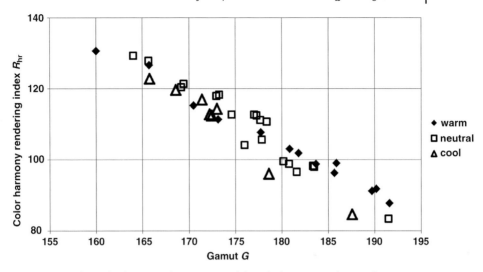

**Figure 6.34** Relationship between color gamut G and the color harmony rendering index $R_{hr}$. See also the caption of Figure 6.30.

As can be seen from Figure 6.33, the color rendering index $R_{UCS}$ correlates well with color gamut G ($r^2 = 0.75$) for warm white light sources. There is somewhat less amount of positive correlation between $R_{UCS}$ and G ($r^2 = 0.45$ and $r^2 = 0.33$) for neutral and cool white light sources, respectively. The absolute maximum of gamut *and* color rendering attain two light sources in the warm CCT group. As mentioned in Section 6.1, light sources of both high rendering and high gamut properties exhibited high visual naturalness [38]. Best vividness, however, was achieved by the light sources of somewhat lower color rendering and high gamut. Figure 6.34 shows the relationship between color gamut (G) and the color harmony rendering index $R_{hr}$.

As can be seen from Figure 6.34, there is a significant negative correlation between color gamut (G) and the color harmony rendering index $R_{hr}$ for every CCT group (with $r^2 = 0.99, 0.93$, and 0.96 for warm, neutral, and cool, respectively). Depending on the application, either the color harmony rendering property of the light source or the color gamut under the light source can achieve a maximum value. For both aspects of color quality, a trade-off between the two aspects is necessary. Figure 6.35 shows the relationship between the color rendering index $R_{UCS}$ and the color harmony rendering index $R_{hr}$.

As can be seen from Figure 6.35, there is a negative correlation between color gamut (G) and the color harmony rendering index $R_{hr}$ for every CCT group (with $r^2 = 0.74, 0.65$, and 0.45 for warm, neutral, and cool, respectively). Depending on the application, either the color harmony rendering or the color rendering property of the light source can attain a maximum. For the optimization of both aspects of color quality, a trade-off between the two aspects is necessary.

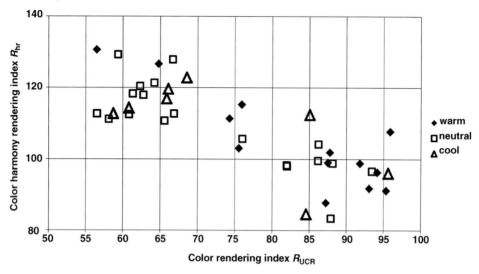

**Figure 6.35** Relationship between the color rendering index $R_{UCS}$ and the color harmony rendering index $R_{hr}$. See also the caption of Figure 6.30.

### 6.5.3
### Whiteness Perception and Light Source Chromaticity

In special applications, an important aspect of color quality may be the degree of perceived whiteness of the reflecting objects of different white tones in the indoor environment. Several different formulas were developed to predict perceived whiteness, for example, the CIE whiteness formula [53] and the so-called C/V whiteness index [54]. The perceived whiteness of CIE daylights reflected from the so-called theoretical perfect reflecting diffuser (an ideal object reflecting 100% of radiation at every wavelength) was predicted by the C/V index. The C/V value increased with the correlated color temperature of daylight, in accordance with visual observations. Normalized to 100 for CIE illuminant D65, the C/V whiteness index of, for example, the 4000 K phase of daylight equaled about 82, while for 12 000 K, the value of 109 was calculated [54].

A straightforward consequence is that the reference illuminants of the CIE color rendering index computation method (see Section 6.1) exhibit different perceived whiteness, according to the above tendency. But light source users do not always prefer a high perceived whiteness (see also Section 5.1.2.4); for example, in the important application of home lighting (at least in Europe), warm white light sources are preferred. Therefore, generally, the objective of spectral design for high light source color quality is not to maximize their whiteness but to avoid any "unusual" tone of the light source (e.g., slightly greenish or purplish) when looking at the light source itself or the white indoor surfaces illuminated by the light source. In today's indoor lighting technology, this aim is achieved by optimizing the relative spectral distribution of the light source so as to come close to the chromaticity of the *reference illuminant* (in the sense of Section 6.1) on the plane of the CIE 1976 uniform chromaticity scale diagram, or $u'$, $v'$ diagram (see Equation 1.9).

As the reference illuminant is either a blackbody (in other words, Planckian) radiator (if the correlated color temperature of the light source is less than 5000 K) or a phase of daylight (if the correlated color temperature of the light source is greater than or equal to 5000 K), this means that the light source shall be located as close as possible to either the Planckian locus or the daylight locus in the $u'$, $v'$ diagram (see Figure 5.5), in order to obtain a suitable white tone preferred by the users at a given correlated color temperature. To achieve this, the chromaticity difference $\Delta C$ (a Euclidean distance on the $u'$, $v'$ diagram) between the light source and its reference illuminant is minimized during the optimization. These chromaticity differences were computed for the example of the 42 light sources listed in Table 6.6. The resulting $\Delta C$ values are listed in Table 6.12.

As can be seen from Table 6.12, $\Delta C$ values range between 0.0000 (MLED9) and 0.0170 (pcLED2). There is a slight decreasing tendency of the $\Delta C$ values with an increasing value of $R_{UCS}$. In today's industrial practice, chromaticity difference values up to $\Delta C = 0.002$ are considered to yield a high-quality white tone while chromaticity difference values up to $\Delta C = 0.005$ are usually considered acceptable. Unfortunately, there is a visual chromaticity mismatch phenomenon between

**Table 6.12** Chromaticity differences ($\Delta C$, Euclidean distances on the $u'$, $v'$ diagram) between the light source and its reference illuminant.

| LS | CCT | $R_{UCS}$ | $\Delta C$ | LS | CCT | $R_{UCS}$ | $\Delta C$ |
| --- | --- | --- | --- | --- | --- | --- | --- |
| pcLED1 | 6344 | 69 | 0.0074 | pcLED6 | 3098 | 74 | 0.0028 |
| pcLED2 | 4579 | 67 | 0.0170 | pcLED7 | 3099 | 76 | 0.0013 |
| pcLED3 | 4870 | 93 | 0.0001 | pcLED8 | 2974 | 76 | 0.0046 |
| FL1 | 4374 | 59 | 0.0032 | pcLED9 | 3348 | 87 | 0.0128 |
| FL2 | 4445 | 88 | 0.0049 | pcLED10 | 4234 | 61 | 0.0004 |
| HMI | 4370 | 86 | 0.0081 | pcLED11 | 4309 | 66 | 0.0048 |
| retrLED | 2683 | 87 | 0.0050 | pcLED12 | 4333 | 67 | 0.0026 |
| pcLED4 | 2640 | 65 | 0.0092 | pcLED13 | 5046 | 56 | 0.0010 |
| pcLED5 | 2797 | 95 | 0.0038 | pcLED14 | 3982 | 61 | 0.0085 |
| FL3 | 2773 | 56 | 0.0032 | pcLED15 | 4061 | 76 | 0.0033 |
| FL4 | 2637 | 88 | 0.0020 | pcLED16 | 4821 | 86 | 0.0017 |
| TUN | 2762 | 96 | 0.0043 | pcLED17 | 5650 | 61 | 0.0005 |
| MLED1 | 2775 | 93 | 0.0007 | pcLED18 | 5575 | 66 | 0.0005 |
| MLED2 | 3042 | 92 | 0.0010 | pcLED19 | 5046 | 56 | 0.0010 |
| MLED3 | 3032 | 94 | 0.0008 | pcLED20 | 5225 | 58 | 0.0063 |
| MLED4 | 4520 | 82 | 0.0007 | pcLED21 | 5043 | 63 | 0.0004 |
| MLED5 | 4541 | 82 | 0.0009 | pcLED22 | 6369 | 59 | 0.0004 |
| MLED6 | 4947 | 88 | 0.0034 | pcLED23 | 4966 | 64 | 0.0106 |
| MLED7 | 6476 | 85 | 0.0019 | pcLED24 | 6540 | 66 | 0.0040 |
| MLED8 | 6451 | 85 | 0.0039 | pcLED25 | 6369 | 59 | 0.0004 |
| MLED9 | 6219 | 96 | 0.0000 | pcLED26 | 5018 | 62 | 0.0069 |

Example of the 42 light sources of Table 6.6. See also the caption of Table 6.6.

narrowband (e.g., white RGB LED light sources) and broadband (like their reference illuminants) spectral power distributions [55] even if the two stimuli have the same $u'$, $v'$ values. Therefore, the value of $\Delta C$ can sometimes yield very surprising visual results (like greenish or purplish tones), especially for the case of white RGB LED light sources. The origin of this phenomenon may be the inaccuracy of the color matching functions of the CIE 1931 standard colorimetric observer (see Section 1.1.1), interindividual variations of color vision, and a nonlinear chromatic adaptation phenomenon [56].

## 6.6
### Effect of Interobserver Variability of Color Vision on the Color Quality of Light Sources

When talking about the color quality of indoor light sources, the interobserver variability of color perception and color cognition among the observers (or light source users) cannot be ignored. One aim of this section is to enumerate those mechanisms of color perception that may contribute to the interobserver variability of perceived color quality or color quality judgments, within the group of observers of normal color vision (Section 6.6.1), while the challenges of improving the color quality for deficient color vision are not dealt with.

In a simple, demonstrative example, the effect of the interobserver variability of color matching functions on the variability of perceived color quality will be shown. In Section 6.6.1, the variability of color matching functions is quantified and the effect of this variability will be followed in a color appearance model. Then, the interobserver variability of chromatic lightness perception, color rendering, color harmony rendering, and color gamut is estimated for a set of different light sources.

The above-mentioned sample calculation, however, is just a demonstration of the possible effect of the variability of one component of color vision (color matching functions). Note that the neural mechanisms of color vision are adaptable enough to counterbalance the innate variability of photoreceptor properties. Hence, the actual variability may be less than the computed one. On the other hand, there may be large variations among subjects at later stages of the evaluation of neural color signals.

Indeed, according to Kuehni, individual observers have a personally set scale of visual color difference assessment that varies widely, "individually easily up to a factor 10 and more" [57]. In turn, as seen in Section 6.5.1, the color rendering index requires a similarity judgment scale ($R$) as its final step of calculation. This $R$-scale exhibits a nonlinear relationship with perceived color differences and its interobserver variability may be less than expected from the large variability of visual color difference assessment.

Authors of this book are not aware of any reliable psychophysical model of these variations. Therefore, in Section 6.6.2, a more pragmatic approach is followed, that is, a direct analysis of the interobserver variability of the experimental results presented previously in Chapter 6 within the groups of observers of those experiments. Note that different observer groups, for example, of different ethnic origin may evaluate certain aspects of color quality, especially color harmony or color preference, in a different way but this is out of the scope of this section. Intercultural differences related to long-term memory colors were discussed in Section 3.4.

The correlation between the color harmony model of Section 6.3 based on Hungarian observers [19] and Ou et al.'s color harmony model based on Chinese observers [21] was only $r^2 = 0.30$, which can be interpreted as a sign of intercultural differences of color harmony assessment.

One issue of Section 6.6.2 concerning the interobserver variability of the continuous visual ratings of the different aspects of color quality (Section 6.4.2) will be analyzed. Besides color perception, *cognitive color* [58] is also subject to changes among the observers. This includes *judgments* about the magnitude of perceived color differences, for example, the rating of color differences on the five-step ordinal similarity rating scale of Section 6.2.1 or on the more continuous six-step scale of Section 6.5.1. The interpersonal variability of latter judgments will also be shown in Section 6.6.2 together with the experimental variability of perceived color differences within the categories of the five-step ordinal rating scale of Section 6.2.1. Finally, in Section 6.6.3, the relevance of interobserver variability for light source design will be discussed.

## 6.6.1
### Variations of Color Vision Mechanisms

Aspects of the variation of color mechanisms include prereceptoral, receptoral, and postreceptoral factors [59]. At the prereceptoral level, the optical density of the macular pigment, lens, and other ocular media plays an important role. At the photoreceptor level, the low-density spectral absorbance functions of the cone visual pigments may vary. At the postreceptoral level, a neural compensation for white point changes due to prereceptoral and receptoral changes takes place. There are interpersonal variations of color difference mechanisms and postreceptoral chromatic adaptation mechanisms [60], as well as changes of unique hues, focal colors, long-term memory colors, color harmony, and color preference judgments.

As a sample calculation, the 10° $r(\lambda)$, $g(\lambda)$, $b(\lambda)$ color matching functions of 12 selected observers of the Stiles and Burch experiment [61] (available, for example, at the web site www.cvrl.org) were transformed into the corresponding 12 $x(\lambda)$, $y(\lambda)$, $z(\lambda)$ color matching functions. In this calculation, five light sources of different relative spectral power distributions illuminated the 34 test color samples of Figure 6.13 and 4 descriptors of color quality (lightness $L^{**}$, gamut $G$, color rendering index $R_{UCS}$, and color harmony rendering index $R_{hr}$) were computed in the same manner as in Section 6.5.2 except that the 12 different sets of $x(\lambda)$, $y(\lambda)$, $z(\lambda)$ functions of the 12 observers were used, instead of the CIE (1931) 2° color matching functions. Figure 6.36 shows the relative spectral power distributions of the five light sources. TUN, FL645, and pcLED5 were taken from Table 6.6. RGB27 and RGB45 represent RGB LED lamps at a CCT of 2700 and 4500 K, respectively. In comparison, Figure 6.36 also shows the $x(\lambda)$, $y(\lambda)$, $z(\lambda)$ color matching functions of the 12 observers.

# 320  6 Improving the Color Quality of Indoor Light Sources

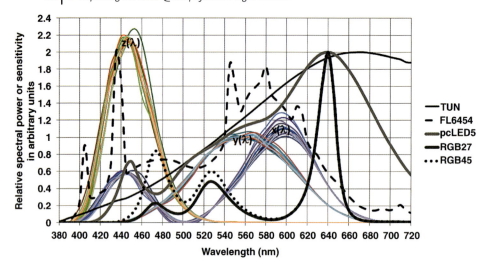

**Figure 6.36** Twelve different sets of x(λ), y(λ), z(λ) functions of the 12 selected observers of the Stiles and Burch experiment [61] compared to the relative spectral power of five light sources: TUN, FL645, and pcLED5 from Table 6.6 as well as RGB27 and RGB45. RGB LED lamps at CCT = 2700 and 4500 K.

The interobserver variability of the quantities $L^{**}$, $G$, $R_{UCS}$, and $R_{hr}$ (see Table 6.13) was quantified by computing the percentage value of their maxima minus their minima among the 12 observers divided by the average.

As can be seen from Table 6.13 and Figure 6.36, the overlapping of the maxima in the spectral power distributions of RGB LED light sources with the spectral ranges of large interindividual variability of the color matching functions causes a large variability of the color rendering index and the color harmony rendering index.

## 6.6.2
### Effect of Variability on Color Quality

In this section, the interobserver variability of the experimental results of this chapter is analyzed without using model predictions. In Section 6.6.2.1, the interobserver variability of the visual ratings of color quality ($V_1$–$V_9$ from Section 6.4.2) is dealt with. In Section 6.6.2.2, the variability of the magnitude of perceived color

**Table 6.13** Interobserver variability of the descriptors of color quality (%) under five different light sources (TUN, FL645, pcLED5, RGB27, and RGB45, see Figure 6.36): lightness $L^{**}$, gamut $G$, color rendering index $R_{UCS}$, and color harmony rendering index $R_{hr}$.

| Light source | TUN | FL645 | pcLED5 | RGB27 | RGB45 |
|---|---|---|---|---|---|
| $L^{**}$ | 0.5 | 0.4 | 0.5 | 0.2 | 0.8 |
| $G$ | 4.2 | 3.9 | 3.7 | 2.4 | 2.4 |
| $R_{UCS}$ | 0.4 | 4.0 | 1.4 | 15.7 | 44.1 |
| $R_{hr}$ | 0.7 | 1.8 | 1.7 | 13.9 | 12.8 |

Variability (%): maximum minus minimum among the 12 observers divided by their average.

## 6.6 Effect of Interobserver Variability of Color Vision on the Color Quality of Light Sources

**Table 6.14** Interobserver variability of the visual scale values of color quality ($V_1$–$V_9$ from Section 6.4.2) in % for the three light sources (INC, CFL, and LED, see Figure 6.18).

| | IOV (%) | | | | | | | | |
|---|---|---|---|---|---|---|---|---|---|
| | $V_1$ | $V_2$ | $V_3$ | $V_4$ | $V_5$ | $V_6$ | $V_7$ | $V_8$ | $V_9$ |
| INC | — | — | — | — | — | — | 23 | 67 | — |
| CFL | 27 | 59 | 48 | 37 | 107 | 57 | 26 | 95 | 74 |
| LED | 32 | 48 | 51 | 55 | 108 | 101 | 28 | 79 | 58 |

| | STD | | | | | | | | |
|---|---|---|---|---|---|---|---|---|---|
| | $V_1$ | $V_2$ | $V_3$ | $V_4$ | $V_5$ | $V_6$ | $V_7$ | $V_8$ | $V_9$ |
| INC | — | — | — | — | — | — | 0.052 | 0.124 | — |
| CFL | 0.058 | 0.096 | 0.084 | 0.083 | 0.229 | 0.117 | 0.056 | 0.177 | 0.152 |
| LED | 0.088 | 0.101 | 0.098 | 0.116 | 0.210 | 0.168 | 0.055 | 0.148 | 0.120 |

IOV: maximum minus minimum among the 30 observers divided by their average (%). INC was a fixed visual reference ($V = 1.000$) for $V_1$–$V_6$ and for $V_9$. $V_1$: brightness; $V_2$: color preference; $V_3$: color harmony; $V_4$: color categories; $V_5$: color gamut; $V_6$: continuous color transitions; $V_7$: color differences; $V_8$: similarity to long-term memory colors; $V_9$: color fidelity (rendering). Standard deviation values of the same data are also shown (STD).

differences (see Figure 6.8) is described and the variability of the color rendering index ($R_{UCS}$) is estimated based on the experimental results. Finally, in Section 6.6.2.3, the interobserver variability of the continuous similarity ratings of Section 6.5.1 is analyzed.

### 6.6.2.1 Variability of the Visual Ratings of Color Quality

For each aspect of color quality ($V_1$–$V_9$) and for each light source (INC, CFL, and LED, see Figure 6.18), different observers obtained different visual scale values. Interpersonal variability (IOV, %) was calculated as the difference between the maximum and the minimum value of the obtained visual scale divided by the mean value of all 30 observers (see Table 6.14). Table 6.14 also shows the corresponding standard deviation values. Figures 6.36–6.39 show the minimum, mean, and maximum visual scales as well as their 95% confidence intervals, for INC, CFL, and LED, respectively.

As can be seen from Table 6.14 and Figures 6.37–6.39, color gamut ($V_5$) exhibits the largest interobserver variability, followed by color transitions ($V_6$) and memory colors ($V_8$). Brightness ($V_1$) and color differences ($V_7$) are less variable. Color preference, color harmony, and color categories exhibit intermediate variability.

### 6.6.2.2 Variability of Perceived Color Differences and the Color Rendering Index

In this section, the interpersonal variability of perceived color differences is analyzed within the categories of the five-step ordinal similarity rating scale of Section 6.2.1 representing excellent (1), good (2), acceptable (3), not acceptable (4), or very bad (5) agreement between the color appearance between the test and the reference light sources. Table 6.15 shows the mean $\Delta E_{UCS}$ values found by eight observers for each

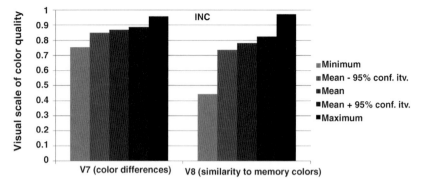

**Figure 6.37** Minimum, mean, and maximum visual scale values of the color quality variables $V_7$ and $V_8$ among 30 observers and 95% confidence intervals of the mean values for the light source INC of Figure 6.18. Note that INC was a fixed visual reference ($V = 1.000$) for $V_1$–$V_6$ and for $V_9$.

category together with their 95% confidence intervals. From these confidence intervals, the interobserver variability of color rendering assessment ($\Delta R_{UCS}$) was estimated by using CRI-CAM02UCS [10].

As can be seen from Table 6.15, the interobserver variability of the color differences in the different categories varies between ±0.19 and ±0.99. These values correspond to a variation of ±1.5 and ±7.9 on the numerical scale of the CRI-CAM02UCS color rendering index [10].

### 6.6.2.3 Variability of Similarity Ratings

The lengths of twice the 95% confidence intervals of the mean continuous similarity ratings $R$ of Figure 6.27 are replotted in Figure 6.40 as a function of $\Delta E_{UCS}$, for the seven different test objects (see Table 6.9) separately. This quantity is considered as a measure of interobserver variability of the similarity ratings. Note that although the $R$ scale was continuous, whole numbers were assigned semantic categories of

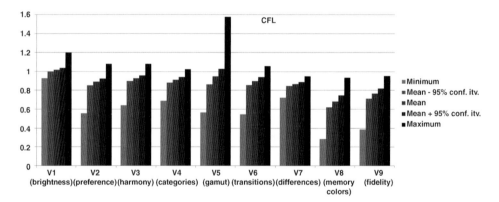

**Figure 6.38** Minimum, mean, and maximum visual scale values of the color quality variables $V_1$–$V_9$ among 30 observers and 95% confidence intervals of the mean values for the light source CFL of Figure 6.18.

### 6.6 Effect of Interobserver Variability of Color Vision on the Color Quality of Light Sources

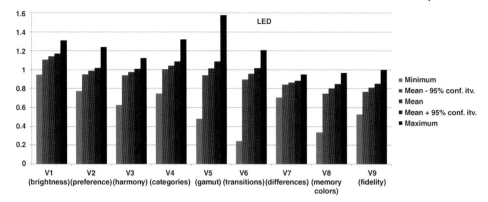

**Figure 6.39** Minimum, mean, and maximum visual scale values of the color quality variables $V_1$–$V_9$ among 30 observers and 95% confidence intervals of the mean values for the light source LED of Figure 6.18.

similarity to the reference color appearance: 1: very good; 2: good; 3: moderate; 4: low; 5: bad; and 6: very bad.

As can be seen from Figure 6.40, the interobserver variability of continuous similarity ratings ranges between 0.3 and 1.0 depending on the magnitude of the color difference between the test light source and the reference light source (see the value of $\Delta E_{UCS}$ on the abscissa of Figure 6.40) and also on the test object. The tendency is that for smaller color differences (e.g., for $\Delta E_{UCS} < 3$), interobserver variability tends to be lower (e.g., less than 0.5 units) and for large color differences (e.g., for $\Delta E_{UCS} > 6$), interobserver variability tends to be higher (e.g., greater than 0.6 units). Note that 1 unit corresponds to a transition between two semantic categories, for example, from "good" to "moderate". For light sources of good color rendering properties, for example, $R_{UCS} > 92$, a typical color difference is equal about one $\Delta E_{UCS}$ unit because $(92 - 100)/(-8.0) = 1.0$. For such a light source, it is likely that two observers of normal color vision will produce the same semantic judgment (e.g., "very good") about the similarity of a color appearance to its reference appearance. For a light source of an inferior color rendering property (e.g., $R_{UCS} = 36$, that is, $\Delta E_{UCS} = 8$), however, it is probable that there will be a semantic disagreement about the color rendering property of the light source, for example, "bad" instead of "low".

**Table 6.15** Mean $\Delta E_{UCS}$ color difference values of eight observers for the ordinal rating categories of Section 6.2.1 (1: excellent; 2: good; 3: acceptable; 4: not acceptable; 5: very bad) together with their 95% confidence intervals.

| Rating category | Mean $\Delta E_{UCS}$ | $\Delta R_{UCS}$ |
| --- | --- | --- |
| 1 | $2.01 \pm 0.19$ | $\pm 1.5$ |
| 2 | $2.37 \pm 0.17$ | $\pm 1.4$ |
| 3 | $3.75 \pm 0.27$ | $\pm 2.1$ |
| 4 | $6.53 \pm 0.46$ | $\pm 3.6$ |
| 5 | $11.28 \pm 0.99$ | $\pm 7.9$ |

$\Delta R_{UCS}$: estimation of the interobserver variability of the color rendering index.

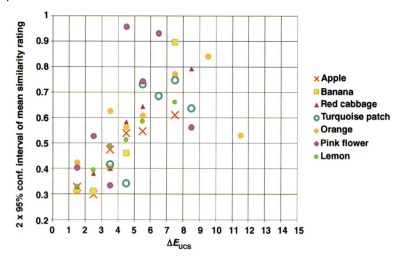

**Figure 6.40** Interobserver variability of similarity ratings R replotted from Figure 6.27. Lengths of twice the 95% confidence intervals of the mean R values as a function of $\Delta E_{UCS}$ (calculated color difference between the test and the reference light sources). The seven test objects can be seen in Table 6.9.

### 6.6.3
### Relevance of Variability for Light Source Design

As seen in Sections 6.6.1 and 6.6.2, the interindividual variability of the different aspects of color quality can be small under certain test light sources but it can be higher for other test light sources. If the interobserver variability of a given aspect of color quality is high (within the group of observers of normal color vision) for a given light source, then the manufacturer cannot expect that this light source will be accepted by all users of normal color vision provided that the light source is optimized for a standard observer.

### Acknowledgments

Authors would like to acknowledge Mrs. Nathalie Krause and Mr. Stefan Brückner (Laboratory of Lighting Technology of the Technische Universität Darmstadt) for constructing and characterizing the experimental setups and performing the visual experiments.

### References

1 CIE 13.3-1995 (1995) *Method of Measuring and Specifying Color Rendering Properties of Light Sources*, Commission Internationale de l'Éclairage.

2 Bodrogi, P., Brückner, S., and Khanh, T.Q. (2010) Dimensions of color quality. 5th European Conference on Color in Graphics, Imaging, and Vision,

CGIV 2010, June 14–17, Joensuu, Finland.
3 CIE 177:2007 (2007) *Color Rendering of White LED Light Sources*, Commission Internationale de l'Éclairage.
4 Bodrogi, P., Brückner, S., and Khanh, T.Q. (2011) Ordinal scale based description of color rendering. *Color Res. Appl.*, **36** (4), 272–285.
5 Guan, S.-S. and Luo, M.R. (1999) A color-difference formula for assessing large color differences. *Color Res. Appl.*, **24** (5), 344–355.
6 Davis, W. and Ohno, Y. (2010) The color quality scale. *Opt. Eng.*, **49** (3), 033602.
7 Luo, MR., Cui, G., and Rigg, B. (2001) The development of the CIE 2000 color-difference formula: CIEDE2000. *Color Res. Appl.*, **26**, 340–350.
8 CIE 159:2004 (2004) *A Color Appearance Model for Color Management Systems: CIECAM02*, Commission Internationale de l'Éclairage.
9 Luo, M.R., Cui, G., and Li, Ch. (2006) Uniform color spaces based on CIECAM02 color appearance model. *Color Res. Appl.*, **31**, 320–330.
10 Luo, M.R. (2011) The quality of light sources. *Color Technol.*, **127**, 75–87.
11 Sándor, N. and Schanda, J. (2006) Visual color rendering based on color difference evaluations. *Lighting Res. Technol.*, **38** (3), 225–239.
12 Bodrogi, P., Csuti, P., Horváth, P., and Schanda, J. (2004) Why does the CIE color rendering index fail for white RGB LED light sources? Proceedings of the CIE Expert Symposium on LED Light Sources: Physical Measurement and Visual and Photobiological Assessment, Tokyo.
13 Borbély, Á., Sámson, Á., and Schanda, J. (2001) The concept of correlated color temperature revisited. *Color Res. Appl*, **26**, 450–457.
14 Fedutina, M. (2010) Color discrimination studies on a LCD monitor. Master thesis, Institute of Printing Science and Technology, Technische Universität Darmstadt (in German).
15 CIE 135 (1999) *CIE Collection 1999: Vision and Color, Physical Measurement of Light and Radiation. 135/2: Color Rendering, Closing Remarks*, Commission Internationale de l'Éclairage.
16 CIE 109-1994 (1994) *A Method of Predicting Corresponding Colors Under Different Chromatic and Illuminance Adaptations*, Commission Internationale de l'Éclairage.
17 Jungnitsch, K., Bieske, K., and Vandahl, C. (2008) Untersuchungen zur Farbwiedergabe in Abhängigkeit vom Lampenspektrum. Tagung Licht 2008, 10–13. September 2008 Ilmenau, Tagungsband, pp. 289–296.
18 CIE 159:2004 (2004) *A Color Appearance Model for Color Management Systems: CIECAM02*, Commission Internationale de l'Éclairage.
19 Szabó, F., Bodrogi, P., and Schanda, J. (2010) Experimental modeling of color harmony. *Color Res. Appl.*, **35** (1), 34–49.
20 Szabó, F., Bodrogi, P., and Schanda, J. (2009) A color harmony rendering index based on predictions of color harmony impression. *Lighting Res. Technol.*, **41**, 165–182.
21 Ou, L.C. and Luo, M.R. (2006) A color harmony model for two-color combinations. *Color Res. Appl.*, **31** (3), 191–204.
22 Goethe, JW. (1970) *Theory of Colors*, The MIT Press, Cambridge, MA (reprinted), translation by C.L. Eastlake (1840) from the German "Farbenlehre", 1810.
23 Chevreul, M.E. (1981) *The Principles of Harmony and Contrast of Colors*, Van Nostrand Reinhold, New York (reprinted), translation by C. Martel (1854) from the French edition, 1839.
24 Van Nostrand, R. (1969) *Munsell: A Grammar of Color*, Reinhold Book Corporation.
25 Nemcsics, A. (1993) *Farbenlehre und Farbendynamik*, Akadémiai Kiadó, Budapest.
26 Judd, D.B. and Wyszecki, G. (1975) *Color in Business, Science and Industry*, 3rd edn, John Wiley & Sons, Inc., New York.
27 Halstead, M.B. (1977) CT Color rendering: past, present, and future, in *Proceedings of AIC Color 77*, Adam Hilger, Bristol, pp. 97–127.
28 Hashimoto, K. and Nayatani, Y. (1994) Visual clarity and feeling of contrast. *Color Res. Appl.*, **19** (3), 171–185.
29 Smet, K., Ryckaert, W.R., Pointer, M.R., Deconinck, G., and Hanselaer, P. (2011)

Color appearance rating of familiar real objects. *Color Res. Appl.*, **36** (3), 192–200.

30 Smet, K.A.G., Ryckaert, W.R., Pointer, M.R., Deconinck, G., and Hanselaer, P. (2010) Memory colors and color quality evaluation of conventional and solid-state lamps. *Opt. Express*, **18**, 26229–26244.

31 Guo, X. and Houser, K.W. (2004) A review of color rendering indices and their application to commercial light sources. *Lighting Res. Technol.*, **36**, 183–199.

32 Judd, D.B. (1967) A flattery index for artificial illuminants. *Illum. Eng.*, **62**, 593–598.

33 Thornton, W.A. (1974) A validation of the color preference index. *J. Illum. Eng. Soc.*, **4**, 48–52.

34 Thornton, W.A. (1972) Color-discrimination index. *J. Opt. Soc. Am.*, **62**, 191–194.

35 Xu, H. (1993) Color rendering capacity and luminous efficiency of a spectrum. *Lighting Res. Technol.*, **25**, 131–132.

36 Fotios, S.A. (1997) The perception of light sources of different color properties. Ph.D. thesis, UMIST, Manchester, UK.

37 Pointer, M.R. (1986) Measuring color rendering – a new approach. *Lighting Res. Technol.*, **18**, 175–184.

38 Rea, M.S. and Freyssinier-Nova, J.P. (2008) Color rendering: a tale of two metrics. *Color Res. Appl.*, **33** (3), 192–202.

39 Hashimoto, K. and Nayatani, Y. (1994) Visual clarity and feeling of contrast. *Color Res. Appl.*, **19** (3), 171–185.

40 Yaguchi, H., Takahashi, Y., and Shiori, S. (2001) A proposal of color rendering index based on categorical color names. International Lighting Congress, Istanbul.

41 Davis, W., Ohno, Y., Davis, W., and Ohno, Y. (2005) Toward an improved color rendering metric. *Proc. SPIE*, **5941**, 59411G.1–59411G.8.

42 Schanda, J. (1985) A combined color preference – color rendering index. *Lighting Res. Technol.*, **17**, 31–34.

43 Jost-Boissard, S., Fontoynont, M., and Blanc-Gonnet, J. (2009) Perceived lighting quality of LED sources for the presentation of fruit and vegetables. *J. Mod. Opt.*, **56** (13), 1420–1432.

44 Mahler, E., Ezrati, J.J., and Viénot, F. (2009) Testing LED lighting for color discrimination and color rendering. *Color Res. Appl.*, **34**, 8–17.

45 Rea, M.S. and Freyssinier, J.P. (2010) Color rendering: beyond pride and prejudice. *Color Res. Appl.*, **35** (6), 401–409.

46 Podobedov, V., Ohno, Y., Miller, C., and Davis, W. (2010) Colorimetric control and calibration of NIST spectrally tunable lighting facility. CIE 2010 Conference on Lighting Quality & Energy Efficiency, March 14–17, Vienna.

47 Schanda, J., Madár, G., Sándor, N., and Szabó, F. (2006) Color rendering – color acceptability. 6th International Lighting Research Symposium on Light and Color, Florida.

48 Nakano, Y., Tahara, H., Suehara, K., Kohda, J., and Yano, T. (2005) Application of multispectral camera to color rendering simulator. Proceedings of AIC Color '05, pp. 1625–1628.

49 Ling, Y., Bodrogi, P., and Khanh, TQ. (2009) Implications of human color constancy for the lighting industry. CIE Light and Lighting Conference, Budapest.

50 Nayatani, Y. (1997) Simple estimation methods for the Helmholtz–Kohlrausch effect. *Color Res. Appl.*, **22**, 385–401.

51 Fairchild, M. and Pirrotta, E. (1991) Predicting the lightness of chromatic object colors using CIELAB. *Color Res. Appl.*, **16** (6), 385–393.

52 Nayatani, Y. and Sakai, H. (2008) An integrated color-appearance model using CIELUV and its applications. *Color Res. Appl.*, **33** (2), 125–134.

53 CIE 015:2004 (2004) *Colorimetry*, 3rd edn, Commission Internationale de l'Éclairage.

54 Katayama, I. and Fairchild, M.D. (2010) Quantitative evaluation of perceived whiteness based on a color vision model. *Color Res. Appl.*, **35** (6), 410–418.

55 Bieske, K., Csuti, P., and Schanda, J. (2006) Colour appearance of metameric lights and possible colorimetric description. CIE Expert Symposium on Visual Appearance, October 19–20, Paris.

56 Oicherman, B., Luo, M.R., Rigg, B., and Robertson, A.R. (2009) Adaptation and colour matching of display and surface colours. *Color Res. Appl.*, **34** (3), 182–193.

57 Kuehni, R.G. (2008) Color difference formulas: an unsatisfactory state of affairs. *Color Res. Appl.*, **33** (4), 324–326.

58 Derefeldt, G., Swartling, T., Berggrund, U., and Bodrogi, P. (2004) Cognitive color. *Color Res. Appl.*, **29** (1), 7–19.

59 CIE 170-1:2006 (2006) *Fundamental Chromaticity Diagram with Physiological Axes – Part 1*, Commission Internationale de l'Éclairage.

60 Oicherman, B., Luo, M.R., Rigg, B., and Robertson, A.R. (2008) Effect of observer metamerism on color matching of display and surface colors. *Color Res. Appl.*, **33**, 346–359.

61 Stiles, W.S. and Burch, J.M. (1959) NPL colour-matching investigation: final report. *Opt. Acta*, **6**, S1–S26.

# 7
# Emerging Visual Technologies

The aim of this chapter is to present selected issues of colorimetry and color science with relevance to the evaluation and optimization of future and emerging self-luminous visual technologies (particularly LED displays) as well as light emitting diode (LED)-based indoor light sources. Section 7.1 deals with flexible displays, laser displays, and LED displays. Special attention is focused on color gamut extension algorithms and the temperature dependence of wide-gamut LED displays.

Section 7.2 describes the potential of applying emerging light sources, especially LED illuminants for indoor applications focusing on tunable LED lamps for accent lighting, the co-optimization of brightness and circadian rhythm, the possibilities of the accentuation of different aspects of color quality, and the use of new phosphor blends. Section 7.2 also describes the implications of the mechanisms of human color constancy for the spectral design of novel light sources. Finally, the content of the book is summarized in Section 7.3.

## 7.1
### Emerging Display Technologies

#### 7.1.1
#### Flexible Displays

Besides their good image quality, flexible display technologies are convenient to use primarily due to their portability [1]. According to Allen's classification [2], flexible displays can be (1) curved but not flexed during use; (2) mildly flexible but not severely flexed (like rolling); and (3) completely flexible like rolling paper or cloth. Technological components of flexible displays include the flexible substrate, backplane electronics, display materials, coating, sealing, and packaging technologies [2].

Flexible displays are manufactured on metal foil, thin glass, or plastic. The choice of the substrate material is essential. Concerning the display technology, electrochromic, LCD, OLED, and electrophoretic technologies can be considered [2]. One of the emerging flexible display technologies is carbon nanotube (CNT) technology that can be used to prepare transparent conducting coatings on flexible substrates.

*Illumination, Color and Imaging: Evaluation and Optimization of Visual Displays*, First Edition.
Peter Bodrogi and Tran Quoc Khanh.
© 2012 Wiley-VCH Verlag GmbH & Co. KGaA. Published 2012 by Wiley-VCH Verlag GmbH & Co. KGaA.

Conventional ITO (indium tin oxide, a transparent conductor) can be substituted by CNTs to be used as driving electrodes in a flexible display that can be mechanically bent without loss of electro-optical performance and flexibility [3].

So-called single-walled carbon nanotubes dispersed in fluorinated rubber increase the stretchability of electric wiring. Using this technology, a stretchable (by 30–50%) active-matrix OLED display with integrated printed elastic conductors, organic transistors, and organic light emitting diodes was constructed. It could be spanned over a hemisphere without mechanical or electrical damage [4].

Polymers are valuable materials for flexible displays because they are transparent, lightweight, flexible, robust, and suitable for mass production. Challenges for this emerging technology include the technologies of encapsulation of OLEDs against water vapor, the choice of the cathode, the electro-optic materials, the thin-film transistors, the transparent conducting anode, the barrier layered polymer substrates, the active materials for OLEDs, LCDs, and OTFTs, and the dielectric and coating materials [1].

It is also possible to print large-area plastic electronic systems on mechanically flexible polymer substrates. Circuit elements can be combined with organic semiconductors to produce active-matrix backplanes for electronic paper with microencapsulated electrophoretic inks. These displays exhibit excellent performance concerning both the transistors and the optical characteristics [5].

Flexible displays enable the user to interact with a computer at an emerging user interface called organic user interface (OUI). In contrast to conventional graphical user interfaces (GUIs), an OUI is a flexible physical surface, a display which is not only an output device but also an input device, at the same time. In an OUI, pointing can be replaced by multi-touch operations and certain functionalities of the computer can be initiated by manipulating the shape of this flexible display. In addition to this, an OUI itself can change its shape to initiate a dialog with the user. A system of multiple displays of different shapes can also be imagined [6].

### 7.1.2
**Laser and LED Displays**

Laser projectors are used in various applications including planetariums and flight simulation. Recent laser projectors with integrated electronics contain two main modules, the RGB laser source and the projection head. These two modules are connected by an optical fiber cable. Individual laser projectors are often combined into a multichannel laser projection system. Concerning their resolution, UXGA ($1600 \times 1200$ pixels) with 12 bits per channel color resolution is available today [7].

Laser displays intrinsically have a large color gamut according to their extremely saturated ("monochromatic") primary colors. To reduce noise and increase image quality, the digital input image signal is often processed by acousto-optical modulators [7]. Due to its high resolution, big depth of focus, large color gamut, and high luminous dynamic range, a laser planetarium display can provide the visual illusion of celestial bodies of very vivid colors virtually floating on the absolutely dark background [7].

Besides holographic displays, an important emerging application is volumetric imaging in which the image appears in a volume. Langhans *et al.* distinguish two basic types of volumetric displays, static volume displays and swept volume displays [8]. Swept volume displays include rotating screens, while in a static volume display, the points of a three-dimensional transparent crystal light up and draw the three-dimensional image. The crystal is doped with optically active ions of rare earths. These ions are excited by two intersecting IR laser beams with different wavelengths and emit visible radiation. Instead of a crystal, special glasses and polymers can also be used [8].

Today's rapid progress of semiconductor laser technology enables the development of laser home theaters available for the consumer market. These home theaters consist of a laser scanning display using MEMS scanners and compact RGB laser sources. The MEMS scanners rotate along two orthogonal axes and the three solid-state laser beams (RGB) are directly modulated by the input signal, while their combined laser beam is projected onto the screen by the scanners [9]. Modulating directly by the laser current is a practicable alternative to the above-mentioned acousto-optical modulation.

LED displays have already been dealt with in Chapter 2, in the context of both projection displays (light sources combined with colored filter mosaics) and direct-view displays (LED mosaics and LED backlights). In this section, some important technological features of LEDs influencing the current development trends of LEDs for emerging display applications are described including the issues of switch-on characteristics, luminous flux, luminous efficacy, junction temperature, lifetime, and dimming behavior.

The development of the components of high-brightness LED light sources includes the optimization of chip structure, finding the appropriate semiconductor materials, enhancing the adhesives for the chip, bond technology, and the arrangement of the electrodes. Another objective is the development of novel phosphor materials to convert the radiation of a blue LED chip into green, yellow, orange, and red.

The luminous output of colored phosphor-converted (pc) LEDs of a given hue tends to be more stable against temperature changes than the luminous output of their chip LED (i.e., a photonic semiconductor junction without phosphor conversion) counterparts of the same hue [10]. The explanation is the superior temperature stability of the blue indium gallium nitride (InGaN) chip LED serving as the basis of the colored phosphor-converted LED compared to other semiconductor LED material systems, especially gallium arsenide phosphide (GaAsP), the basis of red and yellow chip LEDs. Note that, within a set of phosphor-converted LEDs in mass production, to achieve a negligible variation of hue is (among others) an important requirement. This necessitates careful testing and binning of the manufactured product.

One advantage of LEDs is their fast switch-on characteristics. A typical example of a red chip LED is shown in Figure 7.1.

As can be seen from the example of Figure 7.1, the luminous flux of this red LED (lower curve) is delayed 283 ns along the horizontal timescale (concerning the 50% signal levels) after switching on the electronic driving signal (upper curve). The

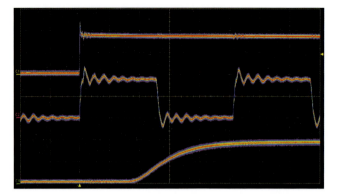

**Figure 7.1** Switch-on characteristics of a red chip LED. Abscissa: timescale, one division corresponds to 100 ns. Yellow triangle below the timescale: outset of the electronic driving signal (upper curve, relative units). Lower curve: relative luminous flux of the LED. Middle curve: synchronization signal (not relevant here). *Source*: Technische Universität Darmstadt.

middle curve is a synchronization signal (not relevant here). Also, there is a luminous rise time of 205 ns. Theoretically, this feature corresponds to a maximum frequency of about 1 MHz for the driving signal without significant loss of luminous flux enabling high frame rates on LED displays.

In practice, much lower pulse frequencies are applied. In a real-life example of an RGB LED driving system of an LED backlight for LCD, RGB converters were operated at the switching frequency of 300 kHz and supplied current pulses at the PWM dimming frequency of 500 Hz [11]. In the same example, there was also a color control feedback system using an RGB photosensor [11]. It should be noted that the response time features of such color feedback systems require special attention [12]. Also, the current, temperature, and aging compensation for the photosensors themselves have to be solved. These issues may make color feedback systems inaccurate or very expensive.

Concerning LED dimming (i.e., controlling the time average of LED luminous flux) behavior, pulse width modulation (PWM) yields a linear relationship between luminous flux and duty cycle (pulse form or percentage of time the LED is switched on at a given PWM frequency) that is advantageous for LED displays [10, 11]. The alternative dimming method, that is, constant current reduction (CCR), is nonlinear. For the case of CCR, the luminous efficacy of the LED increases at lower duty cycle levels, at least for blue LEDs and phosphor-converted blue LEDs. For red chip LEDs, this increase is absent and sometimes a decrease of luminous efficacy can be observed at lower duty cycle levels [10].

Long LED lifetime is also an essential criterion for the emerging display applications. For example, information and signage LED displays work around the clock. Here, the question is how the current and the junction temperature ($T_j$) of the LED influence the lifetime of the LED. Based on measured and extrapolated luminous flux versus in-service time (i.e., the time elapsed since switch-on and continuously

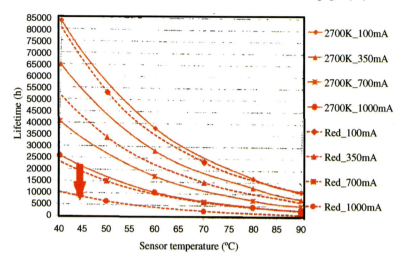

**Figure 7.2** An example of lifetime (L: ordinate, in hour units) versus sensor temperature characteristics ($T_S$, abscissa in °C units) [13]. Orange symbols (continuous curves): a phosphor-converted white LED (CCT = 2700 K). Red symbols (dash curves): a red chip LED. Levels of electric current: 100, 350, 700, and 1000 mA. The red arrow pointing downward indicates the reduced lifetime expectation of the red chip LED compared with the white LED. Reproduced with permission from *Licht*.

burning) characteristics at different junction temperatures, Vinh and Khanh [13] found an exponential relationship predicting lifetime $L$ (defined by the time to 70% reduction of initial luminous flux, measured in units of hours) from junction temperature $T_j$ (measured in Kelvin units) (see Equation 7.1).

$$L = a\, e^{bT_j} \tag{7.1}$$

In Equation 7.1, the parameters $a$ and $b$ depend on the type of LED and on LED driving current [13]. In practical measurements, the temperature of the LED is measured by a sensor placed on a specific position of the LED. This position is the so-called LED thermal pad provided by the manufacturer. There is a linear relationship between sensor temperature $T_S$ and junction temperature $T_j$ [13]. An example of lifetime versus sensor temperature characteristics can be seen in Figure 7.2 for the case of a phosphor-converted white LED (CCT = 2700 K) together with a red chip LED at four different levels of electric current (100, 350, 700, and 1000 mA).

As can be seen from Figure 7.2, the lifetime of the red chip LED is significantly less than the lifetime of the white phosphor-converted LED. For example, at $T_S = 60\,°C$ and $I = 700$ mA, $L = 10\,000$ h for the red chip LED and $L = 17\,000$ h for the white LED. The implication for emerging LED displays is that, for a permanent display, it is essential to select LEDs of similar lifetime characteristics; for example, red chip LEDs should be replaced by the more durable phosphor-converted red LEDs. Note that the placing of the phosphor layer far away from the blue emitting surface – in a so-called remote phosphor LED – reduces the rate of phosphor degradation [14].

## 7.1.3
### Color Gamut Extension for Multi-Primary Displays

The input image for the display is often captured in a standard color gamut. Subsequently, specific algorithms are necessary to guess a preferred color appearance to exploit the full color capability of a wide color gamut display comprising multiple ($n > 3$) primary colors (see Sections 2.3.2 and 5.2.2). For example, Hoshino's algorithm [15] extends CIELAB lightness $L^*$ and chroma $C^*$ simultaneously, depending on lightness level, from the standard gamut to the wide gamut. Instead of CIELAB, however, it is advisable to use CIECAM02 today because the latter color space offers perceptually more uniform color scales. A recent color gamut extension algorithm adapts to the statistical content (color histogram) of the input image instead of a fixed extension method [16].

The key point of these methods is chroma enhancement. But an overall increase of chroma does not always yield a pleasing color appearance. For example, the important skin tones should not be oversaturated as can be seen from the preference score of the hue-dependent chroma boost experiment in the hue range of skin (Figure 3.11). Also, neutral tones (such as white and light gray) should preserve their neutrality. A solution can be a combination of general chroma extension with a mapping algorithm onto the tolerance volumes in the neighborhood of important long-term memory colors. As mentioned in Section 3.4.3, such a mapping imitates the color shifts in human color memory toward long-term memory colors.

Anyway, if different color mapping algorithms are applied in the different regions of color space (e.g., in the neighborhood of the different long-term memory colors), then smooth transitions are necessary between the different regions [17]. Otherwise, spatial color artifacts tend to appear in the output image. Also, gamut extension functions of very high gain factors should be avoided because they lead to spatial artifacts (e.g., false contours) in the output image [16]. An interactive color gamut extension tool was also developed [17]. Using this tool, observers are able to change the colors of an input image according to their choice, depending on the region of CIELAB color space.

For the emerging LED-based wide-gamut displays, an important problem is that the luminous output of the LEDs changes with the temperature of the LED device. If there is a considerable luminance change or chromaticity change of the different colored LEDs due to temperature changes, then these changes may threaten the result of color gamut extension and deteriorate image color quality. Note that, as indicated in Section 7.1.2, a color control feedback system cannot be considered as an ultimate solution. Thus, an important requirement is to maintain the white point and the color gamut of the LED display amid changing temperatures resulting from, for example, the changing thermal environment of the LED display.

To elucidate the temperature dependence of the LED display's luminous output, a demonstrative computational example of a model four-primary wide-gamut LED display with an idealized RGB filter set is shown below. In this example, it is assumed that the model display is operating in color sequential mode (Section 2.3.2). This means that in the time sequence in which a particular LED is emitting, only the

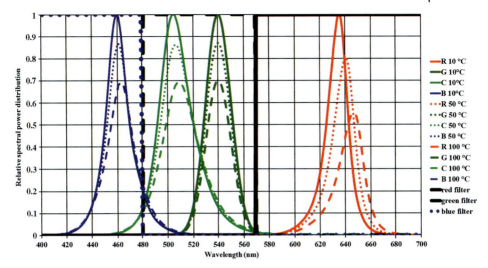

**Figure 7.3** Temperature dependence of the relative spectral power distributions (RSPDs) of four types of chip LEDs: red (R), green (G), cyan (C), and blue (B) in the theoretical model four-primary color sequential (RGCB) LED display. Three sensor temperatures: $T_S = 10$, 50, and 100 °C. For every LED, the maximum of the RSPD is equal to 1 for $T_S = 10$ °C. RSPDs of $T_S = 50$ and 100 °C are scaled relative to $T_S = 10$ °C. R, C, and B curves are based on measurements of typical chip LEDs, while the G curves are simulated curves. Spectral transmissions of model RGB color filters are also shown.

subpixels of the corresponding color filter are transmitting. Particularly, when the green LED is emitting or the cyan LED is emitting, only the green filter is transmitting.

First, the temperature dependence of the LEDs themselves is analyzed. Figure 7.3 shows the temperature dependence of the relative spectral power distributions of four types of chip LEDs, red (R), green (G), cyan (C), and blue (B), for three sensor temperatures ($T_S = 10$, 50, and 100 °C). The R, C, and B characteristics are based on physical measurements of typical chip LEDs, while the G curves result from a theoretical simulation.

In Figure 7.3, the maxima of the relative spectral power distributions were set to unity at $T_S = 10$ °C for every LED, whereas the relative spectral power distributions corresponding to $T_S = 50$ and 100 °C were scaled relative to $T_S = 10$ °C. As can be seen from Figure 7.3, the luminous output of the LEDs decreases with increasing temperature.

There is also a significant shift of the maximum wavelength of the LEDs, especially for the red LED: 645 nm (100 °C) instead of 632 nm (10 °C). The change of the maximum wavelength of the blue LED is 463 nm (100 °C) instead of 459 nm (10 °C), and 508 nm (100 °C) instead of 504 nm (10 °C) for the cyan LED. The change of the maximum wavelength with temperature was neglected for the simulated green LED.

The effect of the above temperature-related changes on the LED display's color performance is shown in Figure 7.4.

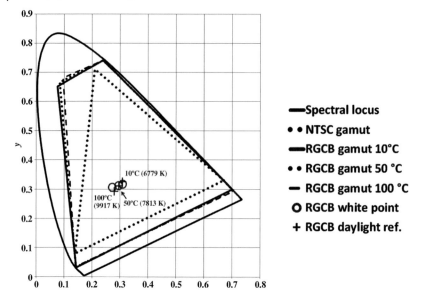

**Figure 7.4** Temperature dependence of the four-primary color sequential (RGCB) model LED display in the CIE x, y chromaticity diagram. The diagram was computed by using the temperature characteristics of the four chip LEDs in Figure 7.3. RGCB display white points, corresponding daylight reference illuminants (CCT in K units), and chromaticity gamut for three sensor temperatures ($T_S$ = 10, 50, and 100 °C). The RGCB display white point was adjusted to the NTSC white point ($x$ = 0.31; $y$ = 0.316) at $T_S$ = 10 °C.

Figure 7.4 shows the shift of the white point with changing temperature, the corresponding daylight reference illuminants, and the chromaticity gamut in a CIE x, y chromaticity diagram for three sensor temperatures, $T_S$ = 10, 50, and 100 °C. The white point was adjusted to the NTSC white point ($x$ = 0.31; $y$ = 0.316) at $T_S$ = 10 °C by weighting the intensities of the four LEDs. As can be seen from Figure 7.4, with increasing temperature, the white point of the display is shifted toward higher color temperatures.

At $T_S$ = 100 °C, there is a well noticeable chromaticity difference ($\Delta uv$ = 0.01) between the white point and its daylight reference; that is, the display's white point is shifted toward the top left of the chromaticity diagram, that is, toward greenish tones. The chromaticity gamut shrinks continuously with increasing temperature. It equals 150.4% of the NTSC gamut at $T_S$ = 10 °C, 149.6% at $T_S$ = 50 °C, and 147.9% of the NTSC chromaticity gamut at $T_S$ = 100 °C.

As pointed out in Section 7.1.2, the lifetime of chip LEDs is generally less than the lifetime of pcLEDs. Hence, for emerging LED displays, it may be interesting to consider the replacement of chip LEDs by pcLEDs. As a further theoretical example, two of the four chip LEDs, the red one and the cyan one, were replaced in the above RGCB model display by a typical red pcLED and a typical cyan pcLED. Figure 7.5 shows the temperature dependence of the relative spectral power distributions of these pcLEDs for three sensor temperatures, $T_S$ = 10, 50, and 100 °C.

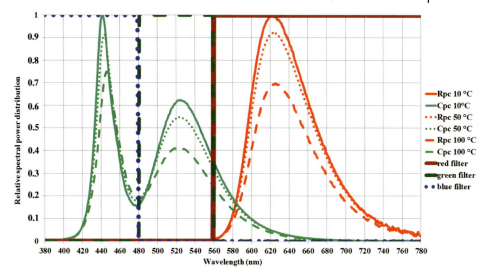

**Figure 7.5** Temperature dependence of the relative spectral power distributions of two types of phosphor-converted LEDs: red (Rpc) and cyan (Cpc) used to replace the red and cyan chip LEDs in the model four-primary color sequential (RGCB) LED display (the green and blue chip LEDs remained the same as in Figure 7.3). Three sensor temperatures: $T_S = 10$, 50, and 100 °C. For every LED, the maximum of the RSPD is equal to 1 for $T_S = 10$ °C. RSPDs of $T_S = 50$ and 100 °C are scaled relative to $T_S = 10$ °C. These curves are based on measurements of typical red and cyan pcLEDs. Spectral transmissions of model RGB color filters are also shown.

As can be seen from Figure 7.5, the luminous output of the pcLEDs decreases with increasing temperature (compare with Figure 7.3). The shift of the maximum wavelength of the cyan and red phosphor components of LED emission toward longer wavelengths (transmitted by the green and red color filters, respectively) is significantly less than that for the case of the corresponding chip LEDs. The amount of the wavelength shift was only 3 nm (626 nm at 100 °C and 623 nm at 10 °C) for the red pcLED. For the cyan pcLED, no wavelength shift could be extracted from the result of this measurement.

Figure 7.6 shows the effect of temperature (imagine that the temperature sensor is placed onto the LED chip) on the model LED display with the replaced (i.e., phosphor-converted) red and cyan LEDs. The CIE $x$, $y$ chromaticity diagram of Figure 7.6 was computed by using the temperature characteristics of the green and blue chip LEDs of Figure 7.3 and the red pcLED and cyan pcLED characteristics of Figure 7.5. The white point was adjusted to the NTSC white point at $T_S = 10$ °C.

As can be seen from Figure 7.6, with increasing temperature, the white point of the display is shifted toward higher color temperatures (7807 K at 100 °C versus 6779 K at 10 °C) but the amount of this shift is less than that for the full chip LED display (9917 K at 100 °C versus 6779 K at 10 °C, compare with Figure 7.4). At $T_S = 100$ °C,

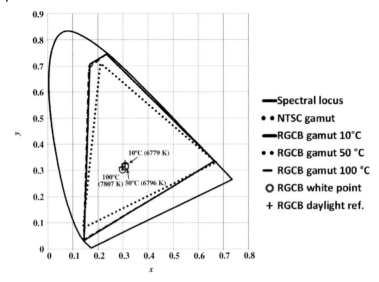

**Figure 7.6** Temperature dependence of the four-primary color sequential (RGCB) model LED display with red and cyan phosphor-converted LEDs (Rpc and Cpc) in the CIE x, y chromaticity diagram. The diagram was computed by using the temperature characteristics of the green and blue chip LEDs of Figure 7.3 and the Rpc and Cpc of Figure 7.5. RGCB display white points, corresponding daylight reference illuminants (CCT in K units), and chromaticity gamut for three sensor temperatures ($T_S = 10$, 50, and 100 °C). The RGCB display white point was adjusted to the NTSC white point ($x = 0.31$; $y = 0.316$) at $T_S = 10$ °C.

there is less chromaticity difference ($\Delta uv = 0.007$) between the white point and its daylight reference than in case of the full chip LED display ($\Delta uv = 0.01$).

For the red pcLED and cyan pcLED display, there is a slight increase of chromaticity gamut with increasing temperature. It equals 122.3% of the NTSC gamut at $T_S = 10$ °C, 122.6% at $T_S = 50$ °C, and 122.9% at $T_S = 100$ °C. The chromaticity gamut of this model display (Figure 7.6, about 123% on average) is less than the chromaticity gamut of the full chip LED display (Figure 7.6, about 149% on average). The reason is that in this example conventional red and cyan phosphors were used that emit at less advantageous wavelengths than their chip LED counterparts.

Observe in Figures 7.3 and 7.5 that the peak wavelength of the cyan phosphor is shifted toward longer wavelengths (on average 525 nm) compared to the cyan chip LED (on average 506 nm). This shift causes a loss of gamut due to the cyan pcLED at the top left of the x, y chromaticity diagram. Similarly, the peak wavelength of the red phosphor is shifted toward shorter wavelengths (on average 625 nm) compared to the cyan chip LED (on average 639 nm). This shift causes a loss of gamut due to the red pcLED at the bottom right of the x, y chromaticity diagram. Consequently, for emerging wide-gamut displays, the optimization of the spectral emission characteristics of phosphors may be an important issue together with the enhancement of their efficiency (today's red phosphors are not efficient).

## 7.2
## Emerging Technologies for Indoor Light Sources

Indoor lighting technology is experiencing a revolution today due to the demand of replacing tungsten lamps by energy-efficient technologies that also provide good color quality. Compact fluorescent lamps are one choice but the problem is that some wavelength ranges are typically underrepresented in their spectra resulting in poor color quality. White pcLEDs represent a more promising technology because a wide choice of phosphors is available today that can be combined with blue LED chips making a very flexible spectral engineering possible. These white pcLEDs are sometimes combined with non-white chip LEDs (red, amber, cyan, or green) or non-white pcLEDs.

Non-white LED combinations can be used for accent lighting or decorative lighting (Section 7.2.1) but today's most important indoor application is general white indoor lighting (Sections 7.2.3 and 7.2.4). The relevance of the electric and temperature stability of the LED light source will be pointed out in this section. Circadian spectral design and its co-optimization with the aspects of color quality (chromatic lightness, color rendering, gamut, and harmony) is an emerging technology described in Section 7.2.2. The implications of human color constancy on spectral design will be dealt with in Section 7.2.5 as an emerging technology.

### 7.2.1
### Tunable LED Lamps for Accent Lighting

General lighting is intended for the general (usually homogeneous) illumination of a room, for example, downward from the ceiling, for example, at an illuminance level of 500–1000 lx. Task lighting focuses a light beam onto a work target to increase its illuminance (up to $10^4$ lx, for example, in emergency medicine) for better visual performance, for example, reading, sewing, crafting, mounting, surgical procedures, or the close visual inspection of a material. Accent lighting, in turn, has an esthetic purpose making the visual environment more beautiful, conspicuous, or attractive. For accent lighting, the dynamic adjustment of the chromaticity of the light source ("tuning") is an important requirement to change the color appearance of the lit environment.

To meet this requirement, LED technology offers not only extended flexibility of spectral design but also many opportunities for energy savings [18]. A field survey of shoppers of retail displays with blue LED accent lighting showed that power could be reduced by 50% without compromising the esthetic value of the display when fluorescent general lighting was eliminated and halogen accent lamps were reduced [18]. Accent lighting can produce specific spatial patterns and color shadings in saturated tones or different white tones. Light sources of accent lighting "enhance the esthetic appeal of certain areas or objects" in the illuminated environment [19], in addition to general lighting that is usually white and spatially homogeneous.

"A luminaire with a color accent is very attractive for certain environments, such as a showroom that displays commercial merchandise, a museum that displays art

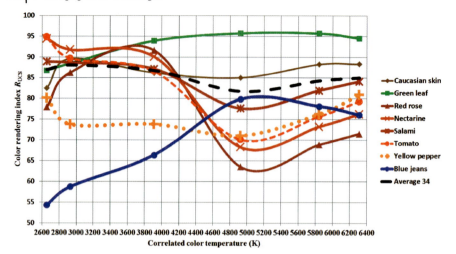

**Figure 7.7** Computational example of a white LED lamp of tunable correlated color temperature (CCT). Ordinate: special color rendering indices $R_{UCS}$ (Section 6.2.2.2) for eight colored objects (see legend) as a function of CCT in K (abscissa).

objects, a hotel or corporate office lobby that provides enhanced illumination to a personnel desk, a performance stage that provides focused illumination on a certain area or a certain performer" [19]. In contrast with non-LED light sources, by the aid of a combination of white LEDs and color LEDs in dedicated luminaires (so-called tunable LED lamps for accent lighting), light output can be changed flexibly and dynamically according to the request of the user. Such accent lighting can be used as a stand-alone light source but also in combination with general lighting.

In the following, a computational example of a tunable white LED lamp is shown. Such a light source can be used for accent lighting because the relative spectral power distribution of the LED lamp can be changed (so-called "tuning") by assigning different weights to its three component LEDs. Thus, the user can tune the correlated color temperature continuously. The example is intended to demonstrate the effect of "tuning" the correlated color temperature on the color appearance of eight important objects of different characteristic hues (Caucasian skin, green leaf, red rose, nectarine, salami, tomato, yellow pepper, and blue jeans). These eight objects were selected from the 34 characteristic indoor objects of Figure 6.13. Figure 7.7 shows the values of the special color rendering indices $R_{UCS}$ (Section 6.2.2.2) for these eight selected objects as a function of the correlated color temperature of the tunable LED lamp.

As can be seen from Figure 7.1, red or orange objects (red rose, nectarine, salami, and tomato) exhibit good color rendering and bluish objects (e.g., the blue jeans) exhibit bad color rendering if the user of the tunable LED lamp sets a lower color temperature, for example, 3000 K. But if the user selects a high color temperature, for example, 5000 K, the opposite situation happens. The point is that, for accent lighting, the correlated color temperature should be selected according to the typical spectral reflectance of the objects in the illuminated scene. For example, for bluish objects, a higher color temperature should be chosen. Note that changing the spatial power distributions of the light source (e.g., introducing colored illumination gradients in the room) can further enhance accent lighting but this issue will not be dealt with here.

## 7.2.2
### Optimization for Brightness and Circadian Rhythm

Visible radiation evokes visual perceptions in the human visual system such as the brightness perception of colored objects illuminated by a light source in an indoor environment. As pointed out in Section 1.3, visible radiation also affects the human circadian clock, a complex subsystem of the human central nervous system [20]. The circadian clock "organizes and orchestrates the timing of all daily biological functions, from complicated physiological systems to single cells" [21]. The hormone melatonin is an important component of the circadian clock: it fosters sleep-in. Exposures to light suppress the production of melatonin by the pineal gland.

Circadian optimization of light source technology is currently emerging and our current knowledge still has to be verified for a reliable application. Specifically, the effect of changing the intensity and relative spectral power distribution of light on nocturnal melatonin suppression has been described and validated and a model of the circadian stimulus (CS) was developed [21]. But it should be kept in mind that "not all light-induced, non-visual responses have the same spectral sensitivity as nocturnal melatonin suppression" [21]. Therefore, care has to be taken to apply these results at this stage of research.

Anyway, in order to show a possibility of circadian optimization as an emerging visual technology in this section, the Rea et al.'s model of the circadian stimulus [21] was applied to the representative set of 42 light sources (Table 6.6). If circadian optimization is applied to the spectral power distribution of an indoor light source, then this optimization should also consider color quality aspects as a supplementary requirement. Nevertheless, this co-optimization should be carried out very carefully because certain applications may require a low level of the circadian stimulus (minimization) together with a high level of a certain aspect of color quality (maximization). This requirement is shown in this section by the example of the relationship between the circadian stimulus of the light source computed by the Rea et al.'s model and four aspects of the color quality of the objects illuminated by the same light source, chromatic lightness, color rendering, gamut, and harmony.

The circadian stimulus is computed in the following way [21]. First, the four contributing retinal signals are computed, that is, the signals of the S-cones, the rods, the intrinsically photosensitive retinal ganglion cells (ipRGCs), and the L + M channel modeled by the CIE 10° photopic photometric observer, $V_{10}(\lambda)$ [22]. Note that the signal of the intrinsically photosensitive retinal ganglion cells also contributes to the controlling of the sustained component of the pupillary reflex [23, 24]. Figure 7.8 shows the spectral sensitivities of these photoreceptors (or their combinations).

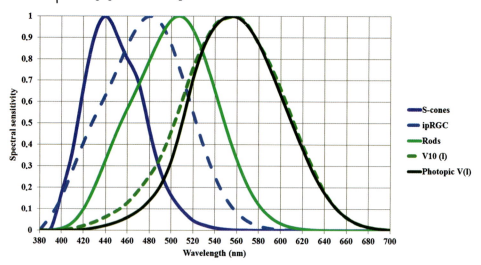

**Figure 7.8** Spectral sensitivities of the photoreceptors (or their combinations) in the Rea et al.'s model [21]. S-cones: $S(\lambda)$; rods: $V'(\lambda)$; intrinsically photosensitive retinal ganglion cells: ipRGC$(\lambda)$; and the L + M channel: $V_{10}(\lambda)$ [22]. These signals contribute to the circadian stimulus according to the Rea et al.'s model [21]. For comparison, the CIE 1924 photopic $V(\lambda)$ function is also shown.

As can be seen from Figure 7.8, the spectral sensitivities of all contributing signals (including $V_{10}(\lambda)$) are shifted toward the blue spectral range compared to the CIE 1924 photopic $V(\lambda)$ function. The S-cone, rod, ipRGC, and L + M signals are computed by multiplying the spectral irradiance at the eye by the corresponding spectral sensitivity function and integrating between 380 and 730 nm.

In the next stage of the model, the S − (L + M) color opponent signal is computed (see Table 1.1). If the S − (L + M) signal is negative, then the circadian stimulus depends only on the ipRGC signal. If S − (L + M) is not negative, then CS depends on the ipRGC signal, the S − (L + M) signal, and the rod signal. For high irradiance levels, the model includes rod saturation. In the final stage of the model, the combination of the above signals is normalized to the photopic illuminance provided by CIE illuminant A (a blackbody radiator at 2856 K) and a signal compression is applied to fit nocturnal human melatonin suppression data [25–27]. The detailed numerical computation procedure of the circadian stimulus is described elsewhere [21] and not repeated here.

In the sample computation of this section, the value of the circadian stimulus [21] was computed as a function of the correlated color temperature of each one of the 42 light sources listed in Table 6.6. For every light source, the illuminance at the eye was set to 700 lx, which is a typical value in a well-illuminated office in the plane of the cornea.

As can be seen from Figure 7.9, the circadian stimulus is an increasing function of color temperature due to the prevailing blue spectral sensitivity of all contributing mechanisms (compare with Figure 7.8). For comparison, two light sources with high

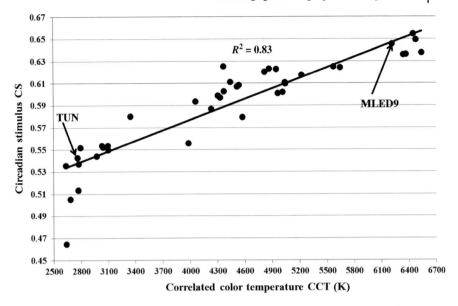

**Figure 7.9** Circadian stimulus [21] as a function of correlated color temperature for the 42 light sources of Table 6.6. For every light source, the illuminance at the eye was set to 700 lx (a typical value in a well-illuminated office in the plane of the cornea). TUN: tungsten halogen lamp; MLED9: a theoretical multicomponent LED light source with high color rendering, $R_{UCS} = 96$.

color rendering, $R_{UCS} = 96$, are labeled in Figure 7.9: MLED9 (a theoretical multi-component LED light source, CCT = 6219 K) and TUN (a tungsten halogen light source, CCT = 2762 K). For MLED9, the value of CS (0.65) is higher than that for TUN (0.54) at the same illuminance level of 700 lx. Therefore, MLED9 has a higher melatonin suppression effect to support better work performance in an office, while TUN has lower melatonin suppression and hence better for relaxing at home, for example, before sleep. This finding implies that illuminance measurement with an illuminance meter is not suitable to evaluate the circadian effect.

As mentioned above, it is interesting to compare the circadian stimuli of these 42 light sources with the aspects of color quality, first with the mean chromatic lightness $L^{**}$ (Equation 6.5) of the 34 characteristic indoor objects (Figure 6.13) under the corresponding light sources. The idea behind this example is that the user of the light source is sitting in a room where the white walls and the ceiling provide the circadian stimulus while the lightness of the typical colored objects is being evaluated. The validity of this idea has to be checked.

Let us first consider the mean lightness ($L^{**}$) of the 34 test objects as a function of correlated color temperature (see Figure 7.10).

As can be seen from Figure 7.10, the correlate of mean object lightness ($L^{**}$) decreases with increasing color temperature, that is, with increasing blue spectral content of the light source. The reason is that the spectral reflectance of this particular set of 34 objects predominates in the middle- and long-wavelength range. Observe

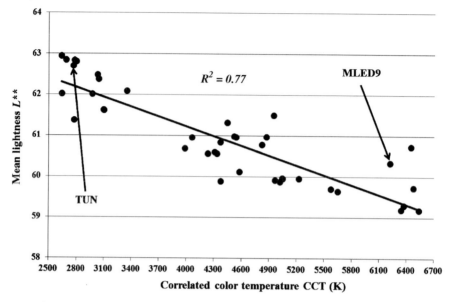

**Figure 7.10** Mean lightness $L^{**}$ (Equation 6.5) of the 34 test objects of Figure 6.13 (characteristic indoor objects) as a function of correlated color temperature for the 42 light sources of Table 6.6.

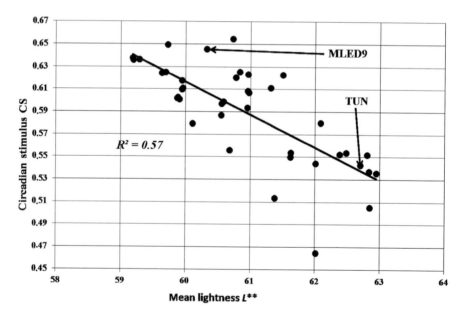

**Figure 7.11** Circadian stimulus [21] as a function of mean lightness $L^{**}$ (Equation 6.5). Mean lightness was calculated for the 34 test objects of Figure 6.13 (characteristic indoor objects). Every point corresponds to one of 42 light sources of Table 6.6. To compute the value of CS, the illuminance at the eye was set to 700 lx for every light source (a typical value in a well-illuminated office in the plane of the cornea).

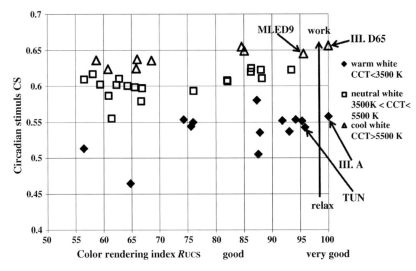

**Figure 7.12** Diagram to select a light source according to the required level of color rendering ($R_{UCS}$) and the user's intent of activity. $R_{UCS}$ was calculated for the 34 test objects of Figure 6.13 (characteristic indoor objects). Compare with Figure 6.31.

in Figure 6.13 that, on average, there is more spectral reflectance above 540 nm than in the bluish spectral range. As can be seen from Figure 7.10, these objects appear on average brighter under TUN (2762 K, $L^{**} = 63$) than under MLED9 (6219 K, $L^{**} = 60$).

For the same set of 42 light sources as in Figures 7.9 and 7.10 and for the same set of 34 colored objects as in Figure 7.10, the value of CS coming from the light source (or from the white walls) is depicted as a function of the mean chromatic lightness $L^{**}$ of the 34 objects under each light source in Figure 7.11.

As can be seen from Figure 7.11, the circadian stimulus [21] decreases with increasing mean lightness as expected from Figures 7.9 and 7.10. The consequence of Figure 7.11 is that, for light sources of low CCT at a fixed illuminance level of, for example, 700 lx (e.g., tungsten light in a Western culture home in the evening), the circadian stimulus is low. Therefore, melatonin is less suppressed and this may result in a more relaxed feeling and better sleep-in. At the same time, especially the important red, orange, and yellow objects in this particular set of 34 typical colored objects appear lighter and this fact tends to enhance the color appearance of the lit environment. In contrast to this, for light sources of high CCT at the same fixed illuminance level (e.g., a cold white LED luminaire in an office during the daytime), the circadian stimulus is high. Therefore, melatonin is more suppressed and this may increase concentration and work performance.

It is also interesting to compare the circadian stimuli of the 42 light sources with the color rendering properties ($R_{UCS}$) of those light sources, together with the representative illuminants A and D65. The value of $R_{UCS}$ was calculated for the 34 test objects of Figure 6.13. Figure 7.12 shows this CS–$R_{UCS}$ diagram (compare with Figure 6.31).

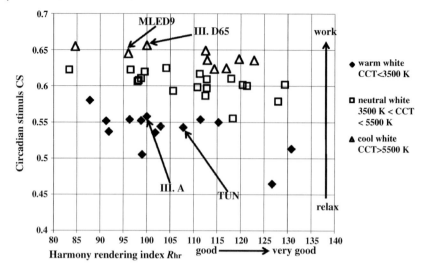

**Figure 7.13** Diagram to select a light source according to the required level of color harmony rendering ($R_{hr}$) and the user's intent of activity. Compare with Figure 6.32.

The CS–$R_{UCS}$ diagram of Figure 7.12 may serve as a possible basis of a future method to select a light source according to the required level of color rendering ($R_{UCS}$) and the user's intent of activity varying between relaxation and work. As can be seen from Figure 7.12, among the light sources of high color rendering, MLED9 or D65 can be used for work while TUN or illuminant A can be used for relaxation. Similar diagrams can be applied for color harmony rendering and for color gamut (see Figures 7.13 and 7.14, respectively).

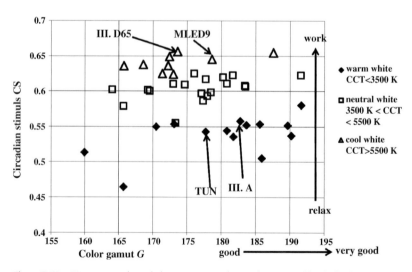

**Figure 7.14** Diagram to select a light source according to the required level of color gamut ($G$) and the user's intent of activity. Compare with Figure 6.33.

On the basis of the CS–$R_{hr}$ and CS–$G$ diagrams of Figures 7.13 and 7.14, a light source can be selected according to the required level of color harmony rendering ($R_{hr}$) or color gamut ($G$) and the user's intent of activity varying between relaxation and work.

## 7.2.3
### Accentuation of Different Aspects of Color Quality

In the sample computation of Section 6.5.2, the relationship among four descriptor quantities of light source color quality (lightness, color gamut, color rendering, and color harmony) was pointed out for the representative set of 42 indoor light sources (Table 6.6) using the 34 characteristic indoor objects of Figure 6.13. In future applications, it may be important to select a light source that accentuates a certain aspect of color quality in a lit indoor scene. In this section, the above set of 42 light sources is examined from this point of view.

Similar to Section 6.5.2, three CCT groups of the 42 test light sources are considered, warm (CCT < 3500 K), neutral (3500 K ≤ CCT < 5500 K), and cool (CCT ≥ 5500 K). In every CCT group, the same four descriptors ($L^{**}$, $G$, $R_{UCS}$, and $R_{hr}$) were considered for every light source as in Section 6.5.2: the mean chromatic lightness of the 34 objects ($L^{**}$), color gamut ($G$), color rendering index ($R_{UCS}$), and the color harmony rendering index ($R_{hr}$). First, the user selects a CCT group (e.g., warm) and then an aspect of color quality is chosen that should be accentuated by the light source (e.g., gamut). For example, the gamut aspect is to be accentuated by the aid of a warm white light source and the user would like to know which one of the warm white light sources should be used to do so.

To alleviate the user's task of light source selection, the descriptor values ($L^{**}$, $G$, $R_{UCS}$, and $R_{hr}$) were transformed within each CCT group in the following way: the mean value of the descriptor for all light sources within the CCT group was subtracted from each descriptor value and then the result was divided by the corresponding standard deviation. Figure 7.15 shows these transformed descriptor values for the light sources in the warm white CCT group.

As can be seen from Figure 7.15, to accentuate lightness with a warm white light source, the user should select FL4, that is, the warm white fluorescent lamp of Table 6.6. To accentuate gamut, MLED1 or pcLED5 should be selected. For color rendering, TUN or pcLED5 and for color harmony, FL3 or pcLED4 should be chosen.

Figure 7.16 shows these transformed descriptor values for the light sources in the neutral white CCT group.

As can be seen from Figure 7.16, to accentuate lightness with a neutral white light source, the user should select MLED6 or FL2. To accentuate gamut, MLED6 should be selected. For color rendering, FL2 or MLED6 and for color harmony, FL1 or pcLED2 should be chosen.

Figure 7.17 shows these transformed descriptor values for the light sources in the cool white CCT group.

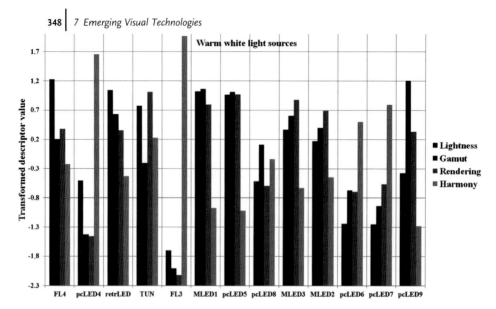

**Figure 7.15** Ordinate: transformed descriptor values (see text) for the following aspects of color quality: mean chromatic lightness ($L^{**}$), color gamut ($G$), color rendering index ($R_{UCS}$), and color harmony rendering index ($R_{hr}$). Abscissa: light sources of the warm white CCT group (CCT < 3500 K) of the 42 light sources of Table 6.6.

As can be seen from Figure 7.17, to accentuate lightness or gamut with a cool white light source, the user should select MLED8. For color rendering, MLED9 and for color harmony, pcLED1 should be chosen.

### 7.2.4
### Using New Phosphor Blends

Phosphor-converted white LED technology is one of today's most promising technologies to replace tungsten lamps for indoor lighting in an energy-efficient manner by providing excellent color quality at the same time [28]. To achieve high color rendering (and other color quality) properties, it is essential to achieve a balanced spectral power distribution and this can be accomplished by using new phosphor blends consisting of multi-phosphors rather than a single yellow phosphor [29, 30]. A so-called 4-pc WLED (i.e., a blue LED chip converted by a mixture of four phosphors) was found to have a general color rendering index of $R_a = 95$ at a correlated color temperature of 4280 K [29].

Following the above idea [29], a computational example of a theoretical 4-pc WLED is shown in this section at four correlated color temperatures between 2900 and 5900 K. The starting point of the computation was to select five sample emission spectra (relative spectral power distributions), a blue LED chip and four phosphors, yellow, orange, green, and red. These sample emission spectra (Figure 7.18) were derived from the spectral measurement of some widely used commercially available phosphor LEDs.

7.2 Emerging Technologies for Indoor Light Sources | 349

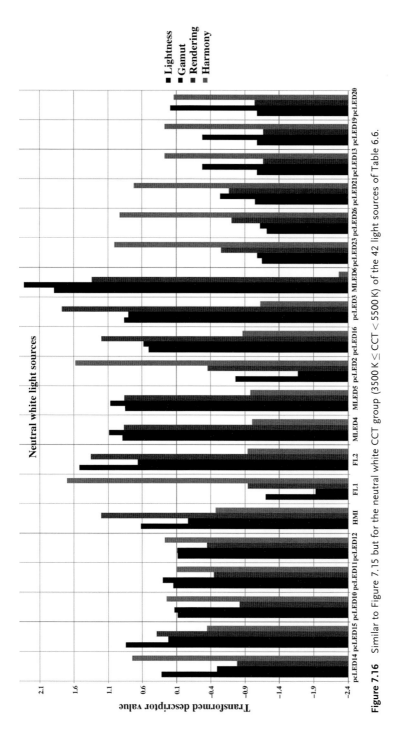

**Figure 7.16** Similar to Figure 7.15 but for the neutral white CCT group (3500 K ≤ CCT < 5500 K) of the 42 light sources of Table 6.6.

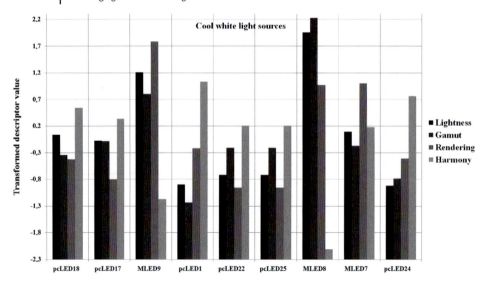

**Figure 7.17** Similar to Figure 7.15 but for the cool white CCT group (CCT ≥ 5500 K) of the 42 light sources of Table 6.6.

A linear combination of the emission spectra of Figure 7.18 was composed. Then, the relative weights of the five components were changed in a limited range (between 0 and 2) in order to move the white point onto the locus of blackbody radiations on the $u, v$ color diagram (if the CCT of the test light source was less than 5000 K) or close to the locus of a phase of daylight (if the CCT of the test light source was equal to 5000 K or more than 5000 K; see Section 6.1, Step 1). In this sample

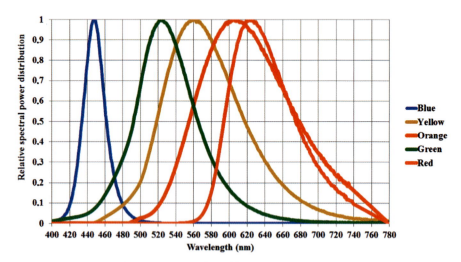

**Figure 7.18** Sample emission spectra derived from the spectral measurement of some widely used commercially available phosphor LEDs: a blue LED chip and four phosphors, yellow, orange, green, and red.

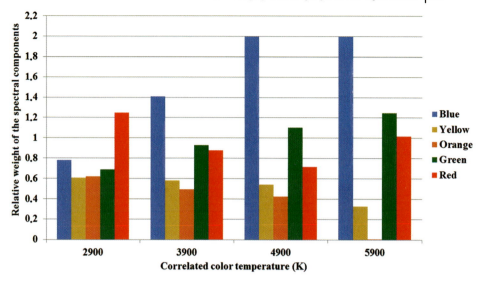

**Figure 7.19** Relative weights of the five spectral components of Figure 7.18 for four correlated color temperatures (CCT = 2900, 3900, 4900, and 5900 K) in four different linear combinations of the five components that achieved accurate white points ($\Delta uv < 10^{-7}$, see text) at these CCTs.

computation, this criterion was achieved accurately ($\Delta uv < 10^{-7}$). Figure 7.19 shows the relative weights of the five spectral components for the four correlated color temperatures.

As can be seen from Figure 7.19, the relative weight of the light from the blue LED chip increases with increasing CCT (as expected). The corresponding relative spectral power distributions of these theoretical 4-pc WLEDs can be seen in Figure 7.20.

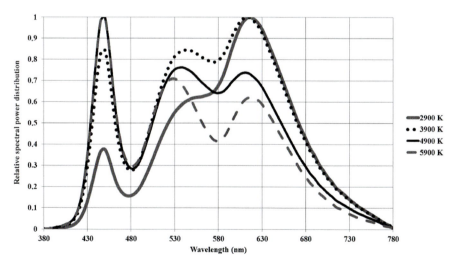

**Figure 7.20** Relative spectral power distributions of the theoretical 4-pc WLEDs. Linear combinations of the component spectra of Figure 7.18 with the weights of Figure 7.19.

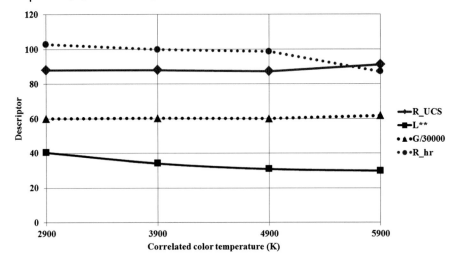

**Figure 7.21** Mean chromatic lightness ($L^{**}$), color gamut ($G$), color rendering index ($R_{UCS}$), and color harmony rendering ($R_{hr}$) of the 4-pc WLEDs of Figure 7.20 as a function of correlated color temperature. The quantities $L^{**}$, $G$, $R_{UCS}$, and $R_{hr}$ are defined in Section 7.2.2.

Figure 7.20 shows how the shape of the relative spectral power distribution of the blue LED and phosphor blend combination changes according to the choice of the CCT. It is also interesting to take a look at the descriptors of chromatic lightness ($L^{**}$), color gamut ($G$), color rendering index ($R_{UCS}$), and color harmony rendering ($R_{hr}$) for these 4-pc WLEDs as a function of the CCT. This is shown in Figure 7.21.

As can be seen from Figure 7.21, the mean lightness of the 34 objects ($L^{**}$) decreases with increasing CCT (about 26%) (for this sample set of colored objects). Gamut ($G$) and the color rendering index ($R_{UCS}$) remain approximately constant (within 4%), while the color harmony rendering property also decreases with increasing CCT (about 15%).

In addition to this computation, it is also interesting to compare these four descriptor values for today's three typical indoor light sources at the correlated color temperature of about 2600 K, that is, the tungsten incandescent lamp (INC, $R_a = 100$), the compact fluorescent lamp (CFL, $R_a = 84$), and the retrofit LED lamp (LED, $R_a = 89$) taken from Table 6.7 and Figure 6.18. Figure 7.22 shows this comparison.

As can be seen from Figure 7.22, mean lightness and gamut do not change among these three light sources significantly. The CFL's color harmony property is better than that of INC and LED. Concerning the mean color rendering index ($R_{UCS}$), however, CFL has a low value ($R_{UCS} = 75$) especially due to three colored indoor objects of Figure 6.13, no. 7 (red rose, $R_{UCS,special} = 31$), no. 21 (tomato, $R_{UCS,special} = 57$), and no. 28 (blue jeans, $R_{UCS,special} = 49$). The reason is that this typical CFL emits little radiation in those important spectral ranges (red or blue, respectively) where these objects exhibit a high spectral reflectance (compare Figure 6.13 with Figure 6.18). Similarly, the lower $R_{UCS}$ value of the typical LED retrofit lamp ($R_{UCS} = 87$) is especially due to two such colored objects, no. 7 (red rose, $R_{UCS,}$

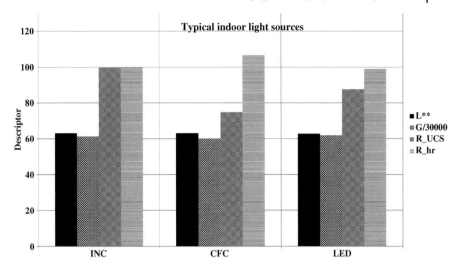

**Figure 7.22** Comparison of the descriptors of mean lightness ($L^{**}$), gamut ($G$), color rendering index ($R_{UCS}$), and color harmony rendering index ($R_{hr}$), for today's three typical indoor light sources at 2600 K, tungsten incandescent lamp (INC), compact fluorescent lamp (CFL), and retrofit LED lamp (LED) (see also Table 6.7 and Figure 6.18.

$_{special} = 72$) and no. 28 (blue jeans, $R_{UCS,special} = 60$). The special color rendering indices ($R_{UCS,special}$) of these three light sources are shown in Table 7.1.

As can be seen from Table 7.1, for CFL, in addition to the above-mentioned objects, several other objects also exhibit $R_{UCS,special}$ values lower than 80, for example, further reddish objects (apple, ham, and salami), two skin tones (Asian and Caucasian), green, yellow, and brownish objects (green and yellow–green leaf, yellow–green pear, banana, yellow shirt, yellow lily, yellow cheese, and bread). One consequence of the above findings for the design of emerging light sources is that special care has to be taken to cover the important ranges of the spectrum to be able to render all important object colors. The other consequence is that the value of the CIE color rendering index ($R_a$) based on the eight desaturated color samples (see Section 6.1, Step 7) may be misleading; for example, for CFL, $R_a = 84$ but there are some very badly rendered common objects, as seen above.

Finally, note that the computational example of 4-pc WLEDs in this section has only a limited validity because the linear combination of the five emission spectra of Figure 7.18 represents only a coarse first approximation of the relative spectral power distribution of a real multi-phosphor device. The reason is that there are multiple reflections, reabsorptions, and reexcitations among the blue LED chip and the phosphors within the device. These effects depend strongly on the geometry of the WLED and can be modeled by ray-tracing computations [31–33] or a physical model [34]. Let us also mention the emerging field of ZnO nanorod white LEDs yielding balanced spectral power distributions and promising color rendering properties with $R_a$ values of up to 98 and all 14 CIE special color rendering indices above the value of 90 [3].

## 7 Emerging Visual Technologies

**Table 7.1** Special color rendering indices ($R_{UCS,special}$) of the three light sources of Table 6.7 and Figure 6.18 for the 34 colored objects of Figure 6.13.

| Type | Object description | No. | INC | CFL | LED |
|---|---|---|---|---|---|
| CIE $R_a$ | | | 100 | 84 | 89 |
| Skin tones | African | 1 | 100 | 83 | 93 |
| | Asian | 2 | 100 | *77* | 89 |
| | Caucasian | 3 | 100 | *78* | 90 |
| Leaves | Green | 4 | 100 | *71* | 88 |
| | Yellow–green | 5 | 100 | *77* | 88 |
| | Lilac | 6 | 100 | 81 | 83 |
| Flowers | Red rose | 7 | 99 | *31* | *72* |
| | Orange gerbera | 8 | 100 | 90 | 92 |
| | Yellow lily | 9 | 100 | *65* | 82 |
| | Pink peony | 10 | 100 | *77* | 86 |
| Fruits-1 | Peach | 11 | 100 | 86 | 94 |
| | Yellow–green pear | 12 | 100 | *77* | 88 |
| | Nectarine | 13 | 100 | *65* | 87 |
| | Apricot | 14 | 100 | *74* | 88 |
| | Granny Smith apple | 15 | 100 | *75* | 87 |
| Food-1 | Yellow cheese | 16 | 100 | *71* | 83 |
| | Salami | 17 | 100 | *69* | 94 |
| | Ham | 18 | 100 | *64* | 92 |
| | Brown bread (outside) | 19 | 100 | *76* | 86 |
| Vegetables | Zucchini | 20 | 100 | *70* | 87 |
| | Tomato | 21 | 99 | *57* | 90 |
| | Yellow pepper | 22 | 100 | *78* | 89 |
| | Cucumber | 23 | 100 | *77* | 88 |
| Fruits-2 | Banana | 24 | 100 | *68* | 83 |
| | Orange | 25 | 100 | 85 | 90 |
| Toy | Smurf | 26 | 100 | 92 | 94 |
| Food-2 | Strawberry yogurt | 27 | 100 | 91 | 96 |
| Textiles | Blue jeans | 28 | 99 | *49* | *60* |
| | Red trousers | 29 | 100 | 85 | 88 |
| | Pink jacket | 30 | 100 | 81 | 90 |
| | Yellow shirt | 31 | 100 | *76* | 83 |
| Books | Blue book | 32 | 100 | *79* | 82 |
| | Red book | 33 | 100 | 81 | 95 |
| | Yellow book | 34 | 100 | 88 | 95 |

Values less than 80 are indicated by bold italic numbers.

### 7.2.5
### Implications of Color Constancy for Light Source Design

For the emerging technologies of indoor light sources, understanding the mechanisms of human color constancy can be important in the future [35]. Color constancy is the ability of human observers to recognize the spectral reflectance of an object's surface color despite changing observation conditions due to the changing spectral properties of the light source illuminating the object. A striking example is switching on warm tungsten light in a room with many different colored objects instead of a cool white fluorescent light source. Although objects exhibit huge chromaticity changes, their color appearance remains astonishingly constant.

Several different mechanisms of the human visual system contribute to the phenomenon of color constancy and researchers of very different fields have been involved in modeling it including psychologists, computer scientists, engineers, and neuroscientists [35]. Color constancy cannot be fully described by considering only the changes of illumination chromaticity (and other crude assumptions about the human visual system), that is, by the application of a straightforward chromatic adaptation transformation like in today's algorithms intended to describe the color rendering property of light sources.

As at the level of human photoreceptor signals, object surface reflectance functions and illuminant spectral power distributions are not uniquely separable [36, 37], the human visual system uses some further clues found in the illuminated scene resulting from the knowledge of previous observations of millions of natural objects under different illumination conditions. Even so, human color constancy is often imperfect and the degree of color constancy depends on the actual visual scene.

For the optimization of today's novel LED lamps, it is important to recognize that their spectral power distributions may differ significantly from the natural broadband illumination under which the human visual systems is able to establish color constancy. These novel spectra may impede the work of human color constancy mechanisms and result in a poor color rendering. Therefore, it would be interesting to apply a unified theory of color constancy to light source spectral design by considering the spectral reflectance characteristics of the most important objects in the given application. Unfortunately, such a theory does not exist at the time of writing this book and it may not emerge in the near future. Nevertheless, it is interesting to review those factors that affect human color constancy with relevance to light source optimization, as will be seen below.

Chromatic adaptation is an important mechanism of color constancy taking place at a lower level of the human visual information processing workflow. Its nonlinearity and cone signal interdependence aspects, however, are often ignored. In the chromatic adaptation transform (CAT) of CIECAM02 (so-called CAT02), only the cone signals of the white point affect the output of the CAT. But there are many other factors such as local contrasts and global contrasts in the illuminated scene as well as the shape and texture of the colored objects.

Local contrast in the illuminated scene causes the visual effect of chromatic induction; for example, a small gray patch in the focus of visual attention appears slightly pinkish inside a large uniform green patch. This effect is affected by global contrasts in the scene, that is, by remote stimuli [35, 38]. It was shown that the color appearance of a test patch was affected not only by the color of its surround but also by the variance of its surrounding colors, that is, by the different contrasts and saturations around the test patch. It was found that the objects appear more vivid against low-contrast gray surrounds than against high-contrast multicolored surrounds [39]. This finding indicates the limitations of quantifying adaptation by a uniform adapting field employed by most of today's CATs.

Relational color constancy is a further important issue relevant to color rendering [35]. This is the constancy of perceived *relations* between the colors of surfaces under changes in illumination [40]. It has been suggested that relational color constancy can be achieved by examining the spatial cone excitation ratios within the scene [41]. For most natural surfaces, the cone excitation ratios between distinct surfaces remain almost invariant under *natural* illuminant changes [41]. The relevance to color rendering is that a test light source of poor color rendering changes the relations between the colors of the surfaces in the illuminated scene compared to the reference light source [35].

Computed spatial cone excitation ratios of a set of colored test objects under the test and reference light sources seem to offer a quantitative color rendering method. According to our knowledge, such a method has not been implemented and compared with subjective color rendering judgments yet – possibly because relational color constancy alone cannot describe the changing color appearance of the surfaces under the illuminant [35].

As mentioned above, *chromatic textures* (color shadings and fine geometric structures visible on the illuminated surfaces) represent a further clue for the higher order mechanisms of the human visual system to establish color constancy. Natural objects often exhibit textures unlike the homogeneous test color patches of most current color rendering computation methods. This is one of the reasons why a tabletop (still life) arrangement of real objects (Figure 6.17) is a better choice to explore light source color quality (in general) than a combination of homogeneous color patches (Figure 6.7).

Due to their colored shadings, natural objects have typical color *distributions* that can be represented, for example, in the CIECAM02 $a_C$, $b_C$ diagram [35, 42]. These color distributions change in a characteristic way if the light source illuminating the objects is changed and the human visual system is able to predict [43] these characteristic changes unless the objects are illuminated by an unusual or incomplete spectral power distribution of a new light source. An example of a set of 12 real objects and 2 light sources (a white phosphor-converted LED light source of good color rendering and an RGB LED light source of bad color rendering) is shown in Figure 7.23. The *XYZ* values of these objects were measured by a high-resolution high-end imaging colorimeter and converted into CIECAM02 $a_C$, $b_C$ values.

As can be seen from Figure 7.23, the *color distribution* of the textures of the 12 objects is *distorted* under the RGB LED light source compared to the white phosphor LED. Therefore, color constancy *breaks down* when the white phosphor LED is switched off and the RGB LED is switched on instead.

As seen above, human color constancy is a very complex phenomenon. Due to the lack of a unified mathematical model at present, today it is not feasible to use these findings for the optimization of light source spectral power distributions. But such

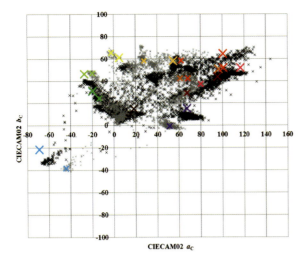

**Figure 7.23** Color distributions of a set of real objects (a yellow pepper, an orange yarn, a red tomato, a green pepper, a red rose, an orange dahlia, a violet table cloth, a cyan table cloth, a yellow–green table cloth, a yellow table cloth, a red table cloth, and a brown table cloth) in a CIECAM02 $a_C$–$b_C$ diagram. Objects are illuminated either by a white phosphor-converted LED light source of good color rendering (small gray crosses) or by an RGB LED light source of bad color rendering (large black crosses). Small (large) colored crosses show the mean $a_C$–$b_C$ values of these objects under the white LED (RGB LED) [42]. Objects were measured by a high-resolution high-end imaging colorimeter. Reprinted with permission of IS&T: The Society for Imaging Science and Technology, sole copyright owners of CGIV 2010 – 5th European Conference on Color in Graphics, Imaging, and Vision.

considerations may serve as the basis of emerging light source technologies – especially for the spectrally more flexible LEDs.

## 7.3
## Summary and Outlook

This book points out how the knowledge about the human visual system – especially color vision – can be used to optimize self-luminous color displays and indoor light sources illuminating a scene of colored objects reflecting the light of the light source. To optimize the appearance of colored objects, the spectral power distribution of the light source has to be co-optimized with the spectral reflectance functions of the objects. The aim of this optimization is to enhance the components of light source color quality (Chapter 6).

The components of light source color quality include color fidelity, color harmony, color gamut, color preference, chromatic lightness of the colored objects, enhancement of color differences, visual clarity, continuity of color transitions, similarity to long-term memory colors, and acceptability, that is, the correspondence of the object's shape and texture to its color under the light source illuminating it.

A set of measured spectral reflectance functions of important indoor objects including different skin tones was shown. This is relevant for the computation of the numerical correlates (indices) of the above aspects of perceived color quality of the light source illuminating the objects. For example, the numerical correlate of color fidelity perception is the color rendering index (Section 6.2). The practical relevance of such indices is that the lighting engineer can optimize the spectral power distribution of a new (planned) light source by changing the technological parameters of the light source, while light source technology permits only a limited spectral optimization because important wavelength ranges can be absent or underrepresented in the spectral power distribution of the light source due to its technological limitations. Hereby, LED chip technology and phosphor technology open up novel horizons for optimization.

Concerning self-luminous displays, colorimetric calibration and characterization methods were described with the aim of displaying the color stimuli accurately. Colorimetric calibration of displays is essential for accurate color image reproduction but also in psychophysical experiments exploring how observers react to a technological feature of the display. More specifically, a psychophysical model fitted to a data set of human visual responses obtained on a display can predict the numerical correlates of human perception or visual cognition correctly only if all color stimuli are *physically accurate* at the stage of collecting the data set. For example, the colorimetric characterization of the PDP monitor in Section 2.4 – including the identification of its technological limitations such as the extent of its spatial inhomogeneity – is relevant for the psychophysical investigation of the color size effect.

By the same token, in the equiluminance legibility experiment (Section 3.3.2), the luminance of the text and its background has to be the same. To achieve this, the results of the CRT characterization methods (Section 2.2.2.1) were used, while it was shown that every display technology has its own tone curve or electro-optical transfer function, different spatial and temporal properties, and also different technological artifacts, that is, visual annoyances arising from the specific hardware or image rendering algorithm.

A *complete* characterization model of a display should describe not only the colorimetry of the display (i.e., predict the *XYZ* values of extended uniform color patches) but also the aspects of *spatial* color, for example, spatial color resolution or the display's ability to show fine continuous color shadings. Factors such as the uniformity of the whole light emitting surface (Section 2.1.5) or the dependence of the color stimulus on viewing angle (Section 2.1.6) are equally important, while it should be noted that the *exact* physical measurement of display *viewing angle* characteristics is a complicated *gonioradiometric* procedure.

Characterization models are often included in the display's hardware (at least partially), for example, in order to convert the native tone curve of the display into a CRT-like tone curve. They (or parts of them) are often implemented as a software method, for example, as an ICC color profile (Section 4.1.2). But usually, what the user of the display needs is not only a simple colorimetric profile enabling the accurate display of *XYZ* values but also a comprehensive image processing algorithm adjusting the color appearance of an original image to the viewing conditions and to the objectives of color reproduction (Section 3.2). Thus, a color profile can include long-term memory-based (Section 3.4.3) or image preference-based (Section 3.6) optimization or even emotional optimization (Section 4.6).

Design methods were presented to obtain a large color gamut with optimum color primaries (Section 5.2) together with a set of multi-primary subpixel architectures (Section 5.4) to optimize spatial color resolution. The reason is that novel *multi-primary* self-luminous display technologies appeared on the market working with novel color subpixel architectures with more than three primary colors and sometimes also with a large dynamic luminance range. These technologies include color multi-primary LCD technology with dynamic LED backlights (Section 2.3), large plasma displays (Section 2.2.2.2), and LED projectors (Section 2.2.2.5). According to their large color gamut, all natural objects (even those exhibiting very saturated colors especially red, purple, and orange) can be displayed accurately.

Color gamut *extension* algorithms (Section 7.1.3) are also used in the color profile because multi-primary displays often exhibit a more extended color gamut than the color gamut of the original image. A good example is a four-primary system exhibiting a color gamut wider than the one of five- or six-primary displays (Section 5.2.2). Color image rendering methods should also be able to display the image on new multi-primary subpixel architectures (Section 5.4). Although the seven-primary architecture of Figure 5.19 covers fewer colors than the four-primary architecture, it is able to render spatial color distributions optimally.

An interesting future research direction is comparing the importance of color gamut with the importance of spatial color representations on multi-primary subpixel architectures. A subject of future research can be the optimization of image rendering algorithms displaying an existing color image file on a complex colored subpixel mosaic to avoid any spatial color artifact. It is not completely solved how to extend the color gamut without color fringe artifacts (Section 5.3) in those image areas containing delicate continuous color transitions.

Dynamic LED backlights enable high dynamic range (HDR) imaging (Section 2.3.3). The appearance of highlights enhances the immersion of the user in the scene and hence the emotional effect although the scattered light in the human eye reduces the technologically achievable dynamic luminance range. In association with this, for an optimum image compression of HDR images, the meaningful lower limit of the luminance range in the vicinity of high contrast should be investigated in the future.

Modern displays often represent large (i.e., spatially extended) color stimuli. A model was presented to describe the specific characteristics of their color appearance (Section 2.4). Future image processing procedures should contain a transformation for such large color stimuli as an option for the case when the image covers a large viewing angle, that is, $>30°$. Visual ergonomic principles (Section 3.1) were also described for computer-controlled displays together with a method of applying color contrast to enhance visual search performance in the presence of numerous colored objects on the user interface (Section 3.3).

Cognitive color, preferred color, and emotional color were also subject of this book. Long-term memory colors were found to be important to enhance the perceived color quality of color displays (Section 3.4.3). The strength of visually evoked emotions related to cinematographic motion pictures was computed by a mathematical model based on the technical parameters of the video sequence (Section 4.6).

Future comprehensive image processing algorithms should include these aspects (memory colors, color preference, and visually evoked emotions). The original colors of the image can be transformed toward long-term memory colors or preferred colors or in the direction of strengthened emotions by considering not only the color signals of the *image* but also their local and global contrasts. Memory and preferred colors and visually evoked emotions exhibit large interpersonal variability including variations with age and country of origin. These variations are partially unexplored and represent interesting future research directions. A further field of research concerns mesopic color appearance because important applications take place in the mesopic (twilight) range of vision, for example, digital or analog cinema including home cinema or interior car lighting. Mesopic color image appearance models have to be further developed for the image processing chain in these applications

## Acknowledgments

Authors would like to acknowledge the support of the following coworkers of the Laboratory of Lighting Technology of the Technische Universität Darmstadt: Mr. Marvin Böll, Mr. Stefan Brückner, Ms. Nathalie Krause, Mr. Wjatscheslaw Pepler, and Mr. Quang Vinh Trinh.

## References

1 Choi, M.C., Kim, Y., and Ha, C.S. (2008) Polymers for flexible displays: from material selection to device applications. *Prog. Polym. Sci.*, **33**, 581–630.

2 Allen, K.J. (2005) Reel to real: prospects for flexible displays (invited paper). *Proc. IEEE*, **93** (8), 1394–1399.

3 King, R.C.Y. and Roussel, F. (2007) Transparent carbon nanotube-based driving electrodes for liquid crystal dispersion display devices. *Appl. Phys. A*, **86**, 159–163.

4 Sekitani, T., Nakajima, H., Maeda, H., Fukushima, T., Aida, T., Hata, K., and Someya, T. (2009) Stretchable active-matrix organic light-emitting diode display using printable elastic conductors. *Nat. Mater.*, **8**, 494–499.

5 Rogers, J.A., Bao, Z., Baldwin, K., Dodabalapur, A., Crone, B., Raju, V.R., Kuck, V., Katz, H., Amundson, K., Ewing, J., and Drzaic, P. (2001) Paper-like electronic displays: large-area rubber-stamped plastic sheets of electronics and microencapsulated electrophoretic inks. *Proc. Natl. Acad. Sci. USA*, **98** (9), 4835–4840.

6 Vertegaal, R. and Poupyrev, I. (2008) Organic user interfaces. *Commun. ACM*, **51** (6), 26–30.

7 Deter, A. (2006) 2nd generation of laser display technology. Innovation Special Planetariums 6, Carl Zeiss, pp. 26–27.

8 Langhans, K., Guill, C., Rieper, E., Oltmann, K., and Bahr, D. (2003) Solid

Felix: a static volume 3D-laser display. *IS&T Rep.*, **18** (1), 1–16.

9 Ko, Y.C., Cho, J.W., Mun, Y.K., Jeong, H.G., Choi, W.K., Kim, J.W., Park, Y.H., Yoo, J.B., and Lee, J.H. (2006) Eye-type scanning mirror with dual vertical combs for laser display. *Sens. Actuators A*, **126**, 218–226.

10 Brückner, S. and Khanh, T.Q. (2011) Dimmung von Hochleistung-LEDs, Nutzen, Methoden und lichttechnische Folgen (Dimming of high-brightness LEDs: benefits, methods and consequences for lighting engineering. *Licht*, **3**, 44–49.

11 Lee, S.Y., Kwon, J.W., Kim, H.S., Choi, M.S., and Byun, K.S. (2006) New design and application of high efficiency LED driving system for RGB-LED backlight in LCD display. Proceedings of the 37th IEEE Power Electronics Specialists Conference (PESC'06), Jeju, Korea, pp. 1–5.

12 Muthu, S., Schuurmans, F.J., and Pashley, M.D. (2002) Red, green, and blue LED based white light generation: issues and control. Proceedings of the Industry Applications Conference, vol. 1, pp. 327–333.

13 Vinh, T.Q. and Khanh, T.Q. (2011) Gefährliche mischung, wirkungen von Strom und Temperatur auf die LED-lebensdauer (A dangerous mixture: effect of current and temperature on the lifetime of LEDs). *Licht*, **11–12**, 76–80.

14 Hunt, C.E., Quintero, J., and Carreras, J. (2011) Appearance degradation and chromatic shift in energy efficient lighting devices. Proceedings of the 19th Color and Imaging Conference, San Jose, CA, pp. 71–75.

15 Hoshino, T. (1991) A preferred color reproduction method for the HDTV digital still image system. Proceedings of the IS&T Symposium on Electronic Photography, pp. 27–32.

16 Xie, Y. and Klompenhouwer, A.M. (2011) Colour image enhancement. U.S. Patent Application No. 2011/0110588 A1.

17 Kang, B.H. and Cho, M.S. (1999) Methods of colour gamut extension algorithm development using experimental data. Proceedings of 1999 IEEE Tencon, pp. 352–355.

18 Freyssinier, J.P., Frering, D., Taylor, J., Narendran, N., and Rizzo, P. (2006) Reducing lighting energy use in retail display windows. Sixth International Conference on Solid State Lighting, *Proc. SPIE*, **6337**, 63371L.

19 Zulim, D., Lydecker, S.H., King, L.C., and Hinnefeld, J.D. (2007) Networked architectural lighting with customizable color accents. U.S. Patent Application No. 2007/0285921 A1.

20 Gall, D. and Bieske, K. (2004) Definition and measurement of circadian radiometric quantities. Proceedings of CIE Symposium '04 on Light and Health, Commission Internationale de l'Éclairage, Vienna, pp. 129–132.

21 Rea, M.S., Figueiro, M.G., Bierman, A., and Bullough, J.D. (2010) Circadian light. *J. Circadian Rhythms*, **8** (2), 1–10.

22 CIE 165:2005 (2005) *CIE 10 Degree Photopic Photometric Observer*, Commission Internationale de l'Éclairage.

23 Viénot, F., Bailacq, S., and Le Rohellec, J. (2010) The effect of controlled photopigment excitations on pupil aperture. *Ophthalmic Physiol. Opt.*, **30** (5), 484–491.

24 Gamlin, P.D.R., Mc Dougal, D.H., Pokorny, J., Smith, V.C., Yau, K.W., and Dacey, D.M. (2007) Human and macaque pupil responses driven by melanopsin containing retinal ganglion cells. *Vis. Res.*, **47**, 946–954.

25 Rea, M.S., Bullough, J.D., and Figueiro, M.G. (2002) Phototransduction for human melatonin suppression. *J. Pineal Res.*, **32**, 209–213.

26 Brainard, G.C., Hanifin, J.P., Greeson, J.M., Byrne, B., Glickman, G., Gerner, E., and Rollag, M.D. (2001) Action spectrum for melatonin regulation in humans: evidence for a novel circadian photoreceptor. *J. Neurosci.*, **21**, 6405–6412.

27 Thapan, K., Arendt, J., and Skene, D.J. (2001) An action spectrum for melatonin suppression: evidence for a novel non-rod, non-cone photoreceptor system in humans. *J. Physiol.*, **535**, 261–267.

28 Khanh, T.Q. (2010) LED – a technology for quality and energy efficiency. Proceedings of CIE 2010 "Lighting Quality & Energy Efficiency", CIE x035:2010.

29 Xie, R.J., Hirosaki, N., Sakuma, K., and Kimura, N. (2008) White light-emitting

diodes (LEDs) using (oxy)nitride phosphors. *J. Phys. D*, **41**, 144013.
30 Winkler, H., Barnekow, P., Benker, A., Petry, R., Tews, S., and Vosgroene, T. (2008) Inorganic phosphors for LED applications. IMID/IDMC/ASIA Display 2008 Digest.
31 Yamada, K., Imai, Y., and Ishii, K. (2003) Optical simulation of light source devices composed of blue LEDs and YAG phosphor. *J. Light Vis. Environ.*, **27** (2), 10–14.
32 Zhu, Y. and Narendran, N. (2008) Optimizing the performance of remote phosphor LEDs. *J. Light Vis. Environ.*, **32** (2), 65–69.
33 Won, Y.H., Jang, H.S., Cho, K.W., Song, W.S., Jeon, D.Y., and Kwon, H.K. (2009) Effect of phosphor geometry on the luminous efficiency of high-power white light-emitting diodes with excellent color rendering property. *Opt. Lett.*, **34** (1), 1–3.
34 Zhu, Y. and Narendran, N. (2010) Investigation of remote-phosphor white light-emitting diodes with multi-phosphor layers. *Jpn. J. Appl. Phys.*, **49**, 100203.
35 Ling, Y., Bodrogi, P., and Khanh, T.Q. (2009) Implications of human colour constancy for the lighting industry. CIE Light and Lighting Conference with Special Emphasis on LEDs and Solid State Lighting, May 2009, Budapest, Hungary.
36 Hurlbert, A.C. (1998) Computational models of colour constancy, in *Perceptual Constancy: Why Things Look as They Do* (eds V. Walsh and J. Kulikowski), Cambridge University Press, pp. 283–322.
37 Smithson, H.E. (2005) Sensory, computational and cognitive components of human colour constancy. *Philos. Trans. R. Soc. B*, **360** (1458), 1329–1346.
38 Shevell, S.K. and Wei, J. (1998) Chromatic induction: border contrast or adaptation to surrounding light? *Vis. Res.*, **38** (11), 1561–1566.
39 Brown, R.O. and MacLeod, D.I. (1997) Color appearance depends on the variance of surround colors. *Curr. Biol.*, **7** (11), 844–849.
40 Nascimento, S.M.C., de Almeida, V.M.N., Fiadeiro, P.T., and Foster, D.H. (2004) Minimum-variance cone-excitation ratios and the limits of relational colour constancy. *Vis. Neurosci.*, **21**, 337–340.
41 Foster, D.H., Nascimento, S.M.C., Craven, B.J., Linnell, K.J., Cornelissen, F.W., and Brenner, E. (1997) Four issues concerning colour constancy and relational colour constancy. *Vis. Res.*, **37** (10), 1341–1345.
42 Bodrogi, P., Brückner, S., and Khanh F T. Q. (2010) Dimensions of light source colour quality. Proceedings of CGIV 2010, 5th European Conference on Colour in Graphics, Imaging, and Vision, Joensuu, Finland, pp. 155–159.
43 Ling, Y., Vurro, M., and Hurlbert, A. (2008) Surface chromaticity distributions of natural objects under changing illumination. Proceedings of CGIV 2008, 4th European Conference on Colour in Graphics, Imaging, and Vision, Terrassa, Spain, pp. 263–267.

# Index

## a

accent lighting  339, 340
acceptable *P/W* ratios  247
achromatic channel  10
achromatic contrast  7, 10, 124, 129
– sensitivity  9–11
achromatic signal  3
acousto-optical modulator  176
adaptation  101, 280
adapted reference  49
adapting field luminance  188
adaptive brightness intensifier method  80
additive mixture  27, 35
additivity  35
addressability  266, 267
Adobe RGB  251
adopted reference white  49
aged and young observers  146
ambient light  46
– reflections  20
AM-LCD  45, 46, 61–66
AM-LCD monitor  68
analog film camera  161
angle-of-view dependence  45, 51
angular size  100
angular velocity  101
antagonistic receptive fields  197
anthropometric results  99
anti-aliasing  259
AOM (acousto-optical modulator)  176
artificial light sources  273
attractiveness  298
attributes of perceived color  13
augmented reality  67

## b

background  97
background luminance  101

backlight  72, 77, 78, 80
backlight luminance compensation (BLC)  80
backlight sources  72, 75, 76
balanced harmony rendering  293
Bayer pattern  262
Bezier spline  149
bit depth  199
black-corrected matrix  64
black emission  56
black level emission  57
black point  79
bleaching field  103
blue channel  216, 219
blur  147
brightness  12, 48, 298, 311

## c

calculated color differences ($\Delta E_{calc}$)  280
camera  19
categorical color names  294
categorical identification  107
CCD filtering mosaic  262
CCFLs  76
CCT. *See* correlated color temperature (CCT)
ceramic lamps  223
channel independence  62
channel interdependence  35, 37, 39–41, 56, 61, 68
characterization models  26, 58
characters  97
chart  179
"checkerboard" arrangement  260
CHF formulas  288, 289
CH formula  288
chip LEDs  333, 335, 338, 339
chip LEDs display  337

*Illumination, Color and Imaging: Evaluation and Optimization of Visual Displays*, First Edition.
Peter Bodrogi and Tran Quoc Khanh.
© 2012 Wiley-VCH Verlag GmbH & Co. KGaA. Published 2012 by Wiley-VCH Verlag GmbH & Co. KGaA.

## Index

chroma  12, 13, 15
chroma enhancement transform  154
chroma extension  334
chromatic adaptation  355
chromatic adaptation formula  283
chromatic brightness  120
chromatic channels  8, 10
chromatic contrast  7, 9, 125
chromatic contrast sensitivity  9
– functions  11
chromatic gratings  124
chromaticity contrasts  107, 111, 116, 120–123, 125, 126, 129, 131, 133
chromaticity contrast thresholds  131
chromaticity coordinates  6
chromaticity differences  46, 62, 65, 112, 113, 218, 251
chromaticity error  270
chromaticity gamut  239, 338
chromatic lightness $(L^{**})$  343, 347, 348, 352
chromatic signals  3
chromatic textures  356
chromostereopsis  107
CIECAM02  49, 81, 82, 88, 106, 144, 240, 243, 289
CIECAM02 $a_C$, $b_C$ diagram  356, 357
CIECAM02 $a_c$-$b_c$ plane  255
CIECAM02 color appearance model  15
CIECAM02-dark version  190
CIECAM02-dim version  190
CIECAM02 $J$ value  147
CIECAM02 $J(Y)$ function  69
CIECAM02-LCD  17, 280
CIECAM02 model  48, 50
CIECAM02-SCD  17, 280
– color space  243
CIECAM02-UCS  17, 280, 281
CIE 1931 chromaticity diagram  6
CIE (x, y) chromaticity diagram  6
CIE 1964 color differences  275
CIE colorimetry  14
CIE color rendering indices ($R_a$)  274, 276, 277
CIEDE2000  17, 280
CIELAB  14, 49, 50, 89, 90, 284
CIELAB chroma ($C^*_{ab}$)  14
CIELAB color differences  31, 58, 118
CIELAB color space  14
CIELAB differences  43
CIELAB hue angle ($h_{ab}$)  14
CIE 1976 lightness  14
CIELUV  14, 49, 50
CIELUV chroma ($C^*_{uv}$)  15
CIELUV color space  15
CIELUV hue angle ($h_{uv}$)  15
CIE 1924 photopic $V(\lambda)$ function  342
CIE 1931 standard colorimetric observer  4, 82
CIE 1964 standard colorimetric observer  4
CIE standard photometric observer  3
CIE 1931 tristimulus values $X$, $Y$, $Z$  275
CIE 1976 UCS chromaticity differences  43
CIE 1960 UCS coordinates $u$, $v$  275
CIE 1976 uniform chromaticity scale diagram (UCS diagram)  14
CIE 1976 uniform color spaces  14
CIE 1964 $U^*$, $V^*$, $W^*$ values  275
CIE $x$, $y$ chromaticity diagram  336, 338
cinema motion picture films  174
circadian behavior  21
circadian clock  341
circadian optimization  341
circadian rhythm  21, 341
circadian stimulus (CS)  341–346
Circle 32 test  295
ClearType™  259, 260
clipping effect  69
closed-view HMD  67
CMOS technology  175
cognitive color  17, 319
cognitive hints  103
cognitive representation of the scene  143
cold cathode fluorescent lamps (CCFLs)  72
color accent  339
color acceptability  295
color algorithms  162
color appearance  12, 25, 48, 81, 87, 88, 143, 190, 274
– models  15
– reproduction  106
color appearance space  238, 250, 252
color artifacts  97
color balance preference  245
color breakup (CBU) artifact  71
color categories  17, 299
color channels  27, 35, 40
– crosstalk  40
– independence  35
color circles  300, 301
color cognition  17
color constancy  354, 355
color differences  17, 32–34, 37, 38, 59, 60, 65, 70, 88, 274, 280
– $\Delta E^*_{ab}$  39, 60
– formulas  280, 281
– magnitude  17
– perception  15
– scales (RCRI)  305

Index | 365

color discrimination  294
color discrimination index (CDI)  275
color distortions  26, 33
color distributions  356, 357
colored objects  273, 274
color ergonomics  98, 107, 121, 124
color error  34
color fidelity  20, 276, 293, 302
color fidelity index ($R_{UCS}$)  312
color filters  78, 79
– mosaic  20, 79
– wheel  70
color fringe artifact (CFA)  237, 258, 261, 268
color fringe error  47, 261
colorfulness  12, 48
color gamut (G)  18, 21, 27, 46, 68, 73–77, 237, 238, 245, 253, 254, 256, 257, 299, 311–315, 346–348, 352, 353
color gamut area  252
color gamut differences  191
color gamut extension  334
color gamut mapping methods  238
color gamut projections  255
color gamut volume  244
color gradations  70
color harmony  17, 286, 287, 294, 298, 299
color harmony rendering  289, 314, 346, 352
color harmony rendering index ($R_{hr}$)  289, 290, 312, 315, 347, 348, 353
color image descriptors  151
color image preference  19, 142–144, 149
– scores  151
color image quality  238
color image rendering methods  262, 268, 269
color image transforms  144
colorimetric characterization  18, 25, 26, 51, 54, 56, 57, 61, 62
– models  27
colorimetric errors  26, 28
colorimetric reproduction  106
colorimetry  4, 48
color management  18
– workflow  106
color mapping algorithms  334
color matching functions  4
color memory  19, 33, 47, 57, 60, 107, 353
color monitors  19
color mottle  70
color names  33, 137, 138
color perceptions  1, 4, 12
color preference  20, 294, 303
color preference index  275
color primaries  238, 239, 249, 267, 269, 270
color primary optimization  243

color primary systems  253
color primary systems (sRGB, NTSC)  251
color process  162
color quality  20, 134, 141, 273, 275, 286, 293, 295–297, 303, 320–322, 341, 347, 348
– factors  303
– four-factor model  302
– of light sources  318
– scale  285, 295
– variables $V_1$-$V_9$  323
color quality scale (CQS)  285, 295
color rendering  20, 223, 273–275, 282, 291, 292, 294, 310, 322, 340, 343, 346, 347
color rendering index ($R_{UCS}$)  282, 313, 314, 321, 323, 345, 348, 352, 353
– difference  284
color rendering scales  305
color resolution  18, 27, 70
color samples  275
color scheme  122
color sequential
– LED displays  78
– mode  74
color size effect  19, 81, 86, 87, 89, 90
– model  88
color space  12, 13, 178, 334
color stimulus  4
color temperature  341
color tracking  57, 60, 62, 64–66
color transitions  294, 299
color vision  18
– mechanisms  319
color wheel  79
combinations of watercolors  299
combined fidelity-preference indices  294
comfort, 98
command length  114–117
compact fluorescent lamp (CFL)  353
complex indoor scenes  304, 311
compression algorithms  9
compression efficiency  210
compression factors  205
compression methods  205
computer workplace  19
cone mosaic  2
cone opponency  8
cones  2
conspicuity  103, 104, 107, 108, 118, 119, 121
conspicuity function  120
conspicuity levels  111, 118, 120–122
constant current reduction (CCR)  332
constrained body postures  99
constrained postures  99

continuous judgments  307
contrast  30, 113
contrast level  127
contrast preferences  123, 133
contrast ratio  9
contrast sensitivity (CS)  2, 6, 9, 19, 123, 196, 216
– chart  10
– function  196
cool white  345, 346
cool white fluorescent lamps  278
correlated color temperature (CCT)  274–277, 317, 340, 343, 344, 351
– groups  347
– preference  147
corresponding colors  179
cost function (p)  239, 250
cost surface  250
coverage  250, 252, 256
coverage computation  243
CQS  285
CQS scale  295
CRI-CAM02UCS  285
CRI-CAM02UCS color rendering index  322
critical flicker frequency (CFF)  10, 101
CRT  51, 53, 54
CRT characterization models  53
CRT display  33, 41, 43, 242
CRT gamma  51
CRT projectors  72
CRT tone curve  64
crystalline lens  134
CS. See contrast sensitivity (CS)
cubic interpolation  220
cultural differences  143
cutoff voltage  54
C/V whiteness index  316
cycles per degree (cpd)  9

## d

dark adaptation  101, 103
dark viewing condition  69
data conversion  220
dataflow quantity  216
data transformation  220
daylight  15, 273
degree of similarity  280
descriptor input-output function  153, 154
detectability  104
detection threshold  102
deuteranope  18
device-independent  28
– color system  163
– format  19

diffuse illumination  223
digital cinema  163
digital color value  35
digital intermediate  205
digital master  205
digital projectors  199
digital signage displays  20
direct glare  102
direct-view CRT displays  72
discomfort glare  47
displaced box filter RGB decimation  259
display backlighting  20
display black  29, 30, 47, 64, 65
– luminance  36
– point  250
– radiation  62
display characterization  25, 35
– models  25
display ergonomics  98
display fill factor  40
display light source  72
display setting  54
display tristimulus values  55
display white point preference ($p_W$)  246
distort the harmony  293
distractor objects  118
DLP projector  71
DMD  51, 68, 73
DMD projectors  70–72, 74, 79
3D optimization  239
Double-chamber viewing booth  277
double opponent  7
downsampling  215
DVD resolution  215
dynamic range (D)  29, 30, 68, 76, 174
dynamic watermark  217

## e

E.B.U. chromaticities  74
edge-type backlight  75
effect of variability on color quality  320
elderly observers  147, 152
electronic color signals  25
ELL formula  118
ELL scale  109, 110
emerging self-luminous visual technologies  329
emotional color  17
emotions  17, 19
equiluminance legibility (ELL)  108
equiluminous  113
ergonomic color  120
– design  107, 108, 110
ergonomic design  98, 122

ergonomic furniture   99
ergonomic guidelines   97, 98
ergonomic principles   98
ergonomics   19
ergonomic standard   111
esthetic image   106
Euclidean differences
– in CIECAM02   280
Euclidean distances   15
extended color gamut   258
eye fatigue   105
eyestrain   97

## f
familiar objects   134, 135
Farnsworth's D-15 test   298
fidelity   303
fields of attention   103
fill factors   41, 42
film density   176
film laboratory   162
film scanner   174
filter mosaic   76, 262
finest spatial details   261
Fitts law   98
five-primary system   251
fixation dwell time   103
flare   70
flexible display   329
flexible substrates   329
flicker   9
flicker artifact   12, 80
flower shape   267
fluorescent general lighting   339
focusing range   101
four primaries   251
four-primary color sequential (RGCB)
– LED display   335, 337
– model   338
four-primary display   76, 257
four-primary system   251
frame rate   205
full chip LED display   337
full screen   29

## g
gain and offset controls   27, 53
gain and offset setting   32
gain control   54
gain/offset settings   28, 29
gain setting   29
gamma   64
– problem   28
gamut   349, 350

gamut area-based index (FCI)   294
gamut boundary   255
gamut color primaries   251
gamut mapping   238
gamut optimization   74
ganglion cells   7, 342
general color rendering index ($R_a$)   274–276, 284
general lighting   303, 339
global contrast   30
– enhancement   147–149
global lightness contrast enhancement   144
global lightness transform   153
GOG model   56, 58, 59
GOG (gain-offset-gamma) model   55
graphics card   34
graphic user interface   98
grayscale   69
grayscale color difference anchor   278
green spatial resolution   261
grouping   121

## h
hard clip method   193
harmonic color   17
harmonic color combinations   33
harmony rendering   291–293
Harmony rendering index $R_{hr}$   346
HDR appearance   106
HDR illusion   80
HDR image   80
HDTV resolution, 219
head-mounted displays (HMDs)   19, 51, 67
head-up display (HUD)   51, 67
Helmholtz-Kohlrausch effect   119, 248, 311
hexagonal architectures   265, 266
hexagonal pixel grid   267
hexagonal RGB subpixel architecture   263
hexagon architectures   268
hexagon patterns   267
Hick-Hyman law   98
high aperture ratio   261
high dynamic range (HDR)
– displays   80
– imaging   18, 79, 106
high pressures   74
HMI discharge lamp   225
HMI reference light source   278
holographic displays   331
horizontal addressability   267
horizontal MTF   266
(HSV) color space   138
HUD projection   67
hue   12, 13

hue angle   15
hue angle differences ($\Delta h_{ab}$)   16
hue-dependent chroma boost   144, 151
hue-dependent chroma boost transform (CH),   150
hue-dependent chroma preference   152
hue differences ($\Delta H^*_{ab}$)   16
hue transforms   144
human color constancy   354
human eye optics   197
human perception   221
human visual system   2, 178, 216
hybrid spatial-temporal color synthesis   78
hyperbolic tangent function   63

*i*

iCAM   49, 238
ICC color management   178
ideal descriptor function   153, 154
illegal copies   215
illuminating spectral content   242
image color appearance (iCAM)   106, 259
image compression   19, 205
image cues   135
image descriptor ($\zeta^{out}$)   151–153
image differences   218
image preference   143, 155
image processing transforms   151
image quality   18, 205
image stabilization time   53
image transforms   144
– parameter   148
immersive   19
– condition   87
– scene   84
– stimuli   89
immersive colors   88
– stimulus   82–84, 88
immersive self-luminous condition   89
immersive self-luminous stimuli   89
imperfect color constancy   283
impression of harmony   293
improve color image quality   134
independence of the color channels   35
index ($R_{UCS}$)   347
indoor illumination   20
indoor light sources   273
indoor productions   223
information content   104
information-intensive mobile displays   104
infotainment   19
intercultural differences   134, 139
interindividual differences   281
interindividual variability   18

interobserver variability   318, 320, 322–324
interpersonal variability   321
intrinsically photosensitive retinal   342
intrinsically photosensitive retinal ganglion cells (ipRGCs)   22, 341
invalid decisions   303
ipRGC($\lambda$)   342
ISO international ergonomic standard   98
isotropy   260, 263, 268
ITO   330
ITU recommendation   218

*j*

jitter   55
JND   1 218
JPEG   9
judgment scales   304
junction temperature ($T_j$)   332, 333
just noticeable difference   218

*l*

Lambertian emission   51
Lambertian light sources   45
Lambertian radiator   61
Lambertian surface   70
large gamut   76
large-gamut primary colors   250
large self-luminous displays   82
large stimuli   85
large viewing angles   19
– displays   81, 84
laser displays   330
laser film recorder   176
laser home theaters   331
laser projectors   330
laser radiation   176
lateral inhibition   197
LCD characterization models   65
LCD light leakage   79
LCD projectors   51, 68, 70
LCDs   44, 45, 51, 60, 66, 68, 79
LCD tone curves   64
L-cones   3
leakage   57, 65
LED backlighting   79
LED-based projector   74
LED dimming   332
LED displays   331, 336, 338
LED driving current   333
LED lamps   340
LED lifetime   332
LED projectors   20
LEDs   77, 80

legal colors   240
legibility   97, 104, 107, 111
lifetime   333
light guide plate   75
lightness ($L^{**}$)   12, 13, 15, 48, 311, 344, 349, 350, 353
– correlate   312, 313
– formula   311
lightness contrast enhancement   144
lightness differences   15
light source   21
– chromaticity   316
– color quality   293
– design   324
liquid crystal   76
local contrast   355
– enhancement   147, 148
local dimming   72, 79, 80
logical pixel   259, 269
long-term color memory   136, 137
long-term memory   139, 273
long-term memory colors   17, 33, 134–136, 140, 242, 294, 301, 334
look-up table (LUT)   51, 62
low-pass function   10
LuAG   74
luminance   9, 11, 29
luminance balance   102
luminance channel   3, 219
luminance contrast   100, 107–109, 111, 112, 121, 125, 127
luminance contrast thresholds   131
luminance outline   120
luminance range   101
luminance ratio   37
luminance resolution   261
luminous efficiency   3
LUT   63–65

## m

MacAdam ellipses   15
Macbeth ColorChecker® chart   179, 279, 285, 295
macular pigment   2, 4
mapping algorithm   334
masking model   65
matching colors   4
matrices ($P_0$, $M_0$, and $\chi_0$)   37
matrices M, $\chi$, or T   57
matrices P, M, and $\chi$   38
matrices P, M, $\chi$, $P_0$, $M_0$, and $\chi_0$   38
matrix ($\chi$)   36, 38, 54
matrix M   38
matrix P   39, 41, 54, 57, 58

matrix $P_0$   39, 41
matrix T   56, 57, 59
MCDM(mean color difference from the mean)   85
M-cones   3
mean similarity ratings   308
measure of coverage ($p_C^{-1}$)   243
mechanical vibrations   104
melatonin suppression effect   343
memory color   134, 138, 139
– reproduction   107
memory colors   30, 141
memory-related color phenomena   17
MEMS scanners   331
mercury pressures   73
metameric   242
metameric color stimuli   245
Michelson contrast   9, 126
minimum CFA   259
minimum color fringe artifact   259
mixed adaptation   50
mixed color sequential (MCS) algorithm   76
mode of color appearance   49
modulation (M)   260
modulation transfer function   260
Moiré pattern   47
monochrome sensor   175
motion pictures   205
motor performance   98
MPEG   9
MTFs   262, 263, 266, 267
multi-color primaries   237
multi-component LED light source   343
multiprimary   259
multi-primary architectures   264
multi-primary color displays   238, 257
multi-primary displays   237, 245, 334
multi-primary mosaics   237
multi-primary subpixel architectures   257, 262, 264, 269
multi-primary subpixel image rendering   264, 265
multi-primary subpixels   258
multispectral imaging   106

## n

native tone curve   27
naturalness   141
natural objects   242
natural spectral reflectances   240
nausea   105
near-eye displays (NEDs)   105
negative film   161
neighboring subpixels   265

neural compensation 319
neutral white 345, 346
NIST 285
NIST color quality scale 294
NIST color set 279
NIST's spectrally tunable lighting facility 295
nocturnal human melatonin
    suppression 342
noise pattern 215
non-rgb sequence 259
NTSC 239
NTSC color gamut 76, 77
NTSC color primaries 74
NTSC gamut 78, 79, 336
NTSC primaries 239
NTSC triangle 75, 77
NTSC white point 338
number of color primaries 245
number of subpixels 265
numerical correlate of perceived
    saturation 15
numerical correlates 1, 12

*o*

objective methods 206
objectives of color image reproduction 105
oblique effect 260
observer preference curve 145
ocular media 2, 4
off-center cell 7
offset knob 54
offset light 46
offset setting 29
OLED displays 71
OLEDs 51, 71, 72, 330
on-center cell 7
optimal color primary sets 251, 255, 256
optimal surface colors 240
optimization principles 238
optimization procedure 249
optimization workflow 21
optimum color gamuts 252
optimum color primary sets 251–254
optimum coverage 251
optimum 4P gamut 256
optimum preference parameter ($p_{opt}$) 146
optimum wide 250
ordinal ratings 286
– scale 280
organic user interface (OUI) 330
original image 206
outdoor film production 223
out-of-color-gamut error term $E_{col}$ 269
out-of-gamut colors 193, 238

out-of-gamut indicator 194
overcoloring 97
overcrowding the display 108

*p*

parallel search 112
parameter ($p_{opt}$) 146
pcLEDs 73, 75, 276, 336–338
4-pc WLEDs 348, 351, 352
PDPs 51, 55–60, 82, 83–86, 88
peak red, green, and blue 41
peak white 35, 36
peak white luminance 6, 29, 30, 43
PenTile Matrix™ 262
– subpixel architectures 263
perceived chromaticity differences 15
perceived colors 12
– differences 278, 321
– harmony 289
perceived similarity ($Z_p$) 280, 281
perceived whiteness 316
perception and cognition 17
perceptual differences 12
perceptually not uniform 15
perceptually uniform 12, 15
perfect reflecting diffuser 14
phosphor blends 348
phosphor chromaticities 28
phosphor constancy 35, 53, 54
phosphor-converted LEDs (pcLEDs) 73, 74, 331, 333, 337
phosphor-converted white LED
    technology 348
phosphor matrices 27, 35
photoreceptor mosaics 18
photoreceptors 2, 342
photoreceptor structure 2
pixel architectures 237
pixel faults 47
pixel rendering 257
Planckian radiator 317
Pointer's database 243
postreceptoral mechanisms 18
predicted color harmony values
    ($CHF_{i,test}$) 290
preference-based color image
    enhancement 151
preference differences 133
preference functions 147
preference scores 143, 145
preferred appearance 17
preferred chromaticity contrasts 126, 129, 131
preferred colors 17, 33

– image reproduction  107
preferred contrast  123, 125, 127
preferred global contrast  147
preferred hue  150
preferred local contrast  147
preferred luminance contrast  131
preferred white point  144
primary colors  20, 238, 240, 265
principal components of color
– quality  294, 303
projector light sources  72, 73
projectors  20, 68
protanope  18
prototypical colors  141
pseudo-technological constraint ($p_T$)  247, 250, 252
PSNR metric  238
PSNR values  206
psychological relationship $R_p(\Delta E_{UCS})$  305, 306, 310
3P system  255
5P system  255
6P system  255
4P systems  256
pulse frequencies  332
pulse width modulation (PWM)  332
pupil area  9, 11
pupil diameter  197
purple line  6
P/W criterion  249
PWM. *See* pulse width modulation (PWM)
P/W ratio ($p_{p/w}$)  247, 248, 250, 255, 265
P/W values  249

*q*
Q(*f*) functions  41, 42
quantization artifact  47
quantization efficiency  244
quasi-monochromatic radiations  6

*r*
rainbow artifact  71
rapid adaptation  102
raster artifact  47
RCRI  286
readability  104
Rea *et al.*'s model  341, 342
real-world colors  240
receptive fields  7, 8
receptoral changes  319
recognition safety  218
reddish objects  242
redefine color rendering index  285
red–green opponent channel  3

red or green channel  216
reference display  33
reference illuminant  273, 275
reference light source  274, 277, 283
reference white  12, 14, 49
reflecting color samples  5
reflections  47
refocusing time  101
regions of visual attention  102
relational color constancy  356
relative background luminance  188
relative tristimulus values  6
remainder $XYZ_{rem}$  269
remote phosphor LED  333
rendering  349, 350
reproducing total appearance  106
reproduction of highlights  106, 107
resolution  174
retinal contrast  133
retinal illuminance  9, 11
retinal mosaic  21
retrofit LED lamp  353
R(*f*) functions  41, 42
RGB backlight  77
RGB color  337
– filters  76
– space  220
*r, g, b* digital color counts  26
RGB filters  73, 77
RGB LEDs  72, 73, 75, 79, 276
– backlights  76, 78, 79
– driving system  332
– illuminators  74, 75
– light source  74
– mosaic displays  72
RGB stripe  257, 259, 260, 262, 263, 265, 267, 268
– architecture  258
*RGB* values  26, 27, 29, 31, 35, 36, 38, 40, 41, 51, 56, 61, 63, 65, 66, 77, 80, 282
RGCB display  338
RGCB gamut  336
robustness  218
rods  2, 342
rotating filter  72
– wheels  74
rotational symmetry  265
roundness  249

*s*
saccade  103
saturation  12, 48
SBI  285
scalable user interface  105

Schanda's combined preference rendering index   294
scheme of colors   122
S-cones   3, 342
screen position   28
SDS test method   219
search commands   111, 113, 114, 118, 121
search performance   111
search target   112, 118
search time   112, 115–118
search zones   114, 115, 117
see-through head-mounted display   67
segmentation   121
self-luminous objects   6
semantic differential scales   296
semiconductor sensors   199
sensor pixels   162
sensor temperatures   333, 338
setting no.   32
settings   27, 30, 33, 34
seven-primary architecture   265, 268
seven-primary hexagon architecture   267
sharp vision   102
short-term color memory   135
short-term memory colors   135
sigmoid function   147
signal-to-noise ratio   199
similarity   307
– to memory colors   301
– ratings   309, 310, 322–324
simultaneous color contrasts   107, 112, 124
sinusoid gratings   124, 125
six-primary architecture   264
six-primary system   251
small and large color stimuli   81
small color differences   17, 300, 301
small-field tritanopia   2
SOCS database   242, 243
soft clip method   193
software ergonomics   111
solid-state laser   176
space of display color primary sets   250
spatial color distributions   237
spatial cone excitation ratios   356
spatial contrasts   7
spatial frequencies   6, 8–11, 123–125, 127, 129, 133, 147, 154, 197, 209, 221
– characteristics   7
spatial independence   39
spatial interdependence   40, 41, 70
spatial low-pass prefiltering   259
spatial nonuniformity   43, 71
spatial opponency   8
spatial power distributions   341

spatial resolution   18, 21
spatial structures   6, 7, 222
spatial-temporal characteristics   195
spatial uniformity   43, 44, 79
special color rendering indices ($R_{UCS,special}$)   274, 340, 354
spectral design   339
spectral locus   6
spectral power distributions (SPDs)   73, 77, 273, 276, 297, 335, 337, 351, 352
spectral reflectance   340, 345
– curves   240
– functions   240–242, 290
spectral sensitivities   4, 175, 342
spectral transmission   73
spherical aberration   197
spotlight   223
sRGB. *See* standard display color space (sRGB)
S-shaped tone curve   64
standard colorimetric characterization model   28
standard display color space (sRGB)   27, 28, 31, 33, 34, 35, 58, 59, 60, 65, 238
– color gamut   243
– model   28
– standard tone curve   31
– tone curves   28, 29, 32, 33
– white point   34
static mask   217
Stiles and Burch experiment   319, 320
still life   296
– arrangement   297
stroke width   101, 267
subjective evaluation index   285
subjective method   206
subpixel luminance error term ($E_{lum}$)   270
subpixels   77, 257, 260, 266, 267, 268
– architectures   21, 257–259, 261
– color image rendering   262
– layouts   266
– mosaic   72, 237
– rendering   18, 258, 270
suitable white tone   317
superposition   220
surround   50, 188
– ratio   50
switch-on characteristics   331, 332
system-independent color space   163
Szabó et al.'s model   287

*t*

tabletop arrangement   296
tanh functions   63

Index | 373

target colors   237, 238
– gamut   240, 242, 243
– set   237, 252
– sets   240
task lighting   339
TCS. *See* test color (TCS)
technological constraints   246
temperature dependence of four-primary color sequential (RGCB) model   336
temperatures   338
– changes   334
– dependence   335, 337, 338
temporal contrast sensitivity   9–11
temporal frequencies   9, 10
test color samples   274, 282
test image   206
test light source   273–275, 277
test objects   290
thermal environment   334
three-color combinations   286, 288
three-dimensional optimization   237
three primary   237
– architectures   262
– mosaics   237
– system   251
threshold   9, 125
threshold chromaticity contrasts   126, 129
threshold contrast   123, 127, 196
tolerable color difference   284
tolerance ellipsoids   284
tolerance volume   141
tone curves   28, 29, 31, 33–35, 51, 58–62, 64–66, 68, 72, 144, 147
– models   27, 55
tone mapping operators   80
total appearance   105, 106
total color differences   15
total immersion   280
trichromatic   2
tristimulus values   25, 26
tritanope   18
troland (Td)   9
tubular fluorescent lamp   223
tunable correlated color temperature   340
tunable LED lamps   339
tunable white LED lamp   340
tungsten halogen   276, 343
tungsten incandescent lamp (INC)   353
tungsten light   223, 273
TV resolution   215
twilight range   30
two-color combinations   286, 288, 289
two-color harmony formulas (CHF)   288
two-dimensional fast Fourier transform (2D FFT)   154
two-primary crosstalk model   65

**u**

UHP (ultrahigh pressure)   72, 75
– discharge lamps   73, 74
under color removal   65
uniform color spaces   17
unusual tone of the light source   316
use of color   107
user acceptance   273
user interfaces   98, 108, 113, 122
users characteristics   21
$u'$, $v'$ diagram   14, 15
– chromaticity   252, 258
$U^* V^* W^*$ color space   274, 283

**v**

$\Delta E^*ab$ value   32
$RGB$ values   55
variability of retinal mosaics   18
viewing angle   45
viewing conditions   139, 179
– parameters   48, 50
viewing direction
– dependence   66
– nonuniformity   46, 61
– uniformity   45, 53, 60
viewing distances   101, 102
viewing environment   49
viewing parameters   280
virtual primaries ($p_V$)   250
virtual reality   67
virtual reality-induced symptoms and effects (VRISE)   67
visibility   104
visual acuity   9, 100, 101
visual artifacts   25, 43
visual attention   97, 103
visual clarity   294
visual color difference ($z_{\Delta Evis}$)   280, 281, 318
visual comfort   109, 111
visual complexity   103
visual display   1
visual ergonomics   17, 19, 100, 123
visual experience   178
visual experiment   178
visual image quality   222
visual information   98
visually acceptable lightness   249
visually acceptable $P/W$ ratios   248, 253
visually evoked emotions   19

visually nonuniform   284
visual mechanisms   21
visual overload   97
visual performance   108
visual preference   142
visual psychophysics   98
visual ratings   320
visual scaling of color harmony   287
visual search   107, 108, 110, 112, 113, 120
visual similarity   280
$V_{10}(\lambda)$   342
$V(\lambda)$ function   3
volumetric displays   331
von Kries transformation   274, 275, 283

## w

Wallis filter   147
Ware and Cowan conversion factor (WCCF) formula   109
warm-up   46, 47
warm white   345, 346
warm white phosphor LED   276
watermarking   9, 214, 215
wavelet-based   205
wearable computing   67
white acceptance ellipse   246
white LEDs   72, 282, 333, 340
– lamp   340
– light sources   74
whiteness perception   316
white pcLEDs   339
white point ($p_W$)   28, 34, 79, 145, 245, 250, 336, 337, 338
– illuminant D65   179
– light source   188
– luminance   188
– monitor   186
– preference   145, 146
white region in the chromaticity diagram   245
white tolerance ellipses   246
white tone   20
wide color gamut   240
wide-gamut CCFL backlights   79
wide-gamut color primaries   237
wide-gamut displays   238, 334
wide-gamut LED display   334
wide viewing angle (WAV) LCD   61
workflow in cinema film and TV production   161
working memory   104
work performance   98
work psychology   98

## x

xenon   75
xenon discharge lamp   72
$x(\lambda)$   4
$x_{10}(\lambda)$   4
$x(\lambda), y(\lambda), z(\lambda)$ color matching functions   319
x, y chromaticity diagram   239
XYZ tristimulus values   4, 5
XYZ values   26, 56, 65

## y

YAG   74
YCbCr color space   220
yellow–blue opponent channel   3
yellow pigment   2
$y(\lambda)$   4
$y_{10}(\lambda)$   4
young and elderly observers   123, 133, 142, 143
young observers   152

## z

$z(\lambda)$   4
$z_{10}(\lambda)$   4
z-scores   280